CLINICAL ATLAS OF POLYSOMNOGRAPHY

CLINICAL ATLAS OF POLYSOMNOGRAPHY

By

Ravi Gupta, MD, PhD

S. R. Pandi-Perumal, MSc

Ahmed S. BaHammam, MD, FRCP, FCCP

Apple Academic Press Inc.
3333 Mistwell Crescent
Oakville, ON L6L 0A2 Canada

Apple Academic Press Inc.
9 Spinnaker Way
Waretown, NJ 08758 USA

© 2018 by Apple Academic Press, Inc.
Exclusive worldwide distribution by CRC Press, a member of Taylor & Francis Group
No claim to original U.S. Government works
Printed in the United States of America on acid-free paper
International Standard Book Number-13: 978-1-77188-663-5 (Hardcover)
International Standard Book Number-13: 978-0-203-71151-4 (eBook)

All rights reserved. No part of this work may be reprinted or reproduced or utilized in any form or by any electric, mechanical or other means, now known or hereafter invented, including photocopying and recording, or in any information storage or retrieval system, without permission in writing from the publisher or its distributor, except in the case of brief excerpts or quotations for use in reviews or critical articles.

This book contains information obtained from authentic and highly regarded sources. Reprinted material is quoted with permission and sources are indicated. Copyright for individual articles remains with the authors as indicated. A wide variety of references are listed. Reasonable efforts have been made to publish reliable data and information, but the authors, editors, and the publisher cannot assume responsibility for the validity of all materials or the consequences of their use. The authors, editors, and the publisher have attempted to trace the copyright holders of all material reproduced in this publication and apologize to copyright holders if permission to publish in this form has not been obtained. If any copyright material has not been acknowledged, please write and let us know so we may rectify in any future reprint.

Trademark Notice: Registered trademark of products or corporate names are used only for explanation and identification without intent to infringe.

Library and Archives Canada Cataloguing in Publication

Gupta, Ravi, 1975-, author
Clinical atlas of polysomnography / by Ravi Gupta, MD, PhD, S.R. Pandi-Perumal, MSc, Ahmed S. BaHammam, MD, FRCP, FCCP.
Includes bibliographical references and index.
Issued in print and electronic formats.
ISBN 978-1-77188-663-5 (hardcover).--ISBN 978-0-203-71151-4 (PDF)
1. Polysomnography--Atlases. I. Pandi-Perumal, S. R., author II. BaHammam, Ahmed, author III. Title.
RC547.G87 2018   616.8'498   C2017-907416-4   C2017-907417-2

Library of Congress Cataloging-in-Publication Data

Names: Gupta, Ravi, MD, author. | Pandi-Perumal, S. R., author. | BaHammam, Ahmed, author.
Title: Clinical atlas of polysomnography / Ravi Gupta, S.R. Pandi-Perumal, Ahmed S. BaHammam.
Description: Toronto ; New Jersey : Apple Academic Press, 2018. | Includes bibliographical references and index.
Identifiers: LCCN 2017055531 (print) | LCCN 2017056540 (ebook) | ISBN 9780203711514 (ebook) | ISBN 9781771886635 (hardcover : alk. paper)
Subjects: | MESH: Sleep--physiology | Sleep Wake Disorders | Polysomnography--methods
Classification: LCC QP425 (ebook) | LCC QP425 (print) | NLM WL 108 | DDC 612.8/21--dc23
LC record available at https://lccn.loc.gov/2017055531

Apple Academic Press also publishes its books in a variety of electronic formats. Some content that appears in print may not be available in electronic format. For information about Apple Academic Press products, visit our website at **www.appleacademicpress.com** and the CRC Press website at **www.crcpress.com**

DISCLAIMER:

Every effort has been made by the authors and publishers to contact the copyright holders to obtain their permission for the reproduction of borrowed material. Regrettably, it remains possible that this process is incomplete. Thus, if any copyrights have been overlooked, the publisher will ensure correction at the first opportunity for a subsequent reprint of this volume.

> The front cover illustration was created by using the Starry Night (original was an oil on canvas by the Dutch post-impressionist painter Vincent van Gogh, 1889) with a superimposed picture of a polysomnographic tracing.

CONTENTS

Dedication vii

About the Authors ix

List of Abbreviations xiii

Forewords .... xvii-xx

Preface xxi

Credits and Acknowledgments xxv

1. Normal Sleep 1
2. Common Sleep Disorders 21
3. Introduction to Polysomnography 37
4. Basic Concepts of Polysomnography Channels 53
5. Preparations for the Sleep Study 115
6. Placement of Leads for the Sleep Study 121
7. Starting and Closing the Study 145
8. Calibration and Biocalibration 179
9. Minimal Recording Parameters and Extended Montage 197
10. Montages 221

11.	The Concept of Epochs	235
12.	Artifacts	253
13.	Scoring of Data in Adults	269
14.	Use of Video Polysomnography	387
15.	Use of Sleep Histogram	391
16.	Polysomnography in Children: Scoring Rules	405
17.	Test Protocols	417
18.	Documentation	423
19.	Troubleshooting	431
20.	Manual Titration with Positive Airway Pressure	439
21.	Writing an Informative Report	447
22.	Guidelines for Supplemental Oxygen	453
23.	Infection Control	459
24.	Sleep Lab Management	463
25.	Financial Viability for a Sleep Lab	499
Glossary		507
Index		509

DEDICATION

To our families...

for their abundant support, patience,

understanding, and everlasting love and affection.

ABOUT THE AUTHORS

Ravi Gupta, MD, PhD, MAMS, Certified Sleep Physician (World Sleep Federation), is presently working as Professor in the Department of Psychiatry & Sleep Clinic, Himalayan Institute of Medical Sciences, Dehradun, India. He has contributed chapters to various high profile academic volumes. He has also authored a book "Psychiatry for Beginners," and has more than 100 published articles in various peer-reviewed journals to his credit. He has received numerous awards and fellowships, including the Young Psychiatrist Award from the Indian Psychiatric Society (2008), fellowships from the World Psychiatric Association (2008) and from the Japanese Society of Psychiatry and Neurology (2009), and membership of the Indian Academy of Medical Sciences, New Delhi (2015) to name a few. In 2010, he received a Mini-Fellowship from the American Academy of Sleep Medicine. This fellowship provided him a chance to refine his knowledge and skills in the area of Sleep Medicine under able guidance of Dr. Jim Walker and Dr. Robert Farney at LDS Hospital, Salt Lake City, Utah. Continuing with the legacy of his mentors, he is engaged in imparting knowledge of Sleep Medicine to physicians from various streams through lectures, workshops, and training programs in his present institute.

S. R. Pandi-Perumal, MSc, is the President and Chief Executive Officer of Somnogen Canada Inc, a Canadian Corporation. He is a well-recognized sleep researcher, both nationally and internationally, and has authored many publications. His general area of research interest includes sleep and biological rhythms. He is a well-known editor in the field of sleep medicine and has edited over 20 volumes dealing with various sleep-related topics. He received an honorable mention in the New York Times in 2004. The India International Friendship Society awarded him the prestigious Bharat Gaurav award on January 12, 2013. The Bharat Gaurav Award is usually recognition of a person who has made achievements in a particular field, which in turn will put a positive impact on the society at large.

Ahmed S. BaHammam, MD, FRCP (Lon), FRCP (Edin), ABIM, EDIC, FACP, FCCP, is a tenured Professor of Medicine at King Saud University (KSU). Recognized as a leading expert in the field of sleep medicine, both nationally and internationally, Prof. BaHammam established the first academic sleep disorders center in Saudi Arabia and the first Sleep Medicine Fellowship program and subsequently chaired the committee that established the Saudi regulation for accreditation of sleep medicine specialists and sleep technologists. As a founding member of the Saudi Thoracic Society and the Saudi Sleep Medicine community, Prof. BaHammam has organized over 20 workshops on polysomnography. His advanced polysomnography courses have been conducted in several countries in the Middle East. For his contribution to sleep medicine, he was awarded the Lifetime Achievement Award (2016) by King Saud University. He has published more than 200 scientific articles. He is a member of the editorial board of several international medical journals.

LIST OF ABBREVIATIONS

10–20	ten–twenty (electrode placement system)	CPAP	continuous positive airway pressure
5-HT	serotonin	CSA	central sleep apnea
AASM	American Academy of Sleep Medicine	CSB	Cheyne-Stokes breathing
		CVD	cardiovascular disease
AC	alternate current	Ca^{++}	calcium ion
AV	atrio-ventricular	Cl^-	chloride ion
ABG	arterial blood gas analysis	CO_2	carbon dioxide
ABIM	American Board of Internal Medicine	CMR	common mode rejection
		CMRR	common mode rejection ratio
ALMA	alternate leg muscle activity	DC	direct current
ASV	adaptive servo-ventilation	DMH	dorsomedial hypothalamus
COPD	chronic obstructive pulmonary disease	ECG	electrocardiogram (or EKG)
		EDS	excessive daytime sleepiness

EEG	electroencephalogram	IRLSSG	International Restless Legs Syndrome Study Group
EFM	excessive fragmentary myoclonus		
EMG	electromyogram	K^+	potassium ion
EOG	electrooculogram	KLS	Kleine-Levine syndrome
EPSP	excitatory postsynaptic potential	LC	locus coeruleus
$ETCO_2$	end-tidal CO_2	LDT	laterodorsal tegmental nuclei
GABA	γ-aminobutyric acid	LFF	low-frequency filter (high pass filter)
GERD	gastroesophageal reflux disease	MSLT	multiple sleep latency test
GI	gastrointestinal (tract)	MWT	maintenance of wakefulness test
Glu	glutamate	N3	deep sleep
H	histamine	Na^+	sodium ion
HF	heart failure	NE	norepinephrine
HFF	high-frequency filter	NREM	non-rapid-eye-movement sleep
HFT	hypnogogic foot tremor	O_2	oxygen (saturation)
HTB	high threshold bursting	OCC	out-of-center
HST	home sleep testing	OHS	obesity hypoventilation syndrome
Hz	hertz; a measure of frequency (cycles per second)	OSA	obstructive sleep apnea
		OSAS	obstructive sleep apnea syndrome
IPSP	inhibitory postsynaptic potential	pH	potential of hydrogen

List of Abbreviations

$PaCO_2$	partial pressure of CO_2 in arterial blood	RIP	respiratory inductance plethysmography
PaO_2	partial pressure of O_2 in arterial blood	RLS	restless legs syndrome
		SA	sino-atrial
PAP	positive airway pressure	SCN	suprachiasmatic nucleus
PDR	posterior dominant rhythm	SDB	sleep disordered breathing
PFT	pulmonary function tests	SNRI	serotonin-norepinephrine reuptake inhibitors
PLMS	periodic leg movements during sleep	SOREM	sleep onset rapid eye movement
PPT	pedunculopontine tegmental nuclei	SOREMPS	sleep onset REM periods
		SRS	Sleep Research Society
PSG	polysomnography	TC	thalamocortical (neurons)
RAS	reticular activating system	$TcCO_2$	transcutaneous CO_2
RBC	red blood cells	VLPO	ventrolateral preoptic nucleus
RBD	REM sleep behavior disorder		
REM	rapid-eye-movement sleep		
RERA	respiratory effort related arousal		
RHT	retinohypothalamic tract		

FOREWORD

Sleep is increasingly recognized, along with a healthy diet and physical activity, as one of the three pillars of sustainable health. Conversely, chronic lack of sleep or disturbed sleep is well known to produce detrimental consequences for both physical and mental health. For instance, epidemiological studies have shown strong associations between sleep loss and sleep disorders, on one hand, and increased risks for cardiovascular and psychiatric disorders, premature cognitive impairments, and even mortality, on the other hand. It may be no surprise then that sleep disorders have become a major public health issue in most modern societies.

Polysomnography is the gold standard for measuring normal sleep and for diagnosing many sleep disorders. It provides the most comprehensive and reliable assessment of multiple physiological parameters during sleep. Many of the discoveries in basic sleep and circadian research and in clinical sleep medicine are the direct results of technological advances, particularly polysomnography. Just to note a few examples, the discovery of rapid-eye movement (rem) sleep, the diagnosis of sleep disorders such as narcolepsy, sleep-related breathing disorders, and REM behavior disorder, as well as the observation of sleep state misperception in some forms of insomnia, could not have been made without polysomnography. Of course, polysomnography is only one tool among several assessment modalities of sleep. Depending on whether the context of assessment is clinical or research, adding a clinical interview, along with subjective sleep diaries, behavioral devices such as actigraphy, and a variety of patient-questionnairesare likely to yield useful and complementary information and provide thecompleteassessment of sleep and sleep-related complaints.

The Clinical Atlas of Polysomnography is truly a comprehensive collection of invaluable information and graphical illustrations about everything that is needed to perform thepolysomnographic assessment. From basic concepts about normal sleep and sleep disorders, technical information about calibration, artifacts, and various montages, to practical recommendations for managing a sleep laboratory, this atlas represents a superb referenceguide. The atlas is filled with useful illustrations of actual records depicting standards for monitoring, artifacts recognition, and key findings associated with different sleep pathologies. A compendium of review questions at the end of each chapter is extremely handy for testing new learning. Although intended primarily for sleep technologists, this atlas will serve as a useful reference for any sleep clinician, researcher, or trainee wanting to learn more about the technical and clinical aspects of evaluating sleep and sleep disorders.

The authors of this atlas, Drs. Ravi Gupta, S.R. Pandi-Perumal, and Ahmed BaHamman, are internationally known leading experts in sleep medicine and sleep research. Theyare to becommended for compiling this essential tool for a better understanding normal sleep and diagnosing many sleep disorders. This atlas is likely to remain a key reference in the field for years to come.

Charles M. Morin, PhD, FRSC, DABSM

President, World Sleep Society

FOREWORD

It is a great pleasure and honor for me to write the foreword to the book *Clinical Atlas of Polysomnography* by Dr. Ravi Gupta, Dr. S. R. Pandi-Perumal, and Dr. Ahmed S. BaHammam. It is not so because I am doing the same as the President of the Indian Society for Sleep Research and Asian Sleep Research Society. The reason is something different. We, the leaders in Indian Society for Sleep Research, are engaged in Sleep Medicine education across the country and across the spectrum for last ten years by organizing courses for physicians and technicians and had envisioned the development of a book on PSG for readers not only in India but also in Asia. We thought of a book that will be easy to understand and comprehensive and that one can begin a career in sleep medicine by simply reading it. Now we do not have to hunt for such a book. This long-cherished desire of the core group has been accomplished. I thank all the authors for their tremendous effort and commitment to bring this unique book for the growth of Sleep Medicine.

Sleep disorders are now recognized as a major public health issue. Investigating sleep is the most challenging medical investigation as it requires background knowledge of physiology and sleep medicine and expertise in sleep technology. *The Clinical Atlas on Polysomnography* has everything that you want to learn or teach on sleep technology. The physiological principles of respiration, gas exchange, the genesis of electrical potentials with color illustrations make it easy to understand for technicians.

The beginners who are starting a career in Sleep Medicine will be immensely benefitted by this book. It is not merely another book on polysomnography. It is something different. It begins with a description of normal sleep, covers everything about polysomnography, and ends with career development, sleep laboratory management, and financial viability of a sleep laboratory. You will find learning objectives at the beginning of each chapter and some review questions to test your knowledge at the end. The essence of a sleep technology book is quality of representative graphs capturing various physiological and pathological events during sleep. This book offers the best. The authors have put extra efforts to put color graphs for easy understanding.

Recent developments in sleep data acquisition systems and analysis software, including automated computing techniques, make the life of sleep technicians ever challenging. This book describes all the major PSG machines available in the market with their software features.

No doubt the book is a stepping stone for starting a career in Sleep Medicine, and the authors provide the blueprint for future revisions and refinement with readers' feedback.

— **Hrudananda Mallick, MD, PhD, FAMS**

Professor, Physiology

Dr. Baldev Singh Sleep laboratory

All India Institute of Medical Sciences, New Delhi

President, Indian Society for Sleep Research

President, Asian Sleep Research Society

PREFACE

The branch of Sleep Medicine came into existence approximately 70 years ago when many physicians who had a keen interest in this area developed themselves as somnologists. Initially, Sleep Medicine was focused towards research; however, as happens in science, with evolution of knowledge, pathological conditions related to sleep (i.e., sleep-disorders) were also recognized. With this advancement of knowledge, the practice and scope of Sleep Medicine extended from bench-side to clinics.

Polysomnography has always been a useful tool for understanding physiological and pathological aspects of sleep. In the past 30 years, sleep laboratories have become places not only for research but also for providing clinical care. The person in command in any sleep laboratory is often a sleep technologist who records and scores the data. He is expected to understand not only the machine and the software that he handles but also to possess at least basic knowledge regarding electrophysiological, technological, clinical, and therapeutic aspects of sleep science.

Recording of data is of paramount importance and, thus, training and certification programs were developed for sleep technologists. Well-structured training programs for sleep technologists also helped to establish a minimum standard of care for patients, as trained sleep technologists must cross the established benchmark of expected knowledge and skills in this area. However, in many parts of the globe, structured training programs for sleep technologists are not available. In those geographical areas, despite the need felt by sleep physicians, sleep laboratories are difficult to be established and run in the absence of trained sleep technologists. This present book is an attempt to fill that void and to provide information that a sleep technologist must possess.

It was important to maintain consistency and comparability of data within and across sleep laboratories. To address this issue, rules for scoring the polysomnographic data were laid down by eminent scientists. The American Academy of Sleep Medicine has been a forerunner in the development of sleep technology courses and scoring rules and, most recently, on April 1, 2017, it produced the *AASM Manual for the Scoring of Sleep and Associated Events*, version 2.4. These guidelines and rules are at present considered as standard globally, and this book follows the framework of same. However, this book is intended to complement, not to substitute, for the AASM scoring manual, as many areas that are covered in the manual are not covered here.

This book provides basic information regarding normal sleep, sleep disorders, and electrophysiology aspects of sleep that are outside of the scope of the AASM scoring manual. The book will guide you through the fundamental aspects of, for example, types of overnight sleep studies, establishing a sleep laboratory, preparing the patient for a sleep evaluation study, placement of electrodes and other signal recording devices, and scientific aspects of recording of data. This book also includes chapters on depicting real-time illustrations of sleep data as captured in the sleep laboratory, and the scoring of recording data. Information regarding common montages, artifacts, and troubleshooting in the sleep

Preface

laboratory will facilitate your journey as a trainee sleep technologist. Graphical one-page representations of overnight recorded data (i.e., histograms) can provide a great deal of useful information. An attempt has been made to explain the interpretation of histograms. It is prudent to summarize the data and observations in a report that is not only comprehensive and informative but also is easy to understand, even by physicians whose primary specialty is not Sleep Medicine. A chapter has been dedicated to explain this in detail. Lastly, we have provided ready-made forms, questionnaires, and documents that can either be used as they are or with some modifications.

We hope that our humble attempt will be welcomed and that readers will find this book useful!

— RG

SRP

ASB

CREDITS AND ACKNOWLEDGMENTS

This volume owes its final shape and form to the assistance and hard work of many talented people. We would like to express our profound gratitude to the many people who have helped and also to some who have contributed without realizing just how helpful they have been.

First of all, the authors would like to express their reverence towards all those scientists, physicians, and sleep technologists whose dedication and contribution have evolved the area of Sleep Medicine and who are still improving our understanding. This book is an attempt to collectively reflect their work.

Our sincere appreciation goes to Prof. Charles Morin and Prof. Hrudananda Mallick who are well known figures in field of sleep medicine. They graciously agreed to write forewords for this book. We know them as great scientists, successful leaders, acclaimed teachers, revered mentors and last but not the least, as humble persons. Their testimonials mean a lot to us. We also would like to thank Prof. Jan Ulfberg, an eminent scientist, a great physician and mentor for many persons in Sleep Medicine for going through the book prior to the print and providing his insights. We are humbled to receive encouraging comments from him. We want to say him a big, "Thank you!".

We gratefully acknowledge our corporate partners, namely Philips Respironics Inc., Cadwell Inc., and Somnomedics LLC, who kindly accepted our request to provide requested illustrations for this volume.

The authors would also like to acknowledge several of our friends and colleagues—Dr. Abhishek Goyal (Associate Professor, Department of Pulmonology, All India Institute of Medical Sciences, Bhopal, India), Dr. Rajanish Sharma (Sleep Physician, Rudraksh Clinic, Jaipur, India), Dr. Supriy Jain (Consultant Cardiologist, Jaipur, India), and Dr. Sourav Das (Sleep Physician, Somnos Sleep Clinic, Kolkata, India)—for providing some illustrations for the book. The authors would also like to acknowledge the technical help provided by the following colleagues: Divinagracia E. Gacuan, RPSGT; Smitha George, RPSGT; and Karen Lorraine Acosta, RPSGT, from the University Sleep Disorders Center at King Saud University, Riyadh, Saudi Arabia. Mr. MM Mathavan from the Himalayan Institute of Medical Sciences, Dehradun, has contributed a chapter "Financial Viability for a Sleep Unit" and we are thankful to him for providing insight on this important issue.

The authors would also like to acknowledge the close co-operation we have received from each other. We think we made a good team, even if we say so ourselves!

No volume can be completed without the untiring efforts of many publishing professionals. Producing a volume such as this is a team effort and we acknowledge with gratitude the work of the editorial department of Apple Academic Press. We are especially indebted to Mr. Ashish Kumar, the President of Apple Academic Press, who was an enthusiastic and instrumental supporter from start to finish. Our

Credits and Acknowledgments xxvii

profound gratitude is offered also to Mr. Rakesh Kumar, Production Editor, whose equally dedicated efforts promoted a smooth completion of this important project. They both provided unflagging dedication, invaluable help, and encouragement. We appreciate their intellectual rigor and personal commitment to our project.

We also thank the Apple Academic Press production department colleagues for their meticulous work. They all gave unstintingly of their time, energy, and enthusiasm. This talented and dedicated team of copy and production editors strengthened, polished, trimmed, and conscientiously checked the text for errors.

Last, but certainly not least, we are most grateful to our wonderful wives, families and friends, who provided love and support too valuable to measure. We owe everything to them; without their support this work would have been not completed. With unwavering optimism and encouragement, they saw the work through from the conception of an idea to the completion of an interesting project.

Without a whole host of dedicated individuals, this volume would never have come to completion. We recognize these people individually and collectively for their contribution. To all these people goes our sincere gratitude. Their willingness to contribute their time and expertise made this work possible, and it is to them that the greatest thanks are due. They made our work possible and pleasurable.

For this, and for so much else, we are ever grateful.

—**Ravi Gupta, India**

S.R. Pandi-Perumal, Canada

Ahmed S. BaHammam, Saudi Arabia

NORMAL SLEEP

LEARNING OBJECTIVES

After reading this chapter, the reader should be able to:

1. Define sleep and changes in sleep across age.
2. Discuss neurobiological mechanisms of sleep including the two-process model.
3. Discuss different stages of sleep along with physiological changes.
4. Discuss the neurobiology of EEG waves seen across different sleep stages.

Contents

1.1	Sleep Across Age 2
1.2	Neurobiology of Sleep..3
1.3	Sleep Architecture 6
1.4	Physiological Changes from Wakefulness to Sleep.......................12
1.5	Neurophysiology of EEG Rhythms from Wakefulness to Sleep ..12
1.6	Atonia During REM Sleep..16
1.7	Concluding Remarks 17
Further Reading 19
Review Questions 19
Answer Key 19

Sleep may be defined as a physiological state of unconsciousness that is reversible, cognitive and perceptual disengagement from and to the environment are seen during this temporary unconsciousness, and from which the arousal is possible in response to any internal or external stimuli. This definition has several important aspects that will help you understand the physiology of sleep. First, it is a state of unconsciousness that is reversible and this reversibility differentiates it from coma. Second, the person is not aware of the environment during sleep, so he does not perceive (although he may sense it) most stimuli during sleep. Perception is different from sensation – when a sensory stimulus is consciously recognized it is considered as perceived. This suggests that a sleeping person cannot take any conscious decision – rather he reacts on instinct. This explains why patients suffering from sleep-walking when aroused forcibly often commit violent acts. The third part of the definition says that a person may be aroused with any internal or external stimuli. Internal stimuli that may arouse a person include pain, anxiety or dyspnea. External stimuli are often auditory or tactile in nature. However, these stimuli must be sufficiently intense to arouse the person, and their intensity also varies across the sleep stages.

1.1 SLEEP ACROSS AGE

Sleep has many important parameters, for example, the total duration of sleep, maintenance of sleep, and at what time of the day we fall asleep. Electrophysiologically, sleep may be divided into two main stages: non-rapid-eye-movement (NREM) sleep and rapid-eye-movement (REM) sleep. NREM sleep can be further divided into three stages: N1 (light sleep), N2, and N3 (deep sleep).

A remarkable variation is seen in all these parameters across age. For example, total sleep time of an infant is around 18–20 hours, which reduces as the child grows and comes to adult timing during mid-adolescence (6–8 hours), after which it remains more or less stable. However, it starts declining in the fourth decade of a person's life. During old age, it reduces to 5–6 hours/day. Regarding continuity, infants have fragmented sleep and they tend to wake up multiple times. It becomes consolidated into a single sleep as the age grows. Children often go to bed early at night, however, during adolescence, they develop a phase delay. Phase delay refers to delayed bedtime and wake time. On the other hand, the phase is considered to be advanced in old age. Aging also influences the electrophysiology of sleep. Children have a high proportion of N3 (deep sleep), which reduces as the person grows old.

1.2 NEUROBIOLOGY OF SLEEP

Sleep is dependent upon two processes: homeostatic and circadian. These processes together decide when we will fall asleep and also the other characteristics of sleep, such as depth, duration, maintenance, and proportion of sleep stages.

The homeostatic (also known as "S") process makes us feel sleepy depending upon the duration of wakefulness. The longer the period of wakefulness, the higher the sleep pressure; in other words, higher are the chances to fall asleep. That is the reason why a person who is awake for a long time falls asleep even during the day. Moreover, this process regulates the proportion of deep sleep (N3 sleep) - the longer the period of wakefulness, the higher the proportion of deep sleep. The homeostatic process is dependent upon two major areas of the brain that are located close to the hypothalamus. The first is the sleep-promoting area – the ventrolateral preoptic nucleus (VLPO), which sends inhibitory signals through the GABAergic neurons to the wake-promoting area of the brain. Wakefulness is dependent upon the monoaminergic nuclei of the brain that are the part of the reticular activating system (RAS). Monoaminergic nuclei include serotonergic neurons (located in dorsal and medial raphe nuclei), noradrenergic neurons (locus coeruleus), histaminergic neurons (tuberomammillary nucleus) and cholinergic neurons (located in pedunculopontine (PPT), laterodorsal tegmental (LDT) nuclei, and basal forebrain nuclei). In addition, there are glutaminergic neurons that are diffusely distributed throughout the RAS. When the VLPO sends GABArgic signals to these neurons, sleep ensues. The activity of both wake-promoting and sleep-promoting areas is modulated through another group of neurons, the hypocretin neurons that are present in the lateral and posterior hypothalamus. Under normal conditions, the hypocretin neurons modulate the activity of the other two areas of the brain in such a manner that only one state of consciousness, that is, wakefulness or sleep prevails (Figure 1.1). Damage to the hypocretin neurons leads to a condition known as narcolepsy. In this condition, a person is not able to maintain either states and experiences bouts of sleepiness during wakefulness.

The circadian process (process "C") is dependent upon environmental light. Certain other factors can also regulate this process, for example, food, emotions, social activity, and exercise. These factors are called "zeitgebers." Among these, light is the strongest factor which can entrain the circadian rhythm. Environmental light falls on the retina and through the retinohypothalamic tract (RHT) (glutaminergic in nature), it sends signals to the suprachiasmatic nucleus (SCN). The SCN one of the nuclei of hypothalamus and acts as the 'master clock' circadian clock. During the dark, this sends signals to the pineal gland to secrete melatonin. Melatonin is a hormone that is released into the blood and circulated to the various organs of the body. The SCN also sends signals to the VLPO through the dorsomedial hypothalamus (DMH) and thus can activate it. In the presence of environmental light, melatonin secretion stops. This is why during the dark, we feel sleepy (Figure 1.2).

To fall asleep, it is important that both the homeostatic and circadian processes are in the same phase. If they are not synchronized, then we may feel difficulty in falling asleep or staying awake. Such a condition leads to the

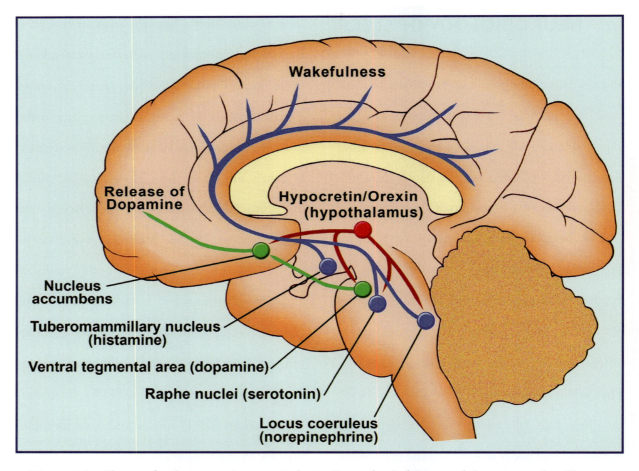

Figure 1.1 Sleep and wake promoting areas in brain: State of wakefulness and sleep are regulated by two areas that exhibit control over each other. Monoaminergic system helps to increase the alertness and forms the part of reticular activating system. These neurons reach cortex and in addition to wakefulness, also responsible for maintaining alertness. GABA neurons from the ventrolateral preoptic area (VLPO) inhibit the monoamine neurons to induce sleep. Hypocretin acts to modulate the activity of both the areas. Laterodorsal tegmental nucleus which is important for wakefulness as well as REM sleep is not shown here.

Normal Sleep

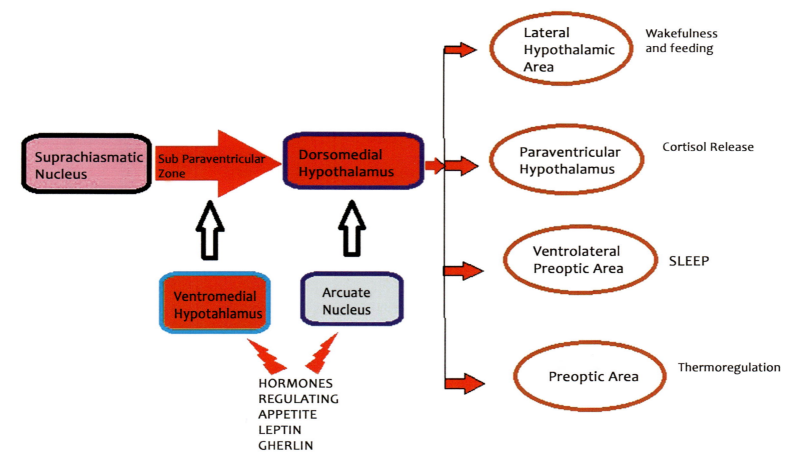

Figure 1.2 Suprachiasmatic nucleus and its connections: Suprachiasmatic nucleus receives signals from retina and passes them to ventrolateral preoptic area (VLPO) through dorsomedial hypothalamus. Dorsomedial hypothalamus has connections with other areas as well that regulate circadian rhythm of other biological functions in addition to sleep.

generation of circadian rhythm sleep disorders (Figure 1.3).

1.3 SLEEP ARCHITECTURE

Grossly, sleep may be divided into two major stages: rapid eye movement sleep (REM) and non-rapid eye movement sleep (NREM), based upon the eye movements, muscle tone, and waveforms in EEG. NREM sleep can further be divided into three stages: N1 sleep (characterized by theta waves in EEG; slow eye movement, and diminution of muscle tone, Figure 1.4), N2 sleep (theta waves, sleep spindles, and K complexes in EEG; absent eye movements and low muscle tone, Figure 1.5) and N3 sleep (more than 20% of epoch has delta waves, Figure 1.6). REM sleep is characterized

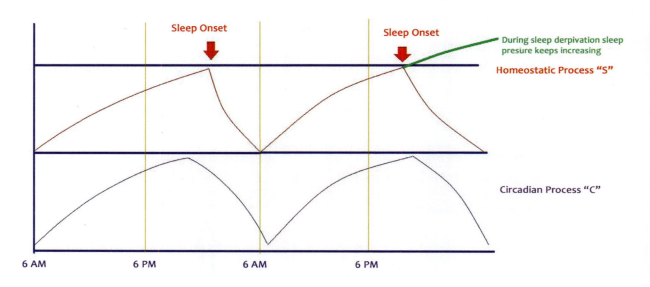

Figure 1.3 Relationship of process C and S in a normal individual: Normally process C and process S overlap with each other so as to increase the propensity to fall asleep at night and propensity to stay awake during the day. Any disruption in this overlapping disrupts the quantity as well as quality of sleep and appears in the form of circadian rhythm sleep disorders.

Normal Sleep

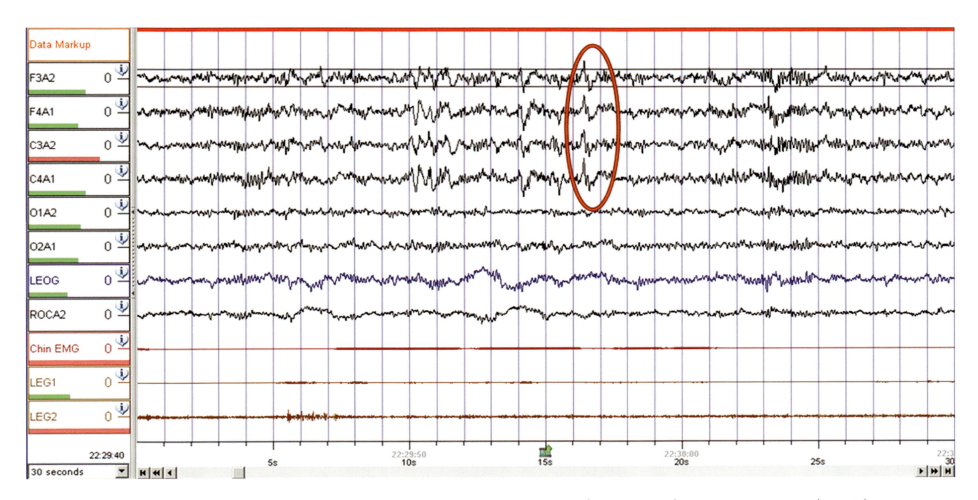

Figure 1.4 30 seconds epoch showing N1 sleep: This epoch shows mix EEG activity (alpha and theta) along with vertex waves (in circle).

8   Clinical Atlas of Polysomnography

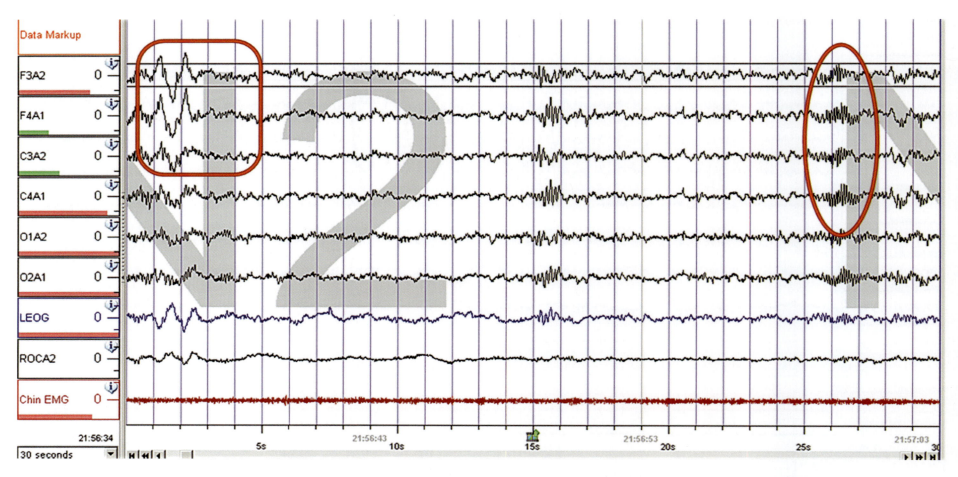

Figure 1.5   30 seconds epoch showing N2 sleep: In this epoch K complex can be seen in the square box and sleep-spindle in circle. Background EEG activity is in theta range.

Normal Sleep

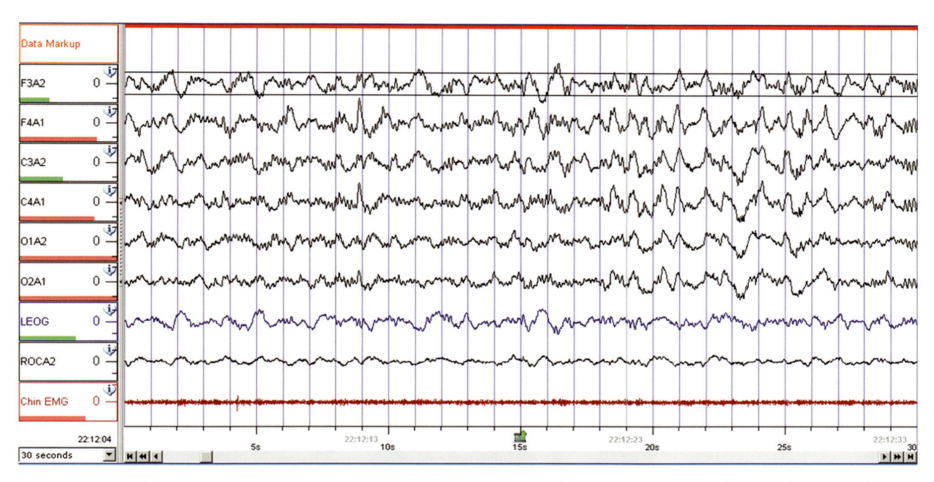

Figure 1.6 30 seconds epoch showing N3 sleep: This epoch show delta activity. Delta waves are also known as slow waves with frequency of 0.5–2 Hz and amplitude of more than 75 mV.

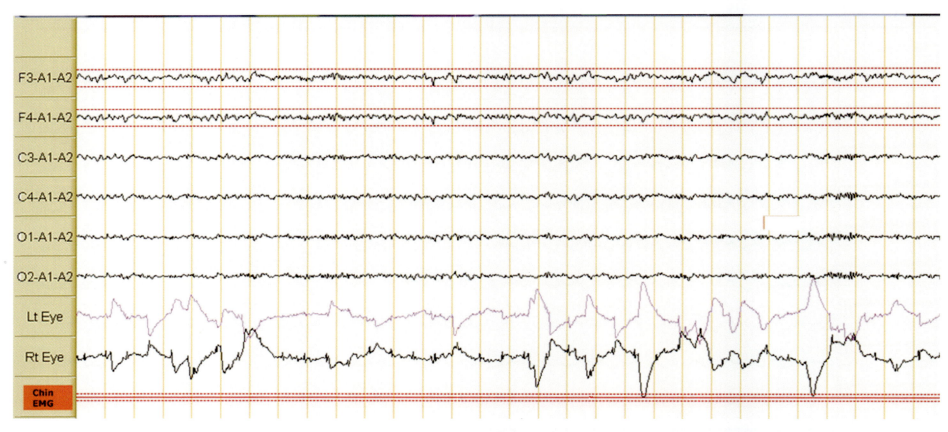

Figure 1.7 30 seconds epoch showing REM sleep: REM sleep is characterized by low voltage mixed frequency activity, rapid eye movements and atonia.

by low-voltage, mixed-frequency activity in EEG, rapid eye movement, and muscle atonia (Figure 1.7).

These stages follow a characteristic pattern in normal persons where NREM and REM sleep keep alternating at an interval of 90–120 min. Sleep starts with NREM sleep, and after approximately 90 min, the first episode of REM sleep appears. This lasts for a few minutes and after this NREM sleep re-starts. This cycle continues throughout the night and during a 7–8 h sleep, an average healthy person has 4–5 cycles of NREM and REM. In addition, as we have already discussed, deep sleep (N3) is regulated by the process "S," so when a person falls asleep after hours of wakefulness, N3 sleep is higher in proportion. Thus, the first half of the night is characterized by a high proportion of N3 sleep. As more time is spent asleep, sleep-pressure lowers, and in the second half of night minimal, if any, N3 sleep is seen. REM sleep is regulated by the process "C" and it is predominantly seen during the second half of the night. Thus, each episode of REM sleep shows progressive lengthening across the night (Figure 1.8).

This has important implications as certain disorders are limited to one sleep stage and they are seen during that part of night. For example, sleep-walking that occurs during N3 sleep is seen more commonly before midnight while REM sleep behavior disorder is more commonly reported close to the morning.

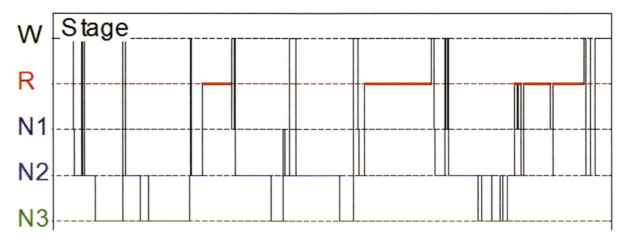

Figure 1.8 Hypnogram: Hypnogram depicts the sleep stages in relation with time. NREM sleep is maximum in the first half of the night, while REM dominates the second half. As the night progresses, duration of N3 periods reduces and that of REM periods increases. NREM-REM has a cycle of around 90–120 min.

1.4 PHYSIOLOGICAL CHANGES FROM WAKEFULNESS TO SLEEP

As the state of consciousness shifts from wakefulness to sleep, and then to different stages of sleep, a number of physiological changes take place inside the body. Many of them are evident during the polysomnography. These changes involve alteration in the brain's electrical activity, muscle tone, respiration, alterations in the cardiac activity, thermoregulation, spontaneous movements, and in the ability to respond to external stimuli. These changes are summarized in Table 1.1.

1.5 NEUROPHYSIOLOGY OF EEG RHYTHMS FROM WAKEFULNESS TO SLEEP

EEG depicts the difference in the summative electrical potentials of the neuronal cells present

Table 1.1 Physiological Changes During Normal Sleep

	Wakefulness		NREM sleep			REM sleep
	Active	Quiet	N1	N2	N3	
Electrical activity of brain	Beta (15–30 Hz) or Gamma rhythm (30–120 Hz) with occasional theta rhythm (4–7 Hz)	Alpha rhythm (8–14 Hz)	Theta rhythm (4–7 Hz)	Theta rhythm (4–7 Hz) with delta	Delta rhythm (<4 Hz)	Low voltage mixed frequency activity similar to wakefulness
Eye movements	Rapid	Rapid	Slow	Absent	Absent	Bursts of rapid eye movements
Muscle tone	Increased	Increased	Reduced	Reduced	Reduced	Minimal/Atonia
Respiration	Irregular rate and variable amplitude depending upon activity	Regular with uniform amplitude	Regular with slowing of rate compared to wakefulness and uniform amplitude	Regular with slowing of rate compared to wakefulness and uniform amplitude	Regular with slowing of rate compared to wakefulness and uniform amplitude	Erratic rate and amplitude
Cardiac activity	Variable within a range	Variable within a range	Regular beats with slowing of rate	Regular beats with slowing of rate	Regular beats with slowing of rate	Irregular heart beats
Thermoregulation	Present	Present	Present	Present	Present	Absent
Spontaneous movements	High	Reduced	Occasional	Occasional	Occasional	Absent
Ability to response to external verbal stimulus	Intact	Intact	Reduced	Reduced	Reduced	Minimal

in the cortex of the brain that lies beneath the electrodes. The cortex is made up of various types of neurons that include pyramidal cells and interneurons, besides glial cells. Pyramidal cells are excitatory in nature and they remain in contact with other cortical neurons through the fibers that they send to other cortical areas (known as association fibers). They also send fibers to the subcortical nuclei and spinal cord (known as projection fibers). These connections are usually reciprocal and thus, other cortical, subcortical nuclei and information coming from the peripheral nervous system (through spinal cord) regulate their activity in a complex manner. In addition, cortical interneurons that are primarily inhibitory in nature also regulate their activity.

During active wakefulness, EEG activity in the gamma or beta band is often visible. Fast-spiking activity of the cortical pyramidal neurons through their interaction with the parvalbumin interneurons generates a gamma rhythm in the EEG. It is thought to represent the synthesis of information as gamma rhythms appear during the sensory stimulation. These pyramidal neurons are glutaminergic and their activity is regulated by the parvalbumin interneurons, which are GABAergic in nature. These interneurons have fast-spiking activity and they rhythmically inhibit and disinhibit pyramidal neurons. Parvalbumin interneurons also get a collateral from the pyramidal cells, thus, excitation of the pyramidal neurons excites the parvalbumin interneurons, which in turn, inhibit the pyramidal neurons.

Beta waves also originate in the cortex and they may either represent a slow gamma activity or it may be possible that the parvalbumin interneurons fire at their natural frequency but some of the pyramidal cells may have a longer refractory period.

Unlike gamma and beta frequency waves, other waveforms of EEG are regulated by the thalamocortical system. Thalamus has relay neurons which receive inputs from the peripheral sensory system and also from the different monoaminergic neurons from the brainstem. These neurons show tonic firing of action potentials. In other words, they show depolarization at regular intervals. On the other hand, when these relay cells are in the state of hyperpolarization, they show bursts of action potential followed by a period of quiescence (Figure 1.9). Burst firing of the thalamocortical relay neurons is important for the generation of various waveforms seen during NREM sleep.

Thalamic relay neurons receive afferents from the periphery as well as the ascending reticular

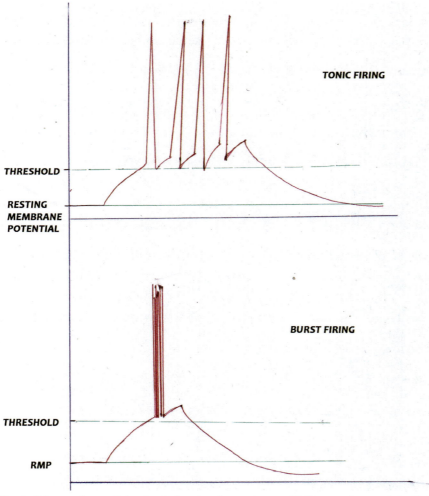

Figure 1.9 Tonic firing vs. burst firing in thalamic relay neurons: Different electrophysiological waves are produced by different electrical activity of thalamic relay neurons. (Adapted from Qiang Zhou, Dwayne W. Godwin, Donald M. O'Malley, Paul R. Adams. Visualization of Calcium Influx Through Channels That Shape the Burst and Tonic Firing Modes of Thalamic Relay Cells. Journal of Neurophysiology Published 1 May 1997 Vol. 77 no. 5, 2816-282.)

activating system. They send their efferent to the thalamic reticular neurons and to the pyramidal cells in the cortex. Cortical pyramidal cells, in turn, send efferents to the relay neurons and the thalamic reticular neurons. It must be noted that thalamic relay neurons and pyramidal cells send excitatory signals, thus, they excite each other and the reticular neurons. Reticular neurons, on the other hand, are inhibitory in nature, and they receive a copy of signals from both the thalamocortical relay neurons and the corticothalamic pyramidal neurons. Reticular neurons, in turn, project to the thalamic relay neurons (Figure 1.10).

Alpha rhythms are seen during quiet wakefulness, especially in the occipital region. Generation of alpha activity is dependent upon the thalamocortical system, in addition to the cholinergic neurons from the brainstem. GABAergic

Normal Sleep

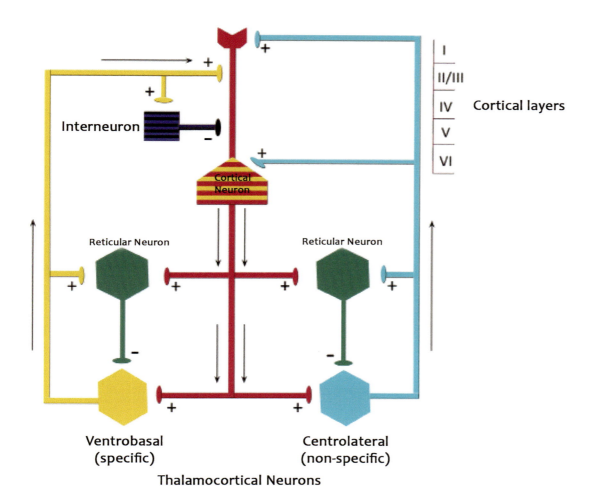

Figure 1.10 Thalamocortical system of neurons: Thalamocortical relay neurons supply to the pyramidal neurons in cortex. Cortical pyramidal neurons send their projection back to relay neurons. Both the neurons also send a copy of their signals to reticular GABAergic neurons, which in turn regulates the activity of relay neurons.

neurons located in the lateral geniculate body periodically silence the high threshold bursting thalamocortical neurons of the occipital area and result in the generation of the alpha rhythm. High threshold bursting (HTB) thalamocortical neurons get depolarized because of cholinergic inputs from the brainstem that acts on the muscarinic receptors present on these neurons. In addition, HTB neurons interact with each other through gap junctions (which is the fastest way of communication between neurons) and, thus, they fire synchronously. Since the activation of muscarinic receptors regulates depolarization of neurons, withdrawal of acetylcholine inputs during the initiation of sleep is depicted as slowing of the EEG to theta rhythms.

Theta rhythms are seen during wakefulness, during N1, N2, and REM sleep. It has been suggested that theta rhythms may represent a slow alpha and, thus, could be regulated by the brainstem cholinergic inputs.

Sleep spindles are 10–14 Hz frequency sinusoidal waves lasting for at least 0.5 seconds (Figure 1.5). These are seen during N2 sleep and are generated by burst firing of the thalamocortical relay neurons and reticular neurons. These sleep spindles are transmitted to the cortex through relay neurons. Connections between various pyramidal cells in the cortex spread them to the widespread cortical areas. Sleep spindles result in hyperpolarization of relay neurons and thus prevent the peripheral sensations from reaching to the cortex. Thus, they help in maintaining sleep.

Delta waves seen during N3 sleep (slow-wave sleep) have their origin both in the thalamus as well as the cortex. Connections between the thalamus and cortex help in synchronization of the two sources of delta waves while connections between various cortical cells help to spread them over widespread cortical areas.

REM sleep is characterized by low voltage mixed frequency activity and sawtooth waves in EEG. REM sleep is modulated by laterodorsal tegmental/ pedunculopontine tegmental (LDT/PPT) cholinergic nuclei situated in the brainstem. These nuclei, depolarize thalamic neurons leading to cortical activation that is important for generation of dreams. These cells also activate GABAergic cells in ventomedial medulla that secrete GABA and glycine on the anterior horn cells in spinal cord leading to profound atonia.

Changes in brain's electrical activity that occur during sleep are depicted in Figure 1.11.

1.6 ATONIA DURING REM SLEEP

Profound muscle atonia develops during REM sleep, leading to multiple ramifications that are clinically important. For example, by producing atonia in upper airway muscles, in particular, genioglossus, it narrows the caliber of the upper airway and increases risks for hypopnea and apnea; when it appears during wakefulness, it manifests as cataplexy, which is a sign of narcolepsy; lastly, absence of atonia during REM sleep leads to the REM sleep behavior disorder.

It is produced by the glycine-mediated (glycineric) inhibition of the anterior horn cells in the spinal cord and the hypoglossal nerve by the REM-on cells of the sub-dorsolateral tegmental nucleus.

Normal Sleep

1.7 CONCLUDING REMARKS

Sleep and wakefulness are regulated by two interconnected areas of the brain, which exert a fine tuning on each other to maintain one of the two states. Sleep is regulated by the circadian as well as the homeostatic processes. Throughout the sleep, neuronal activity keeps changing, which is responsible for the generation of various physiological characteristics that help to determine the different sleep stages.

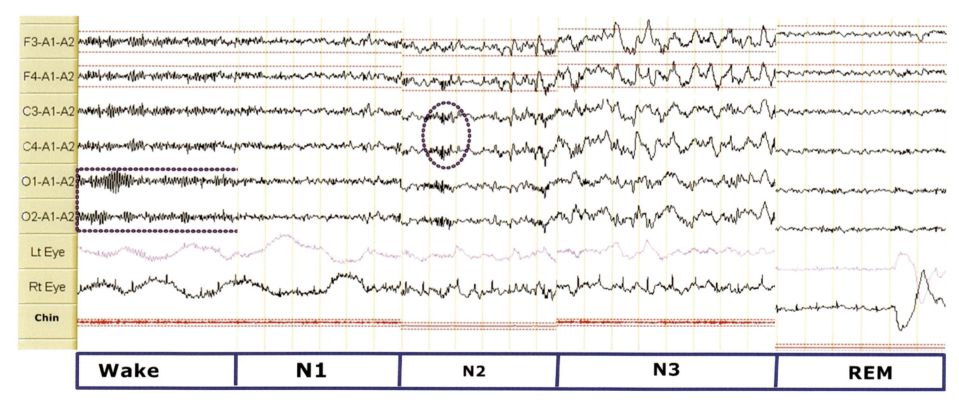

Figure 1.11 EEG, EOG, and chin EMG during various stages of sleep and wakefulness.

FURTHER READING

1. Miner, B., & Kryger, M. H., (2017). Sleep in the Aging Population. *Sleep Med Clin. 12*(1), 31–38.
2. Bathory, E., & Tomopoulos, S., (2017). Sleep Regulation, Physiology and Development, Sleep Duration and Patterns, and Sleep Hygiene in Infants, Toddlers, and Preschool-Age Children. *Curr Probl Pediatr Adolesc Health Care. 47*(2), 29–42.
3. Tarokh, L., Saletin, J. M., & Carskadon, M. A., (2016). Sleep in adolescence: Physiology, cognition and mental health. *Neurosci Biobehav Rev. 70*, 182–188.
4. Schwartz, M. D., & Kilduff, T. S., (2015). The Neurobiology of Sleep and Wakefulness. *Psychiatr Clin North Am. 38*(4), 615–644.

REVIEW QUESTIONS

1. Infant sleep is different from adult sleep as:
    A. first stage to appear is REM sleep
    B. first stage to appear is NREM sleep
    C. first stage to appear is N2 sleep
    D. does not have frequent arousals?
2. As the age progresses, following change is seen in sleep:
    A. sleep duration increases
    B. sleep is dominated by REM sleep
    C. sleep shows frequent arousals
    D. sleep appears shifted to delayed phase
3. Sleep pressure is regulated by:
    A. circadian process
    B. homeostatic process
    C. electrophysiological process
    D. environmental timing
4. Master circadian clock is situated in:
    A. ventro-lateral preoptic nucleus
    B. dorsomedial nucleus
    C. interlaminar nucleus
    D. suprachiasmatic nucleus
5. Low voltage mixed activity is seen during:
    A. REM sleep
    B. quiet wakefulness
    C. active wakefulness
    D. N3 sleep
6. Slowing of EEG during sleep initiation is seen because of:
    A. withdrawal of dopaminergic input
    B. withdrawal of glutaminergic input
    C. withdrawal of cholinergic input
    D. withdrawal of GABAergic input
7. Sleep spindles are generated by:
    A. burst firing of brainstem neurons
    B. burst firing of suprachiasmatic neurons
    C. burst firing of thalamocortical neurons
    D. burst firing of cerebellar neurons
8. Thermoregulation is impaired during:

Normal Sleep

   A. N1 sleep

   B. N2 sleep

   C. N3 sleep

   D. REM sleep

9. Profound atonia is seen during:

   A. quiet Wakefulness

   B. REM sleep

   C. NREM sleep

   D. active wakefulness

10. Patient has irregular breathing during:

    A. REM sleep

    B. N2 sleep

    C. quiet wakefulness

    D. deep sleep

ANSWER KEY

1. A   2. C   3. B   4. D   5. A   6. B

7. C   8. D   9. B   10. A

2

COMMON SLEEP DISORDERS

LEARNING OBJECTIVES

After reading this chapter, the reader should be able to:

1. Discuss common sleep disorders as per the International Classification of Sleep Disorders-3.
2. Diagnose common sleep disorder seen in the sleep laboratory.
3. Understand basic management outlines of common sleep disorders.

Contents

2.1 Obstructive Sleep Apnea (OSA) 22
2.2 Cheyne-Stokes Breathing (CSB) 24
2.3 Obesity Hypoventilation Syndrome (OHS) 26
2.4 Narcolepsy 28
2.5 Restless Leg Syndrome (RLS) 29
Further Reading 31
Review Questions 33
Answer Key 35

In this chapter, we cover the most important sleep disorders that are important from a polysomnographic perspective. For other sleep disorders, readers should refer to a standard textbook of sleep medicine or the International Classification of Sleep Disorders, 3rd edition.

2.1 OBSTRUCTIVE SLEEP APNEA (OSA)

In patients with OSA, the upper airway tends to narrow during sleep, resulting in recurrent closures, which cause repeated apneas despite continued efforts to breathe. This, in turn, results in intermittent hypoxemia and frequent arousals. Intermittent hypoxemia and arousals cause an increase in sympathetic activity, which increases the risk of cardiovascular complications and arrhythmia. Frequent apneas during sleep disturb the sleep architecture, resulting in poor sleep quality and daytime sleepiness.

The obstructive sleep apnea syndrome (OSAS) is defined as five or more abnormal obstructed breathing events per hour of sleep and sleepiness. Studies have reported a prevalence of OSAS in 4% of middle-aged men and 2% of middle-aged women.

2.1.1 OSA Symptoms

Snoring is the most common symptom of OSA. However, not all snorers have OSA. OSA patients may present with choking attacks during sleep, witnessed apnea, nocturia, mouth breathing with dry mouth and throat on awakening, excessive salivation during sleep, excessive sweating, morning headache, nocturnal heart burn, nocturnal palpitation, unrefreshing sleep, and excessive daytime sleepiness (EDS). In some occasions, OSA patients may present with insomnia, particularly in women. In a subgroup of patients, obstructive events may occur mainly during REM sleep, and this is called REM-related OSA. REM-related OSA is seen mainly in women, children and young adults. Usually, patients with this disorder exhibit milder symptoms; however, they may present with nightmares and excessive dreams.

Several questionnaires have been developed to screen for OSA, the most commonly used being the Berlin questionnaire and the STOP-Bang questionnaire.

2.1.2 Risk Factors for OSA

Several risk factors have been associated with OSA, including:

- Obesity: approximately, 70% of OSA patients are obese. Nevertheless, severe OSA can be seen in non-obese subjects with craniofacial abnormalities.
- Increased neck circumference (>17 inches for men and >16 inches in women)
- Gender: OSA is more prevalent in men
- Age: OSA is more prevalent in older people. The risk in women increases significantly post-menopause.

- Craniofacial abnormalities affecting the jaw size: retro- and micrognathia (appears as a small mandible and an overbite).
- Enlarged tonsils and adenoids particularly in children; nevertheless, occasionally it can be seen in adults.

2.1.3 OSA Complications

OSA can lead to EDS, which increases the risk of motor vehicle and other accidents. In addition, there is a great link between OSA and cardiovascular and cerebrovascular complications, such as hypertension, ischemic heart disease, arrhythmias, heart failure pulmonary hypertension, and stroke. New studies have linked OSA with insulin resistance. In addition, OSA can cause depression, and decreased memory and concentration.

2.1.4 OSA Diagnosis

The American Academy of Sleep Medicine (AASM) considers polysomnography to be routinely indicated "Standard" for the diagnosis of sleep disordered breathing. Nevertheless, recent studies have shown that level-III portable studies (home sleep testing) can be used in patients with high clinical likelihood of moderate to severe OSA.

2.1.5 OSA Severity

The severity of OSA is determined by the number of apneas and hypopneas per hour of sleep. That index is called the apnea hypopnea index (AHI). Normal = AHI < 5/hour, mild = 5–15/hour; moderate = 15–30/hour; and severe = >30/hour. Other parameters that may indicate the severity of OSA include desaturation index (the number of desaturations with a 4% (or 3%) drop in SpO_2 compared of baseline/hour of sleep) and time spent with SpO_2 less than 90%.

2.1.6 OSA Treatment

The gold-standard treatment for OSA is the positive airway pressure (PAP) therapy applied non-invasively via an interfacing mask in the form of continuous positive airway pressure (CPAP) or bi-level positive airway pressure (BPAP). The AASM considers PAP therapy as the treatment of choice for mild, moderate, and severe OSA. PAP therapy results in significant improvement in the patient's complaints. Several studies have shown that PAP therapy reduces cardiovascular complications of OSA. PAP therapy side-effects are usually minor and reversible. However, adherence to PAP therapy remains a major obstacle. Therefore, patients should receive a systematic educational program to improve adherence and PAP usage should be monitored objectively and regularly.

2.2 CHEYNE-STOKES BREATHING (CSB)

The AASM defines CSB as a breathing disorder in which there are cyclical fluctuations in breathing, with periods of central apneas or hypopneas that alternate with periods of hyperpnea in a gradual waxing and waning fashion. CSB is seen mostly in patients with heart failure (HF); however, it has also been described in patients recovering from acute pulmonary edema, advanced renal failure, and central nervous system lesions. The presence of CSB in patients with HF is associated with increased morbidity and mortality and impaired quality of life.

2.2.1 CSB Diagnosis

It is usually difficult to differentiate symptoms of HF from symptoms of sleep-related breathing disorders as both problems may have overlapping symptoms, such as poor sleep quality, daytime somnolence or insomnia, paroxysmal nocturnal dyspnea, and easy fatigability. Therefore, a high index of suspicion is required to prevent unjustified delays in diagnosis. Patients with HF and CSB tend usually to have lower body mass index than patients with concomitant OSA. Moreover, CSB has been shown to be more common among males and patients with atrial fibrillation. It is not uncommon to find that patients with HF have co-existing OSA and CSB.

Overnight PSG typically shows cyclical fluctuations in ventilation, with periods of central apnea or hypopnea that alternate with periods of hyperpnea, which predominates during stages N1 and N2 of NREM sleep.

2.2.2 Management of CSB

There is no consensus yet regarding the best management for CSB in patients with heart failure. Management should aim initially to optimize cardiac function. Medical management of HF is not simply needed to improve sleep-related breathing disorders; rather, it is the cornerstone for the management of a dysfunctional heart and thereby increases patient survival.

2.2.2.1 Pharmacological Therapy

Several drugs have been used to treat patients with HF and CSB. Although these drugs frequently reduce sleep-related breathing disorders, they may cause undesirable side-effects and/or interactions with other drugs. Currently, none of these drugs are recommended as a first-line treatment to manage the CSB of patients with HF.

2.2.2.2 Oxygen

The theory behind using oxygen supplementation is that oxygen will offset the hypoxic ventilatory drive and, hence, suppress periodic breathing. Several studies that have examined the role of home oxygen therapy have reported conflicting results and no data are available regarding long-term clinical outcome. For the present time, long-term oxygen therapy is not recommended as a standard treatment for patients with heart failure and CSB.

2.2.2.3 Positive Airway Pressure (PAP) Support

CPAP has been shown to be an effective and safe treatment for patients with acute pulmonary edema; however, its role in patients with CSB is not well-established. A multicenter trial (CAN-PAP) was conducted to evaluate the efficacy of CPAP in reducing mortality and morbidity associated with CSB in patients with HF. The study revealed that CPAP had positive effects on oxygen saturation, left ventricular ejection fraction, and six-minute walk distance, and resulted in a 53% reduction in AHI. However, no significant difference was found in transplant-free survival, the rate of hospitalization, or quality of life. The study group advised against the routine use of CPAP in patients with HF and CSB. However, in the posthoc analysis of the study, patients with residual AHI < 15/hr on CPAP showed improvement in both left ventricular ejection fraction and heart transplant-free survival. Therefore, the investigators recommended that a one month trial of CPAP be continued only if it suppressed AHI significantly. Based on this study, CPAP should not be routinely used to manage CSB in patients with HF.

Adaptive servo-ventilation (ASV) is a new mode of automated pressure support that performs breath-to-breath analysis and delivers ventilatory needs accordingly, in order to prevent the hyperventilation that drives CSB. Numerous observational studies have demonstrated the beneficial effect of ASV in CSB in patients with HF. Nevertheless, randomized long-term controlled trials are needed to determine the long-term clinical efficacy of these devices. ResMed Company issued on May 2015 a serious safety concern during the preliminary primary data analysis from the SERVE-HF clinical trial. The investigators reported an increased risk of cardiovascular death with ASV therapy for patients with symptomatic chronic HF with reduced ejection fraction (≤45%) and moderate to severe central sleep apnea syndrome (CSAS). Based on that, the AASM recently published updated guidelines indicating that ASV targeted to normalize the AHI should not be used for the treatment of CSAS related to HF in adults with an ejection fraction ≤45% and moderate or severe CSA predominant, sleep-disordered breathing.

2.3 OBESITY HYPOVENTILATION SYNDROME (OHS)

To diagnose OHS, the following criteria must be met:

A. The presence of hypoventilation during wakefulness ($PaCO_2 > 45$ mm Hg) as measured by arterial PCO_2, end-tidal PCO_2, or transcutaneous PCO_2.

B. Presence of obesity (BMI > 30 kg/m^2).

C. Hypoventilation is not primarily due to lung parenchymal or airway disease, pulmonary vascular pathology, chest wall disorder (other than mass loading from obesity), medication use, neurologic disorder, muscle weakness, or a known congenital or idiopathic central alveolar hypoventilation syndrome.

It is important to note that OSA often coexists with OHS, in those cases, the diagnosis of both OSA and OHS should be made. About 90% of patients with OHS have coexisting OSA; therefore, symptoms and many of the physical findings of OHS patients are similar to those in patients with OSA, such as excessive daytime sleepiness, snoring, choking during sleep, morning headaches, fatigue, mood disturbance, and impairments of memory or concentration. However, when compared to eucapnic OSA patients, those with OHS tend to complain more often of shortness of breath.

2.3.1 Diagnosis

OHS is a diagnosis of exclusion and, therefore, many diagnostic tests should be carried out to distinguish OHS from other disorders in which hypercapnia is a common finding, such as pulmonary diseases, skeletal restriction, neuromuscular disorders, hypothyroidism or pleural pathology. Tests should include ABG, pulmonary function tests, chest imaging, laboratory tests, electrocardiography (ECG), transthoracic echocardiogram, and polysomnography. ABG sampling is a key test since hypercapnia is a fundamental feature of the disorder and is required to make the diagnosis. Usually, ABG reveals low PaO_2 and a high bicarbonate level, which reflects the chronic nature of the disease.

Pulmonary function tests (PFTs) are essential to exclude other causes of hypercapnia such as chronic pulmonary diseases. Although PFTs can be normal, they usually reveal mild-to-moderate restrictive pattern due to obesity.

Polysomnography in patients with OHS may show oxygen desaturation and hypercapnia during sleep not related to obstructive apneas and hypopneas periods. Hypoventilation is usually more prominent during REM sleep compared to NREM sleep. If $PaCO_2$ can be monitored, it may also demonstrate an increase of more than 10 mm Hg in the $PaCO_2$ level during sleep, compared with levels during wakefulness.

2.3.2 Management of OHS

Untreated OHS is associated with a high mortality rate, a reduced quality of life, and numerous morbidities, including hypertension, pulmonary hypertension, right heart failure, angina, and acute hypercapnic respiratory failure.

Although there are no treatment guidelines for OHS, treatment approaches are based on reversing the underlying pathophysiology of OHS including the reversal of sleep-disordered-breathing, weight reduction, and treatment of comorbid conditions.

2.3.2.1 Weight Loss

Significant weight loss is desirable in patients with OHS and will lead to improvement in pulmonary physiology and function including improvement in alveolar ventilation and nocturnal oxyhemoglobin saturation. However, it is important to realize that weight loss cannot be used as the sole initial treatment.

2.3.2.2 Positive Airway Pressure Therapy (PAP)

Application of positive airway pressure is the mainstay of therapy for OHS. It seems reasonable to start with CPAP knowing that the majority of OHS patients have coexisting OSA. CPAP has been shown to be effective in a group of patients with stable OHS, especially in those with severe OSA. There are no clear guidelines on when to start or switch to bi-level PAP (BPAP); however, BPAP should be strongly considered in patients with OHS without OSA, and in patients with OHS and coexisting OSA, if CPAP is insufficient and hypercapnia persists despite being on long-term CPAP, or if they fail to tolerate CPAP. In addition, BPAP should be used in patients with OHS who experience acute-on-chronic respiratory failure.

Treatment of OHS with PAP improves blood gasses, this improvement could be achieved in 2 to 4 weeks. Therefore, early follow-up is important and should include repeated measurement of ABG with an assessment of adherence to PAP.

2.3.2.3 Oxygen Therapy

Patients with OHS commonly suffer from prolonged episodes of hypoxemia during sleep, in addition to daytime hypoxemia. Therefore, oxygen therapy is needed if hypoxemia persists despite the relief of upper airway obstruction and hypoventilation with PAP therapy, in order to prevent the long-term consequences of hypoxemia on pulmonary vasculature and other vital organs. However, it is important to keep in mind, that treatment with oxygen alone is inadequate and is not recommended as it does not reverse hypoventilation or upper airway obstruction on its own.

2.4 NARCOLEPSY

Narcolepsy is a relatively rare autoimmune disease. It has pentad of clinical features including:

- Irresistible attacks of sleep, which is usually present in all patients. The other feature of narcolepsy is not present in all patients.
- Cataplexy, characterized by sudden bilateral loss of muscle tone brought on by emotions, which can be limited to certain muscles or generalized, resulting in falling down. Full consciousness during cataplexy. Cataplexy is pathognomonic for narcolepsy and is not present in all narcolepsy patients. If cataplexy is present, the patient has narcolepsy type 1. A diagnosis of narcolepsy without cataplexy (Narcolepsy type 2) is appropriate when excessive daytime sleepiness is present with REM phenomenology (hypnogogic hallucinations and sleep paralysis) but without cataplexy.
- Hypnagogic hallucination: Vivid dreams that occur at the transition from wakefulness to sleep (hypnagogic) or from sleep to wakefulness (hypnopompic).
- Sleep paralysis: It is a temporary inability to move or speak that happens when the patient is waking up or falling asleep.

- Interrupted fragmented sleep: Narcolepsy patients may complain of fragmented sleep.

2.4.1 Diagnosis

History gives good clues to diagnose narcolepsy. To confirm the diagnosis, a patient with narcolepsy undergoes an overnight sleep study (PSG), followed by multiple sleep latency test (MSLT). MSLT starts 1.5–2 hours after waking up in the morning. The patient is given 4–5 chances to nap separated by 2 hours. If the patient falls asleep, he is allowed to sleep for 15 min. Sleep latency and sleep onset REM (SOREM) are monitored. The presence of a short sleep latency (<8 min) and two or more SOREM support the diagnosis of narcolepsy.

Periodic leg movement and restless legs syndrome are common among patients with narcolepsy.

2.4.2 Management

The management of patients with narcolepsy and cataplexy aims to improve daytime sleepiness and control cataplexy. For the irresistible attacks of sleep, behavioral therapy and medication are used. Good sleep hygiene, obtaining enough sleep at night, and strategic naps are used. Strategic naps entail getting short naps for a few minutes when circumstances allow. These short naps increase alertness in patients with narcolepsy for 1–2 hr. Additionally, stimulants are used to reduce sleepiness. The first-line treatment is Modafinil. However, Methylphenidate may be used in patients who do not respond to Modafinil.

For cataplexy, the Serotonin Reuptake inhibitors (SSRI) Fluoxetine, or the Serotonin-Norepinephrine Reuptake inhibitors (SNRI) Venlafaxine are used as first-line treatment. For difficult cases, Sodium Oxybate (Xyrem) can be used.

2.5 RESTLESS LEG SYNDROME (RLS)

RLS is a sensory-motor disorder characterized by unpleasant "creepy-crawly" sensations in the lower limbs. Movement of the legs temporarily relieves these symptoms but disrupts the ability to stay asleep during the night, resulting in delayed or fragmented sleep. The major symptoms of RLS are very disturbing sensations in the limbs (98%), and sleep disturbance is often the primary complaint (95%).

The International Restless Legs Syndrome Study Group (IRLSSG) have suggested four diagnostic criteria: (i) an urge to move the legs, usually accompanied or caused by an uncomfortable sensation in the legs; (ii) beginning or worsening of symptoms during periods of rest or inactivity; (iii) partial or total relief of symptoms by movement; and (iv) symptoms that are worse in the evening or night compared to during the day or that occur only in the evening or night (it follows a circadian rhythm). Roughly, 60% of RLS patients are estimated to have a positive family history; furthermore, genetic association studies have linked 5 genes and 10 different alleles to RLS.

The symptoms of RLS follow the circadian fluctuation of dopamine in the substantia nigra and the putamen. RLS patients have lower dopamine and iron levels in the substantia nigra and, therefore, respond to both dopaminergic therapy and iron administration.

2.5.1 Prevalence of RLS

An RLS prevalence of 3.2–12% has been reported in different countries.

2.5.2 Medical Conditions Associated with RLS

RLS can be primary (idiopathic). This entity usually has a genetic predisposition and is seen more in young people and is usually more difficult to treat. However, secondary RLS has been linked to a number of comorbid conditions, such as iron deficiency, renal failure (uremia), diabetes mellitus, and neurological disorders, such as multiple sclerosis and Parkinson's disease, and rheumatologic diseases such rheumatoid arthritis. Moreover, antidepressants use [such as tricyclics and SSRIs (bupropion is an exception and has not been shown to increase symptoms of RLS)], lithium, antihistamines, and dopamine antagonists have an association with RLS. In some patients, RLS may worsen with nicotine, alcohol, or caffeine.

2.5.3 Diagnosis

There are no specific tests for RLS diagnosis. RLS is a clinical diagnosis based on clinical findings. PSG is not required to diagnose RLS. If iron deficiency is suspected, serum ferritin levels should be obtained.

2.5.4 Treatment

The goal of treatment of patients with RLS is to have uninterrupted sleep with minimal sleep latency. In patients with intermittent RLS symptoms that disturb sleep, treatment may be used on an intermittent basis during symptomatic episodes. In these cases, the dopamine agonist carbidopa-levodopa (Sinemet) at bedtime can use as needed. For severe persistent RLS, dopamine agonists, such as pramipexole or ropinirole can be used 1–2 hour before bedtime. However, it is preferred that this class of drugs should be started by a sleep specialist to titrate the proper dose and avoid side effects. The problem with dopamine agonists is that a good proportion of RLS patients may develop tachyphylaxis to the drugs and, hence, augmentation of symptoms. Recently, the calcium channel 2δ ligands gabapentin and pregabalin have been approved for the treatment of RLS. Augmentation is lower with this class of drugs.

Secondary RLS is dependent on the causative conditions, which once managed, RLS can be cured. Therefore, in RLS patients with iron deficiency, pregnancy, and uremia, symptoms may remit after treatment or resolution of these conditions. For patients with ferritin level <112 picomols/L (50 nanograms/mL), iron treatment can be initially started and ferritin levels monitored. If ferritin levels are >112 picomols/L (50 nanograms/mL) and symptoms persist, they can be treated based on their severity.

FURTHER READING

1. Al-Jawder, S. E., & Bahammam, A. S., (2012). Comorbid insomnia in sleep-related breathing disorders: an under-recognized association. *Sleep Breath.* 16(2), 295–304. Epub 2011/03/30.
2. BaHammam, A. S., Al-Shimemeri, S. A., Salama, R. I, & Sharif, M. M., (2013). Clinical and polysomnographic characteristics and response to continuous positive airway pressure therapy in obstructive sleep apnea patients with nightmares. *Sleep Med.* 14(2), 149–154. Epub 2012/09/11.
3. Kushida, C. A., Littner, M. R., Morgenthaler, T., Alessi, C. A., Bailey, D., Coleman, J., Jr., et al., (2005). Practice parameters for the indications for polysomnography and related procedures: an update for 2005. *Sleep.* 28(4), 499–521. Epub 2005/09/21.
4. Oliveira, M. G., Garbuio, S., Treptow, E. C., Polese, J. F., Tufik, S., Nery, L. E., et al., (2014). The use of portable monitoring for sleep apnea diagnosis in adults. *Expert Rev Respir Med.* 8(1), 123–132. Epub 2013/12/07.
5. Epstein, L. J., Kristo, D., Strollo, P. J., Jr., Friedman, N., Malhotra, A., Patil, S. P., et al. (2009). Clinical guideline for the evaluation, management and long-term care of obstructive sleep apnea in adults. *J Clin Sleep Med.* 5(3), 263–276. Epub 2009/12/08.
6. AlDabal, L., & BaHammam, A. S. (2010). Cheyne-Stokes respiration in patients with heart failure. *Lung.* 188(1), 5–14. Epub 2009/12/04.
7. Carmona-Bernal, C., Ruiz-Garcia, A., Villa-Gil, M., Sanchez-Armengol, A., Quintana-Gallego, E., Ortega-Ruiz, F., et al. (2008). Quality of life in patients with congestive heart failure and central sleep apnea. *Sleep Med.* 9(6), 646–651. Epub 2008/01/22.
8. Sin, D. D., Fitzgerald, F., Parker, J. D., Newton, G., Floras, J. S., & Bradley, T. D. (1999). Risk

factors for central and obstructive sleep apnea in 450 men and women with congestive heart failure. *American Journal of Respiratory and Critical Care Medicine.* 160(4), 1101–1106. Epub 1999/10/06.

9. Vital, F. M., Ladeira, M. T. & Atallah, A. N. (2013). Non-invasive positive pressure ventilation (CPAP or bilevel NPPV) for cardiogenic pulmonary oedema. *The Cochrane Database of Systematic Reviews.* 5, CD005351. Epub 2013/06/04.

10. Bradley, T. D., Logan, A. G., Kimoff, R. J., Series, F., Morrison, D., Ferguson, K., et al., (2005). Continuous positive airway pressure for central sleep apnea and heart failure. *The New England Journal of Medicine.* 353(19), 2025–2033. Epub 2005/11/12.

11. Arzt, M., Floras, J. S., Logan, A. G., Kimoff, R. J., Series, F., Morrison, D., et al., (2007). Suppression of central sleep apnea by continuous positive airway pressure and transplant-free survival in heart failure: a post hoc analysis of the Canadian Continuous Positive Airway Pressure for Patients with Central Sleep Apnea and Heart Failure Trial (CANPAP). *Circulation.* 115(25), 3173–3180.

12. Javaheri, S., Brown, L. K., & Randerath, W. J. (2014). Clinical applications of adaptive servoventilation devices: part 2. *Chest.* 146(3), 858–868. Epub 2014/09/03.

13. ResMed. Important medical device warning. 2015 [cited 2015 6/29/2015]; Available from: http://www.thoracic.org.au/imagesDB/wysiwyg/ServeHFDoctorLetter.pdf.

14. Eulenburg, C., Wegscheider, K., Woehrle, H., Angermann, C., d'Ortho, M. P., Erdmann, E., et al., (2016). Mechanisms underlying increased mortality risk in patients with heart failure and reduced ejection fraction randomly assigned to adaptive servoventilation in the SERVE-HF study: results of a secondary multistate modelling analysis. *Lancet Respir Med.* 4(11), 873–881. Epub 2016/11/02.

15. Aurora, R. N., Bista, S. R., Casey, K. R., Chowdhuri, S., Kristo, D. A., Mallea, J. M., et al., (2016). Updated Adaptive Servo-Ventilation Recommendations for the 2012 AASM Guideline: "The Treatment of Central Sleep Apnea Syndromes in Adults: Practice Parameters with an Evidence-Based Literature Review and Meta-Analyses." Journal of clinical sleep medicine. *JCSM: Official Publication of the American Academy of Sleep Medicine.* 12(5), 757–761. Epub 2016/04/20.

16. Gurski, L. A., Knowles, L. M., Basse, P. H., Maranchie, J. K., Watkins, S. C., & Pilch, J., (2015). Relocation of CLIC1 promotes tumor cell invasion and colonization of fibrin. *Mol Cancer Res.* 13(2), 273–280.

17. Al Dabal, L., & Bahammam, A. S., (2009). Obesity hypoventilation syndrome. *Ann Thorac Med.* 4(2), 41–59. Epub 2009/06/30.

18. BaHammam, A. S., (2015). Prevalence, clinical characteristics, and predictors of obesity hypoventilation syndrome in a large sample of Saudi patients with obstructive sleep apnea. *Saudi Med J.* 36(2), 181–189. Epub 2015/02/27.

19. Bahammam, A. S., & Al-Jawder, S. E., (2012). Managing acute respiratory decompensation in

the morbidly obese. *Respirology. 17*(5), 759–771. Epub 2011/11/05.

20. Aaron, S. D., Fergusson, D., Dent, R., Chen, Y., Vandemheen, K. L., & Dales, R. E., (2004). Effect of weight reduction on respiratory function and airway reactivity in obese women. *Chest. 125*(6), 2046–2052. Epub 2004/06/11.

21. Mokhlesi, B. M. (2010). Obesity hypoventilation syndrome: a state-of-the-art review. *Respiratory Care. 55*(10), 1347–1362, discussion 63–5. Epub 2010/09/30.

22. BaHammam, A. M. (2010). Acute ventilatory failure complicating obesity hypoventilation: update on a 'critical care syndrome'. *Curr Opin Pulm Med. 16*(6), 543–551. Epub 2010/09/11.

23. BaHammam, A. S., Alenezi, A. M. M (2006). Narcolepsy in Saudi Arabia. Demographic and clinical perspective of an under-recognized disorder. *Saudi Med J. 27*(9), 1352–1357. Epub 2006/09/05.

24. Bahammam, A., (2007). Periodic leg movements in narcolepsy patients: impact on sleep architecture. *Acta Neurol Scand. 115*(5), 351–355. Epub 2007/05/11.

25. BaHammam, A. S., Pandi-Perumal, S. R., & Neubauer, D. N., (2015). Sodium Oxybate (Xyrem®): A New and Effective Treatment for Narcolepsy with Cataplexy. In: Guglietta A, editor. Drug Treatment of Sleep Disorders: Milestones in Drug Therapy. Switzerland: *Springer International Publishing. p*, 231–248.

26. Walters, A. S. (1995). Toward a better definition of the restless legs syndrome. The International Restless Legs Syndrome Study Group. *Mov Disord. 10*(5), 634–642. Epub 1995/09/01.

27. Allen, R. P., Picchietti, D., Hening, W. A., Trenkwalder, C., Walters, A. S., & Montplaisi, J. (2003). Restless legs syndrome: diagnostic criteria, special considerations, and epidemiology. A report from the restless legs syndrome diagnosis and epidemiology workshop at the National Institutes of Health. *Sleep Med. 4*(2), 101–119. Epub 2003/11/01.

28. Salas, R. E., Gamaldo, C. E., & Allen, R. P., (2010). Update in restless legs syndrome. *Curr Opin Neurol*. doi, 10.1097/WCO.0b013e32833bcdd8.

29. Facheris, M. F., Hicks, A. A., Pramstaller, P. P., & Pichler, I. (2010). Update on the management of restless legs syndrome: existing and emerging treatment options. *Nat Sci Sleep. 2*, 199–212. Epub 2010/01/01.

30. BaHammam, A., Al-Shahrani, K., Al-Zahrani, S., Al-Shammari, A., Al-Amri, N., & Sharif, M., (2011). The prevalence of restless legs syndrome in adult Saudis attending primary health care. *Gen Hosp Psychiatry. 33*(2), 102–106. Epub 2011/05/21.

31. Rottach, K. G., Schaner, B. M., Kirch, M. H., Zivotofsky, A. Z., Teufel, L. M., Gallwitz, T., et al., (2008). Restless legs syndrome as side effect of second generation antidepressants. *J Psychiatr Res. 43*(1), 70–75. Epub 2008/05/13.

REVIEW QUESTIONS

1. OSA occurs because of:
   A. reduction in the calibre of upper airway
   B. reduction in the calibre of lower airway

C. increased calibre of upper airway

D. increased calibre of lower Airway

2. Cheyene Stokes breathing is characterized by:

   A. crescendo-decrescendo pattern of breathing

   B. ataxic breathing

   C. crescendo-decrescendo pattern of breathing with central sleep apnea

   D. no change in breathing pattern with central sleep apnea

3. Untreated OSA has been found to increase chances of:

   A. rheumatoid arthritis

   B. hypothyroidism

   C. diabetes insipidus

   D. cardiac arrhythmias

4. The mainstay of therapy for moderate to severe OSA is:

   A. life style modification

   B. PAP therapy

   C. position therapy

   D. weight management

5. For the diagnosis of Obesity Hypoventilation Syndrome, BMI should be:

   A. at least 18.5

   B. at least 25

   C. at least 30

   D. at least 35.

6. Cataplexy is characterized by:

   A. sudden loss of muscle tone during wakefulness

   B. sudden loss of muscle tone during sleep

   C. sudden loss of muscle tone usually in response to an emotional stimulus during wakefulness

   D. sudden loss of muscle tone where it can't be recovered for hours after a heavy exercise

7. Diagnosis of RLS is based on:

   A. the PSG data

   B. the level of serum ferritin

   C. the neuroimaging

   D. the history

8. Frequent periodic limb movement during sleep:

   A. may be an incidental finding

   B. seen only in cases of RLS

   C. seen only during childhood

   D. always diagnostic of RLS

9. Polysomnography is not useful to diagnose:

   A. Cheyne-Stokes breathing

   B. obesity hypoventilation syndrome

   C. narcolepsy

   D. insomnia

10. Two or more sleep onset REM periods suggest the diagnosis of:

Common Sleep Disorders

A. idiopathic hypersomnia

B. obesity hypoventilation syndrome

C. narcolepsy

D. delayed sleep wake phase disorder

ANSWER KEY

1. A   2. C   3. D   4. B   5. C   6. C

7. D   8. A   9. D   10. C

INTRODUCTION TO POLYSOMNOGRAPHY

LEARNING OBJECTIVES

After reading this chapter, the reader should be able to:

1. Discuss the principle and utility of polysomnography.
2. Understand the basis of differentiating between various polysomnography techniques.
3. Recognize various devices with their advantages and disadvantages.
4. Understand the indications and contraindications of Home Sleep testing.

Contents

3.1 Why Is Polysomnography Required? 39
3.2 Types of Sleep Studies and Sleep Monitoring Devices 40
3.3 Advancement of the Machines 45
3.4 Advantages and Limitations of Various Sleep Studies 45
3.5 Guidelines for the Use of Sleep Studies at Home/Out-of-Center.. 46
3.6 Concluding Remarks 49
Further Reading 49
Review Questions 49
Answer Key 51

For a very long time, the human race has remained interested in the mysteries of sleep. Reference to sleep can be seen in the texts and scriptures across diverse religions. Sleep has been described in ancient Hindu texts, such as Vedas,

Upanishads, and Puranas; the ChristianBible and in the Islamic text, Quran. These books reinforced the need for good sleep and described the omens of bad sleep. Thus, it appears that our ancestors knew about the importance of sleep and the adverse effects of sleep disorders. However, these religious scriptures discuss sleep disorders in a cryptic manner with religious flavor and thus, the ancient knowledge regarding sleep was difficult to decipher and transform to clinical practice.

A sleep study is used to objectively assess changes in physiological parameters that occur during sleep. First, the objective monitoring of sleep became possible after the discovery of the EEG by the German Psychiatrist, Hans Berger, in 1924. He was the first person to record and demonstrate the cortical electrical activity via electrodes applied to the human scalp. He could depict alpha activity during wakefulness and slowing of waves during sleep. A group of researchers from Harvard Medical School and University of Chicago (USA) described the features of NREM sleep between the years 1935–1938. At the same time, researchers kept experimenting with the filters of the EEG to get clearer signals and using different channels to record other bioelectrical potentials, such as electrocardiography, body movements, and respiration along with EEG. However, it took another fifteen years to describe the REM sleep. Two researchers, Eugene Aserinsky and Nathanial Kleitman, developed the electrooculogram and published their findings regarding REM sleep in 1953. Atonia during REM sleep was described by Michael Jouvet in 1959 in cats and, hence, it was proposed that an electromyogram should be recorded during sleep study. Thus, it was recommended to include the EEG, EOG, and EMG channels during polysomnographic recording.

In 1959, the Pickwickian syndrome was first described using the EEG, breathing, and pulse. In 1965, Kulho et al expanded our knowledge regarding the Pickwickian syndrome, using the EEG, and respiratory movements monitoring via a belt, heart rate, and carbon-di-oxide content during expiration. One year later, i.e., in 1966 apnea and sleep fragmentation were described after monitoring the oro-nasal airflow, chest wall movement, and EEG. Thus, you can appreciate that the development of sleep medicine and polysomnography took a long time and both the fields are still evolving. As our knowledge expanded, we kept including electrodes to obtain more data that would help us to understand sleep physiology and the associated pathological changes. It was in 1974, that the term polysomnography (PSG) was used for the first time to describe the simultaneous recording of an EEG, EOG, EMG, and respiratory channels.

Since then, we have witnessed a great development in the technology—size of the machines has reduced to a great extent; specificity and sensitivity of the channels have improved owing to technical advancement; instead of the analog paper and ink-pen based recording, we now have computerized software-based recordings that allow us to get digital signals, and to tailor the recording according to our need.

In this chapter, we will discuss a few basic issues:

- Need for polysomnography;
- Types of polysomnography machines available;
- Advantages and limitations of each of these machines; and
- Guidelines for the use of various devices.

3.1 WHY IS POLYSOMNOGRAPHY REQUIRED?

The functioning of almost all body systems, such as brain, heart, respiratory system, gastro-intestinal system, genitor-urinary system, endocrinal system, and musculoskeletal tone, changes between two states of consciousness, that is, wakefulness and sleep. Even during sleep, the functioning does not remain static and keeps changing, and it is influenced by/gives rise to different sleep stages. For example, during NREM sleep, respiration become stable, heart rate slows and peripheral muscle tone reduces. During REM sleep, respiration and cardiac activity become erratic, peripheral muscle tone is lost and periodic twitches appear in the muscles. There is a noticeable change in the EEG as well. As the person drifts into sleep, EEG activity begins to slow down and certain characteristic waveforms start appearing, for example, vertex waves, sleep spindles, K complexes, delta waves or low-amplitude, mixed-frequency EEG. These EEG waves, in addition to the information from muscle tone and eye channels, help us to determine sleep stages.

At times, these physiological functions get disrupted and give rise to different disease states. This disturbance may be limited to either state of consciousness, that is, wakefulness or sleep, or at times, may be seen during both stages. For example, exercise-induced cardiac ischemia may remain limited to the state of wakefulness. On the other hand, some pathological processes are seen only during sleep. These conditions may, (i) interfere with the initiation or maintenance of sleep, for example, insomnia; (ii) lead to abnormality in one of the physiological functions during sleep, for example, sleep apnea, sleep-related laryngospasm; (iii) be associated with movements during sleep, for example, restless legs syndrome/periodic limb movement disorder, sleep seizure, night terrors, and REM sleep behavior disorder; (iv) lead to excessive sleepiness, for example, narcolepsy, idiopathic hypersomnia, and Kleine-Levine syndrome. Lastly, some pathologies are present during wakefulness but further deteriorate during sleep, for example, non-apnic hypoxemia during sleep in COPD patients. The latter two entities that are seen during sleep interfere with the normal sleep process, and in such a situation, it is termed as sleep disorder.

During polysomnographic recording, we gather data regarding physiological parameters and try to detect any abnormality. Thus, polysomnography provides objective evidence of different pathologies occurring during sleep. In addition, it also helps us in measuring the severity of some sleep disorders (sleep apnea using apnea-hypopnea-index or respiratory disturbance index; periodic limb movement disorder (PLMS) by measuring PLMS index) and treatment of sleep disorders (for obstructive sleep apnea—positive airway pressure titration study). Lastly, polysomnography may be used to measure the effect of treatment in certain conditions, for example, REM-sleep-behavior-disorder and sleep apnea. Besides offering medical help to patients, polysomnography is an extremely useful research tool to understand the physiological changes occurring in various organs of the body, for example, brain, cardiovascular system, respiratory system, muscular system, and upper GI tract during sleep.

3.2 TYPES OF SLEEP STUDIES AND SLEEP MONITORING DEVICES

Depending upon the number of channels that record various physiological parameters and the availability of a qualified sleep technician during the recording, sleep studies may be divided into four major types:

Level IV: This is the elementary machine that contains only one or two channels for the recording of at least one or two respiratory parameters throughout the night. Thus, it records one of the following parameters during sleep—oxygen saturation or respiratory flow, or both signals. The channel is mounted at the appropriate place where it remains throughout the night, and next day, the tracing can be deduced from the data. This machine can be used for screening of obstructive sleep apnea. However, it is not approved by the AASM for diagnosing obstructive sleep apnea.

Level III: These devices have at least a minimum of four channels including, ventilation (at least two channels of respiratory movement or respiratory movement airflow), oxygen saturation respiratory effort, oxygen movement, and airflow), heart rate or ECG, and oxygen saturation (Figure 3.1).

Thus, they have following channels:

- Pulse oximeter
- Nasal airflow
- Chest or abdominal movements
- Electrocardiogram

Level II: This type is also called comprehensive portable polysomnography. It contains all channels that are recorded in the Level I sleep study, except for the video. However, a sleep study is not attended by a sleep technician and can be done at the patient's home. These devices have EEG channels, electrooculogram, and chin electromyogram; in addition to the respiratory monitoring channels and ECG. Thus, the total number of channels is at least seven and may be expended:

- Pulse oximeter.
- Nasal airflow: recorded by either thermistor, pressure transducer or both.
- Chest and abdominal movements: one belt for each.
- Electrocardiogram: At least 2 electrodes that can record any of Leads I, II or III.
- Electro-encephalogram: at least 2 channels of EEG are present. Active electrodes are placed at either of the following positions: frontal, central, and occipital on both sides of the head—right side and left side. They are usually referred to the opposite mastoid area and thus one electrode is placed on the right mastoid and the other on the left mastoid. These channels are sufficient for the determination of sleep stages and arousals.

However, most available monitoring devices provide an option for recording from other areas of the brain, for example, prefrontal and temporal. Thus, devices actually have provision for 24–32 channels of EEG, and thus, they may be used for monitoring sleep-related seizures and parasomnia.

Figure 3.1 Level III devices from various manufacturers.

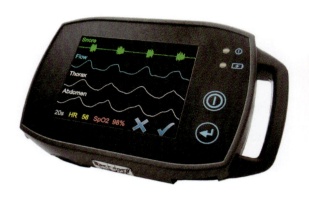

A: Device from Somnomedics.

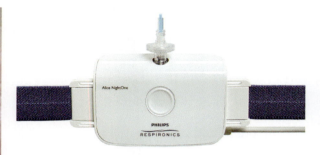

C: Device from Philips Respironics.

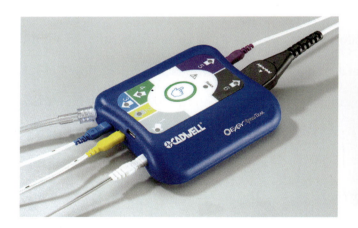

B: Device from Cadwell.

- Electro-occulogram: 2 electrodes; one for the right eye and the other for the left eye. They are usually referred to the mastoid electrode of the opposite side.
- Electromyogram: At least 1 channel for recording chin electromyogram; two are placed on the anterior tibialis muscles of each leg.
- Body position.
- During these studies, Auto-PAP may be used for the titration of PAP pressure in patients with sleep apnea.

Level I: This study is done using devices that are used in type II study. The only difference is that level I sleep study is done in the sleep laboratory and a sleep technologist attends the whole study. However, to gather more information than type II study provides, the following channels are added:

- Video recording that is synchronized to the recording of other data.
- Audio recording that is synchronized to the recording of other data.

In addition to the diagnosis of sleep disorders, this study type allows manual titration of positive airway pressure therapy for patients with sleep-related breathing disorders. Thus, it has a:

- Channel for PAP machines that are utilized during manual titration of PAP in patients with sleep apnea.

The following are the optional channels that may be added depending upon the need:

- Capnograph: End-tidal carbon dioxide or using finger capnograph.
- Channel for pH monitoring in the pharynx: helps in detecting nocturnal gastro-esophageal reflux disease (GERD).
- Esophageal pressure monitor: Its use is limited to the research purposes. It can reliably differentiate between central and obstructive sleep apnea.

Level II: Sleep studies are cumbersome and impractical and, therefore, are used mainly in research.

According to the AASM (2007), portable device must be capable of displaying the raw data for review by the clinician, in order to allow assessment of the quality of the data. Moreover, portable sleep studies are approved for patients with high clinical likelihood of moderate to severe OSA. These devices are not approved to diagnose patients with central sleep apnea, obesity hypoventilation syndrome or other sleep disorders. Additionally, they have not yet been approved for children and elderlies (>65 years) (Figure 3.2).

Figure 3.2 Level I devices from various manufacturers.

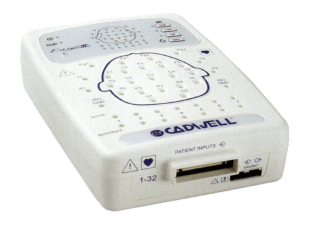

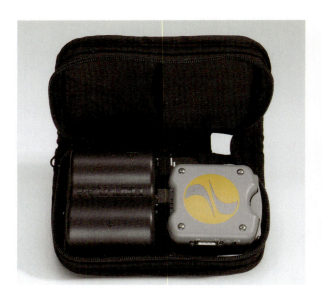

A: Easy III device from Cadwell.

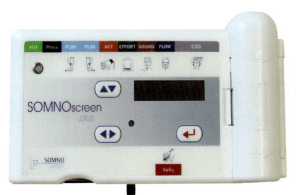

B: Somnoscreen device from Somnomedics.

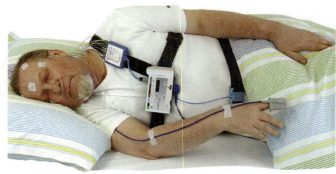

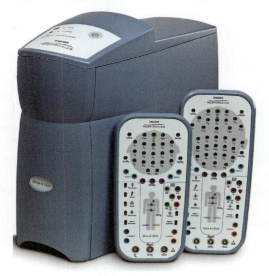

C: Alice 6 from Philips Respironics.

3.3 ADVANCEMENT OF THE MACHINES

In the past few years, considerable advances have taken place with regards to the devices and their software. This has changed the classical definition of the devices. Now some devices fall between type III and type II devices. For example, some of the manufacturers have devices that in addition to the above channels record the snoring or the ambient light, or have 1–2 EEG channels or the body position sensor. In addition, many devices have the facility for generating data from surrogate channels, for example, respiratory movements from the electrocardiogram or snore microphone or snoring from the nasal cannula.

Based on the recent advancement in devices and methods used to capture data, a new classification system was proposed by the AASM in 2011. This system classifies the devices based upon the channels that they have and this has been done specially for the home-sleep-testing. It is known as the SCOPER system where S stands for Sleep, C stands for Cardiovascular parameters, O stands for Oximetry, P stands for Position, E stands for respiratory Effort, and R stands for Respiratory flow. Then, each of these parameters is rated between 1 and 5, depending upon the technique used for the collection of data.

A considerable progress has been made in type II and types I devices as well. Most of the devices that are used for type I and II studies have a cable connection between the head box and the central processing unit for storing the data. With traditional machines, the cable has to be disconnected when the patients want to go to the washroom; if the patient is making any vigorous movement during sleep, as we see in sleep-related seizures and parasomnias, the leads may fall off. Both the conditions result in loss of data as well as limit the mobility of the patient.

Now, devices are available that offer telemetry recording of data. These machines improve the mobility of the patient and the data continues to be recorded even when the patient is mobile. Of course, with vigorous movements, leads may fall off, as happens with traditional monitoring devices (Figure 3.3).

3.4 ADVANTAGES AND LIMITATIONS OF VARIOUS SLEEP STUDIES

So far, it is clear that different types of sleep studies use different kinds of devices and they provide different information. Thus, each of them has limitations and certain advantages. These are depicted in Table 3.1.

3.5 GUIDELINES FOR THE USE OF SLEEP STUDIES AT HOME/OUT-OF-CENTER

The AASM has proposed guidelines for the use of various types of sleep studies. According to the guidelines, Level III studies, which are also known as Home Sleep Testing (HST) or Out-of-center (OCC) studies, can be done in high-risk patients to verify the presence of sleep apnea. However, a negative test does not rule out OSA, and if the clinical suspicion is high, patients should be subjected to level I (in-lab attended) polysomnography. However, when performing an HST, it must be ensured that:

- A trained sleep physician has evaluated the patient comprehensively. Report of the HST/OCC should be made in the background of the clinical evaluation.
- Sleep physician must have access to the raw data generated out of HST/OCC.
- Patient has a high pre-test probability of having OSA.
- Patient is not bale to attend laboratory because of immobility or any critical illness.
- Patient has been educated regarding placement of sensors by a trained sleep physician or sleep technologist.

HST/OCC cannot be used if:

- The patient has congestive heart failure, neuromuscular abnormalities. or severe COPD that may compromise the quality of data.
- The patient has co-morbid or exclusively suffering from sleep disorders other than OSA.

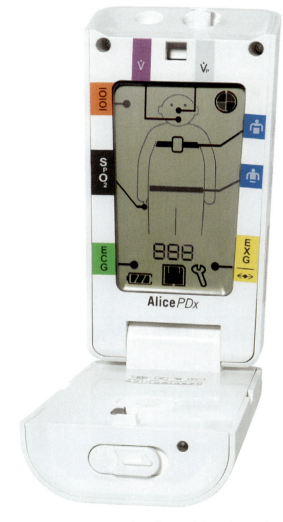

Figure 3.3 Devices that do not fall in classical I-IV level category: Alice PDX from Philips Respironics.

Introduction to Polysomnography

Table 3.1 Types of Sleep Studies and Their Utility

Type of Sleep Study	Advantages	Limitations
Type IV	• Simple to use • Least expensive • Can be done at home	• Calibration not possible • Data quality not ensured as usually patient himself applies the channels • Because of limited channels, artifacts like coughing, movement, talking can not be recognized and may lead to falsification of data • Chances of false negative results if the patient has not slept • Loss of data if the probe gets misplaced during night • Cannot be used for any other condition except for screening of obstructive sleep apnea • Underestimation of the severity of sleep apnea as it is calculated by 'time in bed' rather than 'total sleep time' • Can be used for verification of sleep apnea in high-risk patients
Type III	• Simple to use • Less expensive • Can be done at home • Can pick central sleep apnea and Cheyne-Stokes breathing • Can pick sleep-related systole and other arrhythmias	• Calibration not possible • Data quality not ensured as usually patient himself applies the channels • Because of limited channels, artifacts like coughing, movement, talking can not be recognized and may lead to falsification of data • Chances of false negative results if the patient has not slept • Cannot be used for any other condition except for screening of sleep apnea • Can not provide the optimal information regarding severity as arousals cannot be monitored • Underestimation of the severity of sleep apnea as it is calculated by 'time in bed' rather than 'total sleep time' • Can be used only to 'rule in' OSA but not to 'rule out'. Thus, can be used for 'verification' of sleep apnea in high-risk patients

Table 3.1 (Continued)

Type II	• Less expensive than Type I • Can be done at home • Can pick data regarding sleep maintenance, leg movements during sleep in addition to respiratory and cardiac data	• Loss of data if the probe gets misplaced during night • Can not be reliably used for the diagnosis of parasomnia and sleep related seizures as video and audio channels are not available • Requires a trained technician for hooking up the patient
Type I	• No loss of data as it is attended by a sleep technician • Can be used reliably for the diagnosis of most of the sleep disorders • Can be used to measure sleepiness during the day through multiple sleep latency test/maintenance of wakefulness test • Can be used for the therapeutic purpose, for example, manual titration of PAP devices • Data from a number of parameters is available so artifacts and incidental findings can be recognized • Can be used reliably for research purpose	• Most expensive • Requires a trained technician for hooking up the patient and during PAP titration • Used for 'discovering' the underlying sleep disorders

- The machine does not provide raw data to be verified by a trained sleep technologist/sleep physician.

HST/OCC can be a good technique to:

- Monitor the progress of non-PAP therapies for OSA.

3.6 CONCLUDING REMARKS

To summarize, polysomnography is a good method to assess sleep and sleep disorders. It is useful not only for the clinical diagnosis but for the research as well. Choice of appropriate parameters and technology, along with the availability of trained sleep technician and certified sleep physician, can bring a remarkable change in the quality of life of patients suffering from sleep disorders as well as help in the advancement of knowledge in this field.

FURTHER READING

1. Collop, N. A., Anderson, W. M., Boehlecke, B., Claman, D., Goldberg, R., Gottlieb, D. J., Hudgel, D., Sateia, M., & Schwab, R. (2007). Clinical guidelines for the use of unattended portable monitors in the diagnosis of obstructive sleep apnea in adult patients. *J Clin Sleep Med.* 3(7), 737–747.
2. Littner, M. R. (Ed). (2011). Home portable monitoring for obstructive sleep apnea. *Sleep Medicine Clinics.* 6(3), 261–386.
3. Collop, N. A., Tracy, S. L., Kapur, V., Mehra, R., Kuhlmann, D., Fleishman, S. A., & Ojile, J. M. (2011). Obstructive sleep apnea devices for out-of-center (OOC) testing: technology evaluation. *J Clin Sleep Med.* 7(5), 531–548.
4. Haba-Rubio, J., & Keriger, J. Evaluation Instruments for Sleep Disorders: A brief history of polysomnography and sleep Medicine. In: Introduction to Modern Sleep Technology. Chiang, R. P. Y., & Kand, S. C. (eds.). Springer; Netherlands: 2012, pp. 19–31.
5. Kapur, V. K., Auckley, D. H., Chowdhuri, S., Kuhlmann, D. C., Mehra, R., Ramar, K., & Harrod, C. G. (2017). Clinical Practice Guideline for Diagnostic Testing for Adult Obstructive Sleep Apnea: An American Academy of Sleep Medicine Clinical Practice Guideline. *J Clin Sleep Med.* Jan 31. pii: jc-17-00035.

REVIEW QUESTIONS

1. Gold standard for the sleep study is:
    A. level IV study
    B. level III study
    C. level II study
    D. level I study
2. For screening of obstructive sleep apnea following may be used:
    A. home sleep testing
    B. actigraphy

C. pulmonary function testing

D. peak flow meter

E. pulse oximetry

3. End tidal CO_2 is useful for the diagnosis of:

   A. Cheyene Stokes breathing among children

   B. obesity hypoventilation syndrome among adults

   C. obstructive sleep apnea among children

   D. daytime hyperventilation among children

4. For the patient with suspected parasomnia best sleep study will be:

   A. attended level I polysomnography

   B. home sleep testing

   C. attended level I polysomnography with extended EEG montage and video

   D. attended level I polysomnography with extended EEG montage

5. Home sleep testing is not accurate for assessment of OSA because:

   A. it may give a false negative result or spuriously increased severity of illness

   B. it may give a false negative result or spuriously reduced severity of illness

   C. it may give a false positive result or spuriously reduced severity of illness

   D. it may give a false positive result or spuriously increased severity of illness

6. Polysomnography is required to differentiate between:

   A. obesity hypoventilation syndrome and obstructive sleep apnea

   B. insomnia and hypersomnia

   C. RLS and PLMS

   D. narcolepsy and idiopathic hypersomnia

7. To monitor the progress of PAP therapy following is used:

   A. level I PSG

   B. level II PSG

   C. level III PSG

   D. level IV PSG

8. Telemetry recording of data during PSG is advantageous in:

   A. accurately depicting the EEG waves during parasomnia/seizure

   B. depicting the respiratory waveform during OSA

   C. leg movements during PLMS

   D. ECG during cardiac asystole

9. Synchronized audio-video recording helps in:

   A. diagnosing obesity hypoventilation syndrome

   B. Catathrenia

   C. diagnosing habitual snoring

D. treating seizure disorder

10. Minimum number of EEG channels required to score sleep wake stage:

    A. 4

    B. 3

    C. 2

    D. 1

ANSWER KEY

1. D   2. A   3. C   4. C   5. B   6. D

7. C   8. A   9. B   10. D

4
BASIC CONCEPTS OF POLYSOMNOGRAPHY CHANNELS

LEARNING OBJECTIVES

After reading this chapter, the reader should be able to:

1. Discuss the electrical potentials of cells including resting membrane potential, Inhibitory post synaptic potential, excitatory postsynaptic potential, and action potential.
2. Discuss the mechanisms underlying the generation of various potentials.
3. Discuss mechanism of synaptic and neuromuscular communication.
4. Define electricity, polarity, and the concept of dipole.
5. Understand the principle of use of amplifier and filters during polysomnography recording.

Contents

4.1	Electrical Potentials of the Neurons, Muscles, and Heart	54
4.2	Electrical Concepts	60
4.3	Concepts Related to Digitalized Recordings	77
4.4	Physiology and Recording of Electrical Potentials	85
4.5	Respiratory Data	98
4.6	Body Position	110
4.7	Video Data	110
4.8	Carbon Dioxide Monitoring (PCO_2)	111
Further Reading		112
Review Questions		112
Answer Key		113

6. Discuss the importance of sampling rate and bit resolution
7. Discuss the sources of EEG, EOG, EMG and EKG signals
8. Differentiate between various measures used to monitor airflow and respiratory efforts
9. Understand the concept of oximetry and capnography
10. Discuss the mechanism of body position sensor
11. Discuss the importance of audio-video recording during polysomnography

Since Level I polysomnography is the most informative and gold-standard diagnostic test for sleep disorders, we will focus our discussion on it in this book. Polysomnography records a number of activities, some of them are electrical in nature and recorded as such, for example, EEG, EOG, EMG, and ECG; some of them are mechanical but converted into electrical activity during data acquisition, for example, respiratory flow, chest, and abdominal movements during respiration and body position. Similarly, chemical changes occurring in the body (oxygen saturation and end-tidal CO_2) are converted into electrical signals during the process of recording. These changes are necessary to visualize the data in clear waveforms/numbers for easier interpretation.

In this chapter, the basic concepts for each of these channels will be reviewed. To understand some of the technical aspects of different concepts, one needs to be informed regarding certain physiological functions. Hence, in the initial section, an electrical potential will be discussed with reference to neurons, muscles, and heart.

4.1 ELECTRICAL POTENTIALS OF THE NEURONS, MUSCLES, AND HEART

4.1.1 Resting Membrane Potential

Grossly, we can divide the body into two compartments, intracellular and extracellular. Both compartments are separated by cell membranes. Water and electrolytes are present in our body, both inside the cell and outside the cell, but in different concentrations. The different concentrations across the cell membrane of the three main ions, Na+, K+, and Cl– create a potential that is known as the resting membrane potential. This resting membrane potential is present across all cells such as neurons (unit of the brain and peripheral nervous system) and myocytes (unit of muscles that form various organs of the body including skeletal muscles and heart). The resting membrane potential of neurons is −90 mV, skeletal myocytes are −80 mV, and cardiac myocytes is −85 mV, that is, outside of the membrane is more negative than inside (Figure 4.1).

4.1.2 Action Potential of the Neuron

Different cells inside the body talk to each other to make the organ properly function. For example, neurons talk to each other whenever we learn, discuss, or recall. Similarly, when we make any movement, signals from the motor area of the brain travel to the muscles and communicate with them to contract or relax. Neurons are

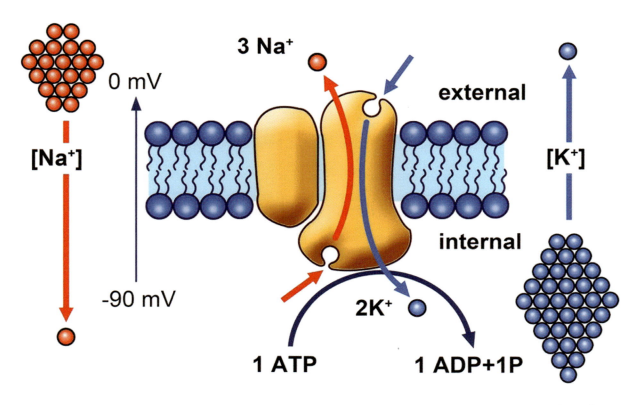

Figure 4.1 Resting membrane potential across cell membrane: Resting membrane potential (RMP) is produced by different concentrations of anions and cations across cell membrane. Each ion produces its potential difference and final RMP is sum of all potentials. In general three ions are most important owing to their concentration and have maximum contribution to the RMP. These are sodium, potassium and chloride ions. A pump known as sodium-potassium pump maintains the concentration of these ions across cell membrane in a narrow range to maintain resting membrane potential.

interconnected through synapses while neurons connect to muscles via motor end plates.

When a neuron gets a signal from the preceding neuron, neurotransmitters are released into the synapse from the axon of the preceding neuron (Figure 4.2A). These neurotransmitters, for example, adrenaline, acetylcholine, glutamate, or GABA get attached to neurotransmitter-specific receptors containing ion channels on the postsynaptic membrane and open them, which allows ions to pass through the postsynaptic neuron and create a local potential. Excitatory neurotransmitters, such as acetylcholine and glutamate open channels that increase the flow of Na+ and Ca++ ions inside the cell and make the resting membrane potential less negative. This change in potential is called as excitatory postsynaptic potential (EPSP). Once the EPSP reaches a threshold value, say +20 mV (in other words, neuronal membrane potential reaches to −70 mV from the resting membrane potential of −90 mV), it opens up the Na+ channels in the axon of neurons that are sensitive to changes in the membrane potential, known as the voltage-sensitive Na+ channels. Then, more Na+ enters inside the neuron and the membrane potential start rising (becomes less negative) with the opening of many voltage-sensitive sodium channels. This change in the neuronal membrane potential is known as the depolarization phase of the action potential. However, the neuron has to return back to its resting membrane potential and to accomplish this, the voltage-sensitive K+ channels open, but with a slight delay compared to the voltage-sensitive Na+ channels. This process allows K+ to move outside cells, thus reducing the positivity inside them, and leads to a fall back of potential towards the resting state. This is known as the repolarization phase of the action potential. However, all these events take place during milliseconds, and the duration of one action potential is between 0.5–1 msec. After the action potential, Na+ and K+ are brought back to their normal concentration across the membrane with the help of the Na+-K+ pump that is present in the cell membrane. It sends three Na+ outside the cell membrane along with bringing two K+ inside the cell to restore the resting membrane potential (Figure 4.2B).

On the other hand, inhibitory neurotransmitters such as GABA act on the GABA channels that allow Cl− to enter inside the neurons and, hence, make the inside of the cell more negative. This is known as the inhibitory postsynaptic potential (IPSP). Like EPSP, it also changes the resting membrane potential towards more negative and by hyperpolarizing the cell, it inhibits neurotransmission.

Basic Concepts of Polysomnography Channels

The action potential that starts in the initial part of the axon traverses till the terminal part of the axon where it brings changes in the permeability of the cell membrane towards Ca++ by activating the voltage-gated Ca++ channels. This is followed by the entry of calcium inside the axon terminal and, in turn, the activation of the vesicles that contain neurotransmitters. They bind with the cell membrane and release their content (neurotransmitter) in the synapse. Released neurotransmitters act on the post-synaptic cell and, thus, the cycle gets repeated.

Thus, the EPSP, IPSP, and action potential change the potential outside the cell. If this process occurs in a small group of neurons, it will produce a difference in the membrane potential of that area relative to surrounding area forming a dipole (see below). These changes can be picked at the surface during an EEG.

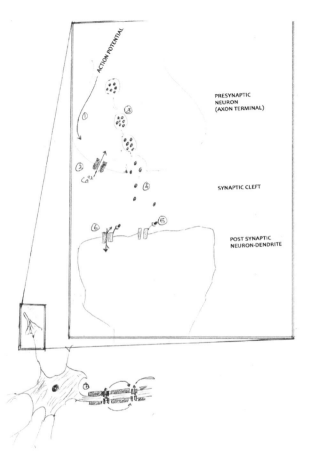

Figure 4.2A Steps in generation of action potential: 1. Action potential in the presynaptic neuron opens the calcium channels in presynaptic axon. 2. Calcium ions move inside presynaptic axon. 3. Entry of calcium ions move vesicles filled with neurotransmitters to the end and makes them fuse with presynaptic membrane. 4 Neurotransmitters released in synaptic cleft. 5. Neurotransmitter attach to receptors on postsynaptic dendrites. 6. Attachment of neurotransmitters to its receptors brings a conformational change and open ion channels. Movement of ions generates local postsynaptic potential that may be either excitatory or inhibitory. 7. Excitatory postsynaptic potential opens channels in axon hillock and generate action potential that travels to end of axon through movement of ions across cell membrane.

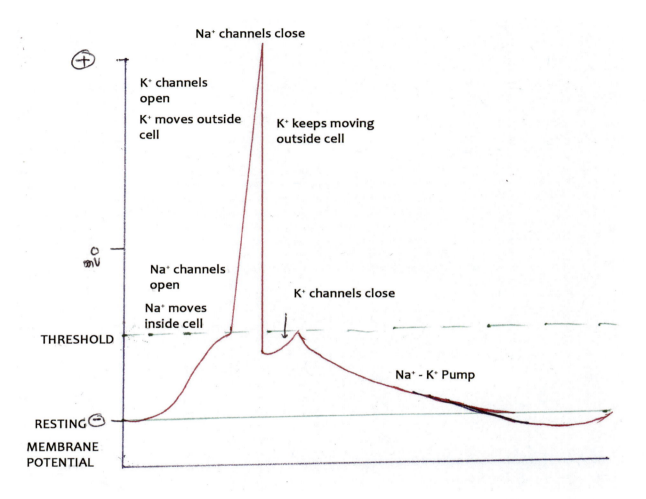

Figure 4.2B Ion changes during action potential: Action potential is best understood as a spike of depolarization (less negativity) that is dependent upon the movement of ions across cell membrane. Various phases of action potential are depicted in this picture.

4.1.3 Communication Between Nerves and Muscles

Muscles are made up of small cells that are known as myocytes. These myocytes contain two types of filaments—actin and myosin that have contractile property. These are supplied by motor nerve fibers and both are connected through a motor end plate. Thus, the motor end plate is made up of an axon of a motor nerve, a synapse, and a myocyte. The axon releases acetylcholine in the synapse when an action potential reaches its end, as discussed above. Acetylcholine gets attached to its receptor on the end plate and opens the acetylcholine-gated channels through which Na+ and Ca++ enter the myocyte. This process changes the resting membrane potential to a relatively more positive potential (to –40 mV from the resting membrane potential of –80 mV). This is known as end plate

Basic Concepts of Polysomnography Channels

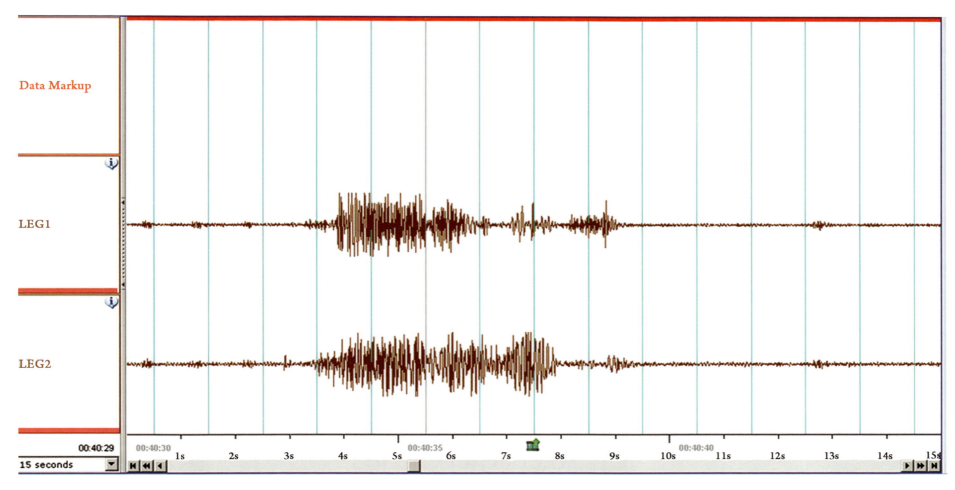

Figure 4.3 Muscle contraction in EMG: A single muscle contraction appears as multiple deflections; each deflection is produced by a small number of motor units.

potential. Similar to nerve fibers, it opens the voltage gates Na+ channels and an action potential is generated. This action potential traverses inside the myocyte through a structure called T-tubules, which contain Ca++ in high concentration. The action potential opens the voltage-sensitive calcium channels in the T-tubules to release Ca++ inside. This Ca++ binds with the actin and myosin to induce contraction.

It must be remembered that a motor (as well as sensory) nerve is made up of multiple nerve fibers, with each nerve fiber representing one axon (or dendrite). In the muscles, one nerve fiber supplies many myocytes ranging from few units to several hundred myocytes. Myocytes that get stimulated by stimulation of one nerve fiber are called a motor unit because they contract together upon stimulation of one nerve fiber. Thus, during contraction of a muscle, multiple motor units work together. Changes in the concentration of ions across the cell membrane produce a localized change in potential that can be picked up through a surface EMG. Since all motor units do not contract together, multiple deflections are seen in an EMG during a single contraction (Figure 4.3).

4.1.4 Potential Generation Across Cardiac Muscles

The mechanism of cardiac muscles contraction is similar to that discussed for skeletal muscles.

4.2 ELECTRICAL CONCEPTS

4.2.1 Basic Concepts of Electricity

Current is represented as (I) in the electrical literature. It can be of two types: alternating current (AC) where the current produces a sine wave with negative and positive polarities, and direct current (DC) where sine waves are not produced. During DC flow, electrons move in a single direction while during an AC current, electrons change their direction with time resulting in oscillations. These oscillations are measured as they occur in a unit time (second) and are expressed as frequency and measured in Hertz (Hz). This is an important concept to understand the filter setting.

Further, the potential difference between two electrodes placed on any charged surface is recorded in Volts (V). Since electrodes are placed over the body, they are not able to assimilate with the underlying surface. Therefore, some resistance is always present during measurement of electrical activity in the underlying area. This resistance is denoted as R and measured in Ohms (Ω). Voltage equals

Basic Concepts of Polysomnography Channels

current multiplied by resistance (V=IR). This means that increasing resistance will lower the potential difference (V) for a given current. It is important to remember this concept while placing leads.

The concept of resistance applies to the DC circuit. For some reasons, that may not appear interesting to medical persons, resistance in an AC circuit is denoted as impedance (Z) despite being measured in Ohms. Hence, for AC circuits the relationship between the three factors is displayed as V = IZ.

4.2.2 Dipole

Dipole refers to two points where one has a negative charge and the other has a positive charge. Positivity of the other is relative to the first. For example, if point A is having -90 mV charge (electrical potential) and point B carries the potential of -70 mV, B is considered positive relative to A (because it has +20 mV difference as compared to A). As the electrical current flows from the negative to the positive side, the direction of current will be from A to B. If we attach a galvanometer to these points, it will show a deflection.

4.2.3 Amplifiers

These devices serve two purposes: amplification and differential discrimination.

As their names suggest, these are the devices that amplify the signals coming to them. This is important because signals generating inside the body are so tiny that it is nearly impossible to detect them. This is quantified as amplifier gain, which is denoted as the ratio between input and output voltage (Vout/Vin). Let us consider an example: an amplifier is set to increase the incoming voltage by a factor of 1000 (or 1 µV of incoming potential is amplified to a signal of 1 mV), then the gain, in this case, would be 1000. It is also expressed in decibels (dB) in some literature. The sensitivity of an amplifier is calculated as the ratio of input voltage to the vertical amplitude of waveform and is measured as µV/mm. In other words, how many micro-volts are covered in an amplitude of 1 mm of the waveform? As expected, the vertical waveform decreases in size with increasing sensitivity. Sensitivity is chosen so as different waveforms appear distinctly in leads, at the same time, waveforms from nearby channels do not overlap with each other (Figure 4.4 A-E).

4.2.3.1 Differential Discrimination

Differential discrimination refers to the capacity of the amplifier to detect the potentials at two inputs and to reject the potential that is identical at two places. The body is a conductive environment

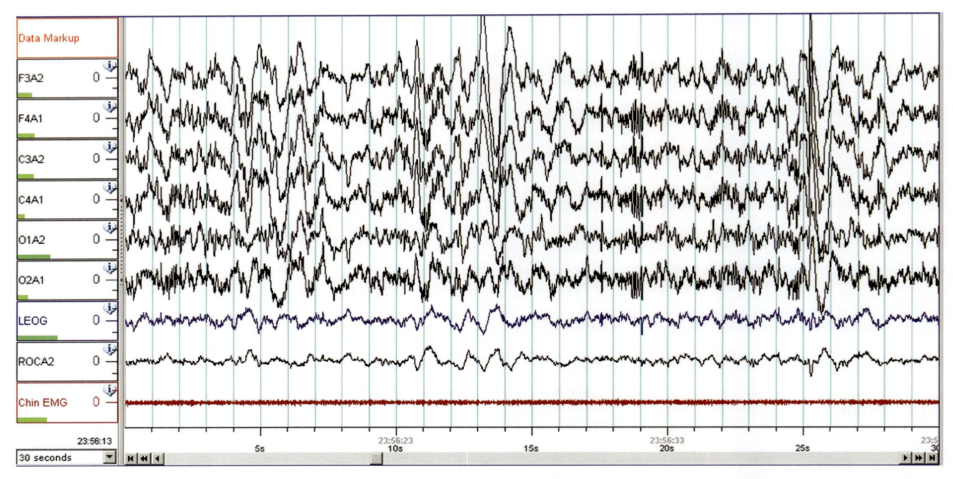

Figure 4.4A Amplitude changes with sensitivity: Amplitude of the waveform in EEG is dependent upon the sensitivity. Lower the sensitivity, higher the amplitude. This can be manually set in an epoch. This epoch shows EEG waves with sensitivity of 7 mV/mm.

Basic Concepts of Polysomnography Channels

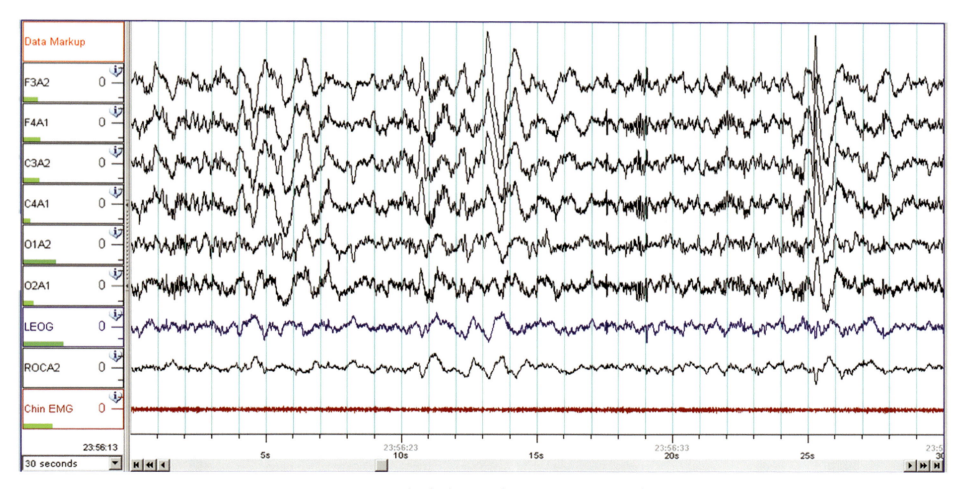

Figure 4.4B Amplitude change with sensitivity set at 10 mV/mm.

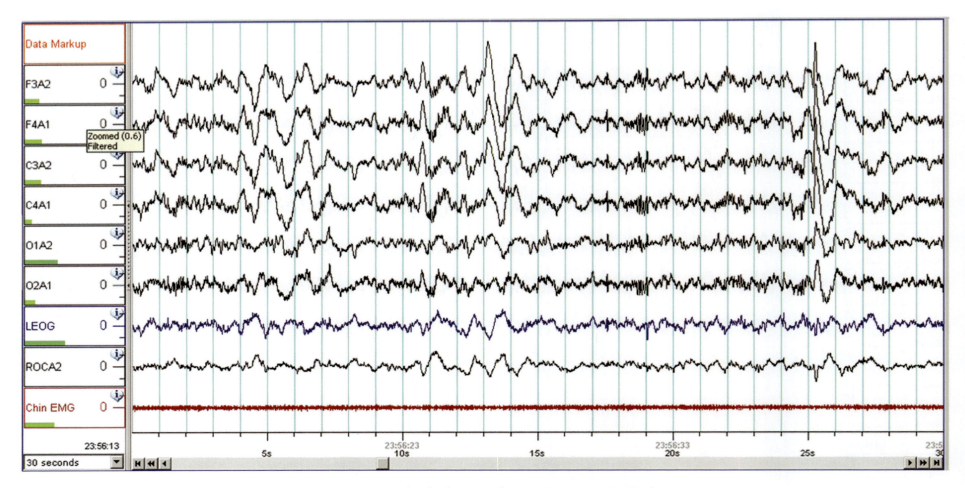

Figure 4.4C Amplitude change with sensitivity set at 12.5 mV/mm.

Basic Concepts of Polysomnography Channels

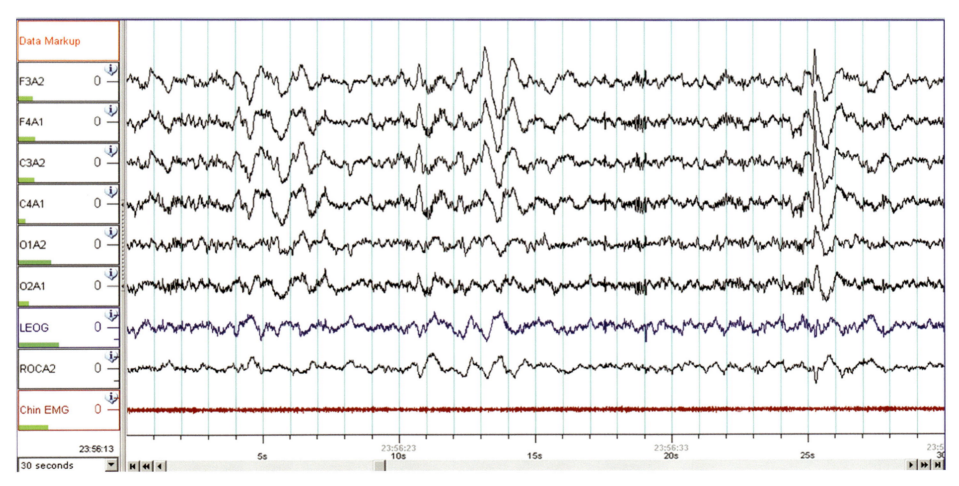

Figure 4.4D Amplitude change with sensitivity set at 15 mV/mm.

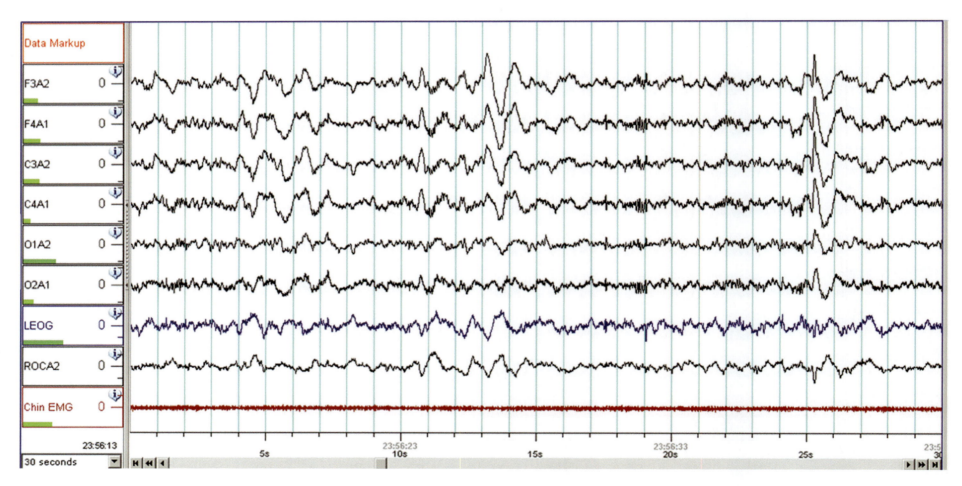

Figure 4.4E Amplitude change with sensitivity set at 17.5 mV/mm.

Basic Concepts of Polysomnography Channels

where a number of organs are functioning and emitting electrical currents. Most of this current is generated in the nerves (somatic or autonomic) or the muscles. For example the electrical activity of the heart is picked up as an ECG and muscles also discharge electrical potentials that can be measured with the help of an EMG, and the brain is continuously working and generating electrical potentials that can be measured as an EEG. Since, salt water is a good conductive medium and is present throughout the body—within and outside the cells, these potentials traverse the whole body (Figure 4.5A). Although their strength (measured as the amplitude of the wave in Volts) decreases as we go farther from the organ that is emitting that current. It results in the appearance of physiological artifacts in the electrical channels of the polysomnography (Figure 4.5B). To remove these artifacts, we use the devices that filter all potentials that are equi-distributed in

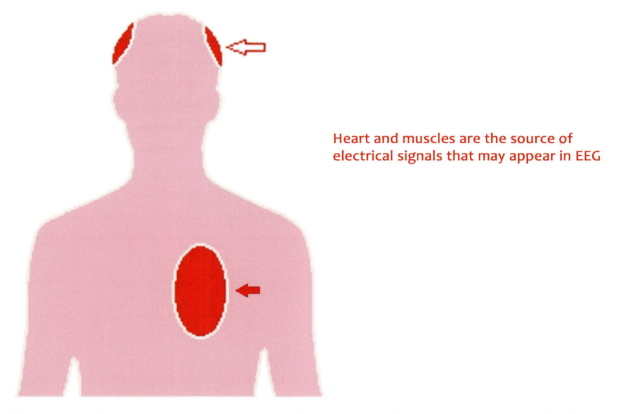

Heart and muscles are the source of electrical signals that may appear in EEG

Figure 4.5A Electrical potentials inside the body: Electrical potentials are generated by almost all cells in the body. However, those of large amplitude traverse to distant areas. These potentials may produce artifacts in EEG. These may be seen as changes in EEG during muscle contraction and EKG artifacts.

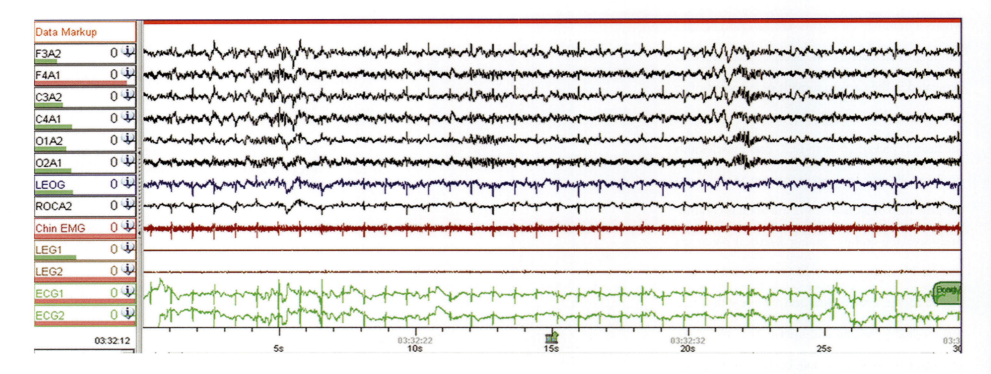

Figure 4.5B ECG artifacts in the EEG derivations: ECG artifacts in the EEG and chin EMG derivations. They can be recognized by their regularity and synchronized appearance with QRS complex in ECG derivations.

all channels. This is done by discriminative amplifiers, which recognize the potentials with similar and opposite polarities. All potentials with similar polarity at two inputs are rejected and that of opposite polarity are allowed to pass through (Figure 4.6). This phenomenon is known as common mode rejection (CMR) and is quantified as the common mode rejection ratio (CMRR), which is the proportion of input voltage to the output voltage of a discriminative amplifier. A good discriminative amplifier will be having a high CMRR, which means that the amplifier will reject the common voltage at two inputs more efficiently. However, the efficiency of a discriminative amplifier depends upon the voltage that is entering at the two inputs, which in turn, is dependent on the electrical activity of the underlying source, in addition to the impedance at the two inputs. Different amplitudes at two inputs will reduce the CMR. In other words, a gross

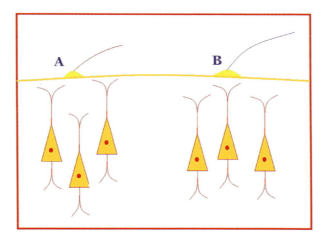 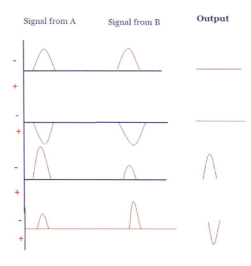

Figure 4.6 Common mode rejection: Potential of same phase are cancelled while that of opposite polarity are added through common mode rejection. Potentials depicted in top two waveforms have same amplitude hence they get canceled. However, potentials depicted in lower two waveforms have same polarity but different amplitudes. Hence, signals from one electrode are more positive or negative than another, and hence, are considered as having different polarity.

discrepancy of impedance at the two inputs will increase the artifacts. Another reason for increased artifacts could be a loose ground electrode, because of which the device will not be able to sense a 'set' reference potential.

4.2.4 Polarity

Polarity refers to the appearance of the final output as either negative or positive. It depends on the relative difference in the potential between two electrodes from which recordings are acquired. If the input to both electrodes is positive then the output is positive; if the input to one is negative and to the other is positive, then the output is the arithmetic sum of the two, and if it is negative in both, then final output is negative (Figure 4.7). Conventionally, positive polarity in the EEG is represented as downward deflection and vice versa.

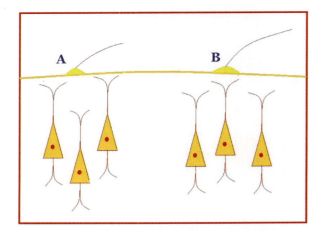
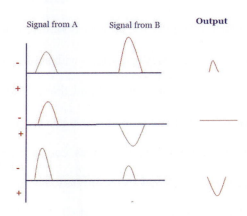

Figure 4.7 Potential difference between two electrodes defines final output: Output potential is determined by difference of the potentials between two points. One point acts as the reference and other as active electrode. In this diagram A is the reference electrode and B is the active electrode.

Depending upon the reference electrode used, channels may be either unipolar (better termed referential) or bipolar. Unipolar electrode refers to the use of a distant electrode as a reference electrode (e.g., scalp electrode referred to an electrode outside an electrically active area viz., mastoid), and bipolar refers to the measurement of the potential difference between two electrodes that are placed on electrically active area, for example, frontal and central. Because of the polarity, the output from unipolar and bipolar electrodes will be different (Figure 4.8 A and B).

4.2.5 Ground

Grounding is done for two purposes: to increase the safety of the patient by grounding the machine and to improve the quality of signal acquisition by grounding the patient.

Since machines are attached to an electrical source for the power and patients are connected to machines through electrodes, any short-circuit inside the machine may allow the current from an external source to enter inside the patient's body and causing the injury. Hence, proper grounding of the machine is of paramount importance.

Another grounding, that is, (patient's grounding) helps in providing a common reference point to the machine to find out the actual potential difference between the two electrodes in question (see common mode rejection). Hence, the ground electrode should be placed at a site where we expect minimal endogenous electrical activity. One such site is the forehead and it is usually placed at the FPz location. Placement of the ground electrode at this location lacks the activity of the cerebral hemispheres. In addition, because of its proximity to the eyes, a ground electrode helps in finding out the unequal impedance.

4.2.6 Filters

As the name implies, filters are used to attenuate the waves of a particular frequency from the tracing. Notch filter helps in attenuating electrical interference of 60 Hz that enters into the channel (Figure 4.9A-C). Once it is turned on, it removes electrical interference from all leads. However, it is important to realize that this filter should not be used routinely as it may obscure important signals. As we have discussed earlier, poor grounding can cause the appearance of 60 Hz artifact in the tracing. If the filter is kept turned on since the outset, this information may be missed and we may not obtain good quality data.

Filters are used to improve the display of waves of physiological interest and to minimize

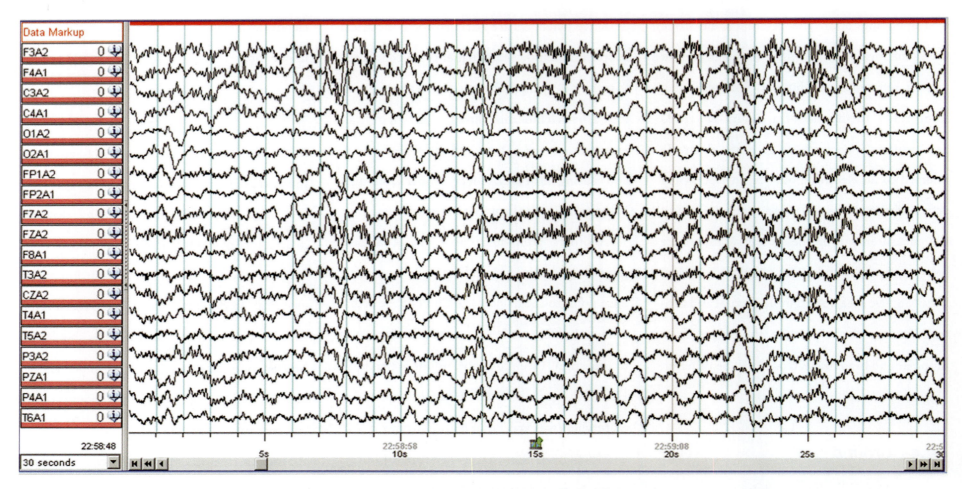

Figure 4.8 Output of same epoch from referential (A) and bipolar (B) montages. A: referential montage all derivations are referred to Auricular electrode of opposite side is shown in this illustration. Compare it with B on next page.

Basic Concepts of Polysomnography Channels

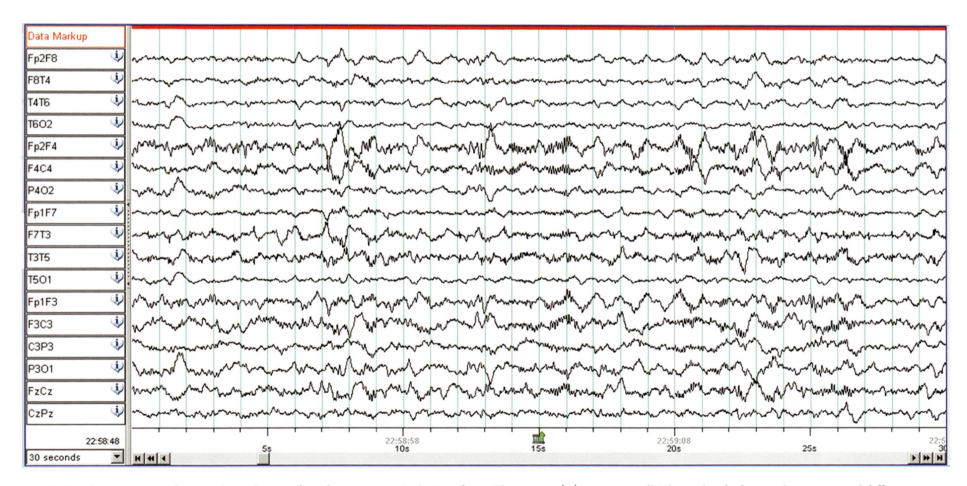

B: Bipolar montage: Adjacent channels are referred anteroposteriorly. In referential montage (A) waves are of high amplitude due to a large potential difference.

74 Clinical Atlas of Polysomnography

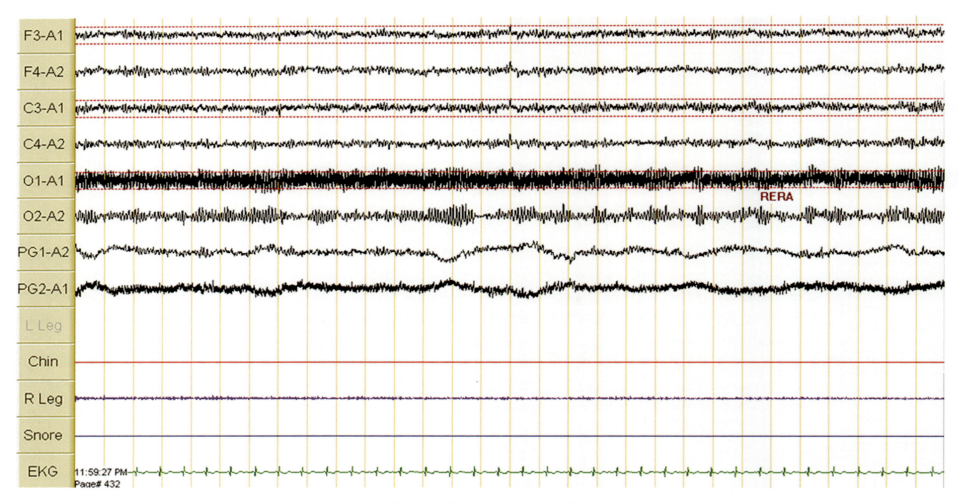

Figure 4.9A Due to high impedance electrical signals appear in O1 and PG2.

Basic Concepts of Polysomnography Channels

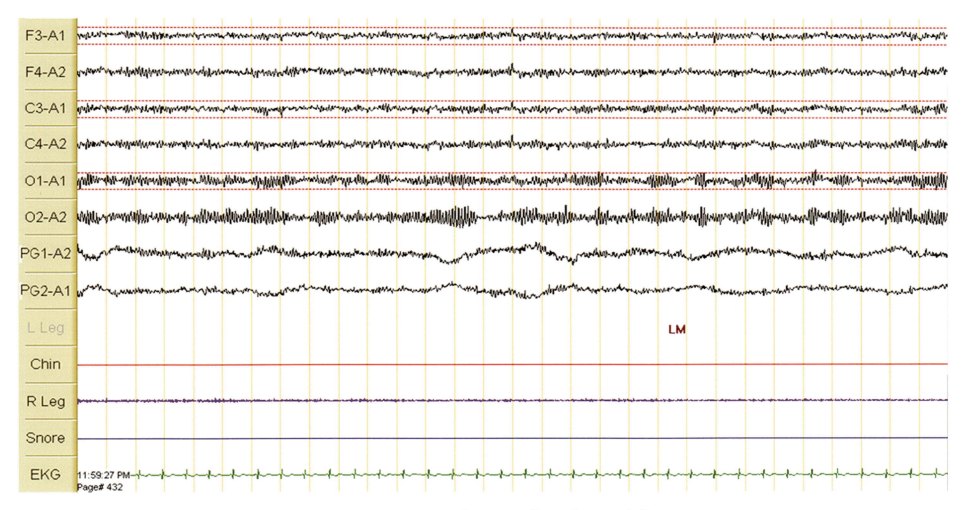

Figure 4.9B Can be improved by applying notch filter

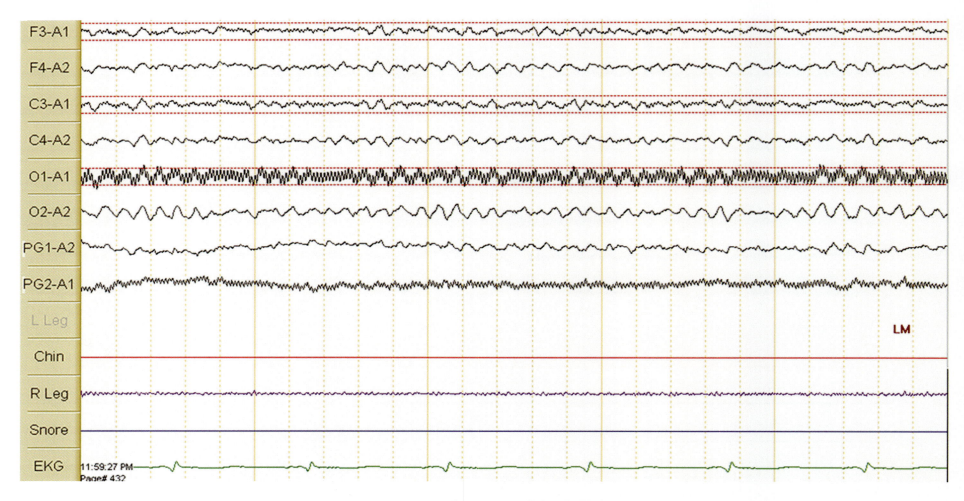

Figure 4.9C Tracing shown in epoch A is shown in 15 seconds.

artifacts. Hence, filter setting varies according to the type of lead under consideration. This can be understood by paying attention to the wave generated during calibration (Figure 4.10). During calibration, a constant current is applied to the filters and is usually of 50 µV. As a result, a sudden deflection followed by a gradual diminution of the wave is seen. The time spent in falling back of the signal to 37% of its maximum amplitude (peak) is known as "time constant," and it has an inverse relationship with low-frequency filters (LFFs) (Figure 4.11). Therefore, increasing the low-frequency filter will reduce the time-constant, and as a result, a waveform will die prematurely. LFFs diminish the low-frequency waves by nearly 30% in order to improve the display of waveforms of frequencies higher than that. That is why it is also known as high-pass filter. For example, in EEG leads, LFF (high-pass filter) is usually kept at 0.35 Hz since we are interested in waves above this frequency, especially for sleep staging. However, in certain cases where the sweating artifact is distorting the tracing, this filter may be set a bit higher, for example, to 1 Hz to minimize the distortion (Figure 4.12 A-D).

Similarly, high-frequency filter (HFF) defines a cut-off frequency where waves having a frequency higher than this cut-off frequency will be attenuated to minimize their display in the tracing. Hence, waveforms of physiological interest are displayed better. For an EEG, it is usually kept at 70 Hz, although the AASM recommends a setting of 35 Hz.

4.3 CONCEPTS RELATED TO DIGITALIZED RECORDINGS

These days most of the sleep laboratories are using computer programs designed to gather, store, and score data. A few other concepts that are unique to the digitalized data are discussed here.

The sampling rate is defined as the frequency at which an analog signal is captured and converted into a digital signal. A low sampling rate (i.e., less frequent acquisition of signal from the body) may induce malformed output (known as aliasing), leading to incorrect reporting (Figure 4.13). Thus, the sampling rate of 200Hz in the EEG means that this signal will be captured 200 times in a second, before converting it to digital format. To obtain a digital signal similar to the analog signal, it is recommended to capture the digital signal at least six times the frequency of the original signal. Aliasing is difficult to recognize, hence, it is advised to keep the recommended settings of filters and signaling rate before starting the recording. In the modern studies, aliasing may also emerge because of poor resolution of the monitor. Monitor Aliasing refers

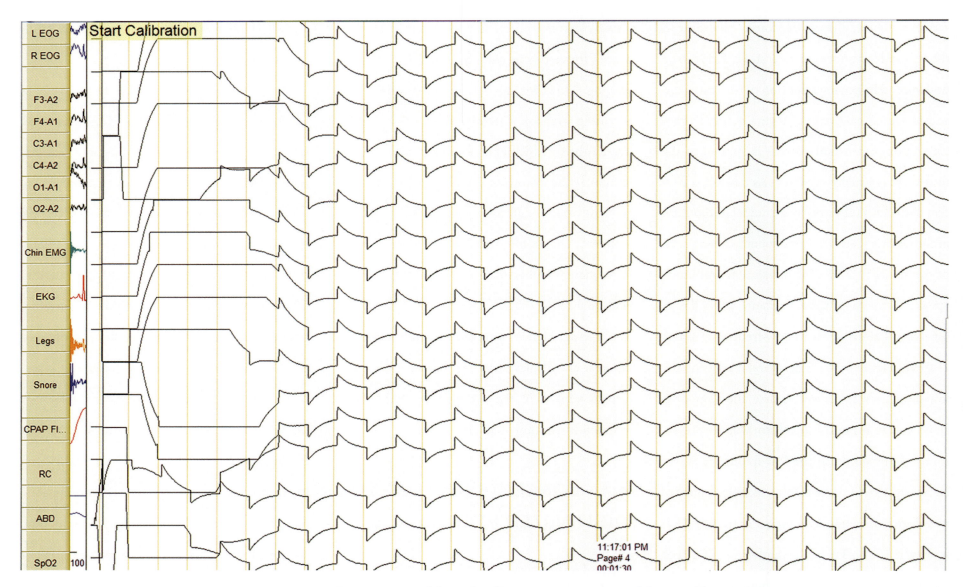

Figure 4.10 Waves generated through calibration: Waves generated during calibration.

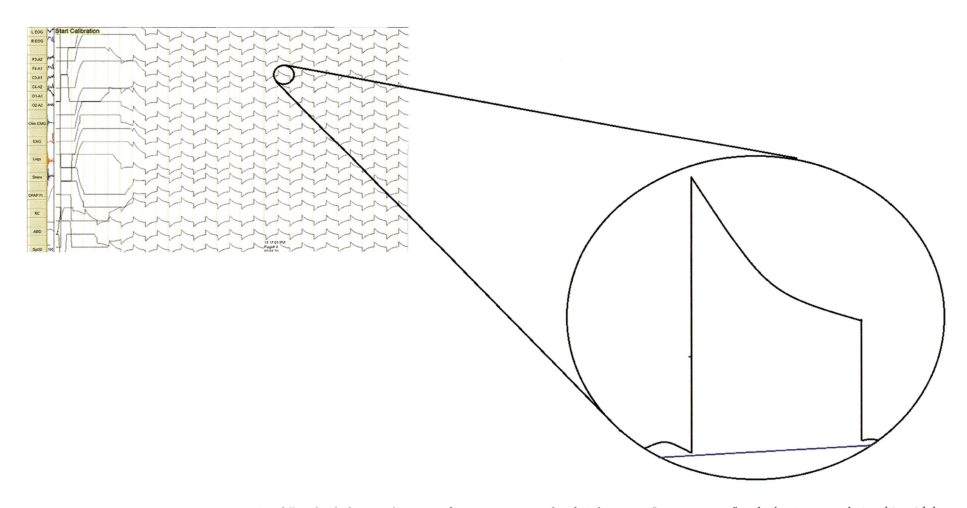

Figure 4.11 Time constant: Time spent in falling back the signal to 37% of its maximum amplitude is known as "time constant" and it has inverse relationship with low frequency filter. This figure has been taken from Figure 4.10.

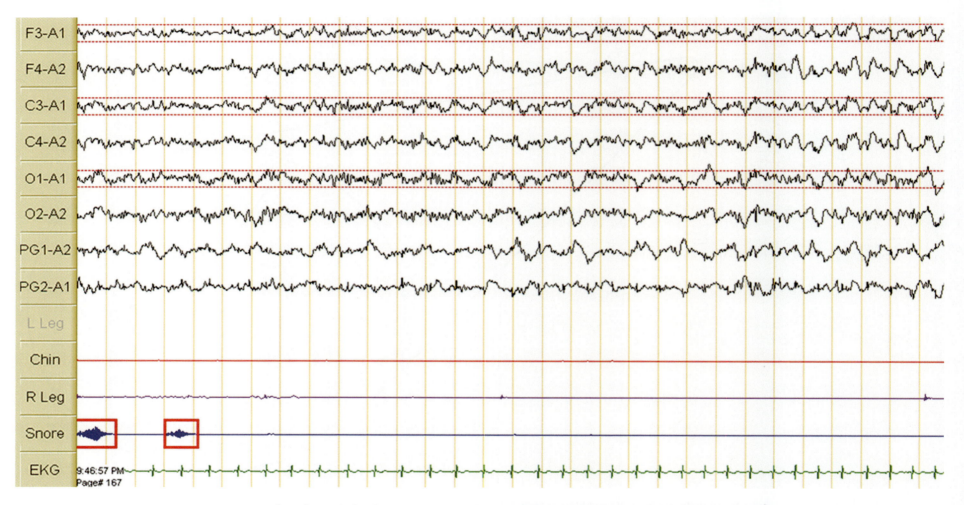

Figure 4.12(A-D) Epochs showing EEG signals using different filters. A: N2 stage with HFF70 LF 0.3.

Basic Concepts of Polysomnography Channels

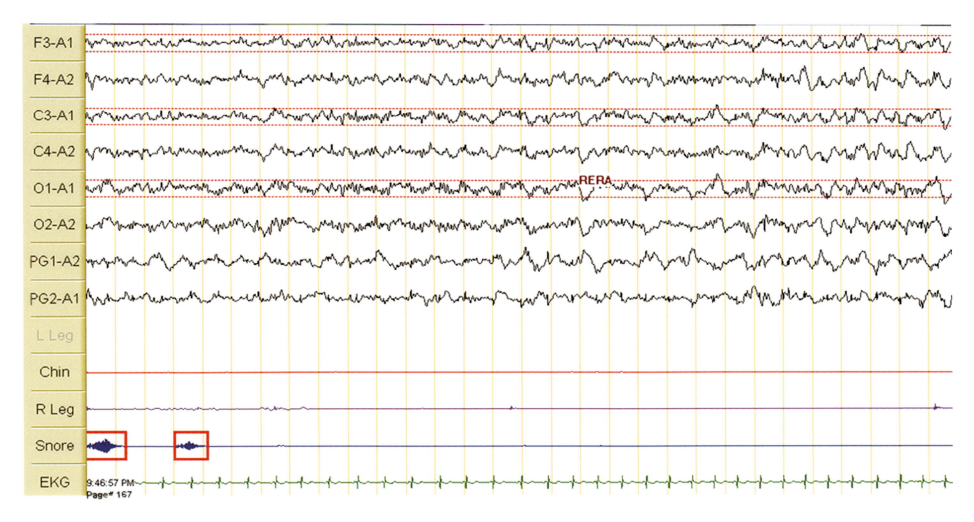

B: Same epoch with HFF 35 LFF 0.3.

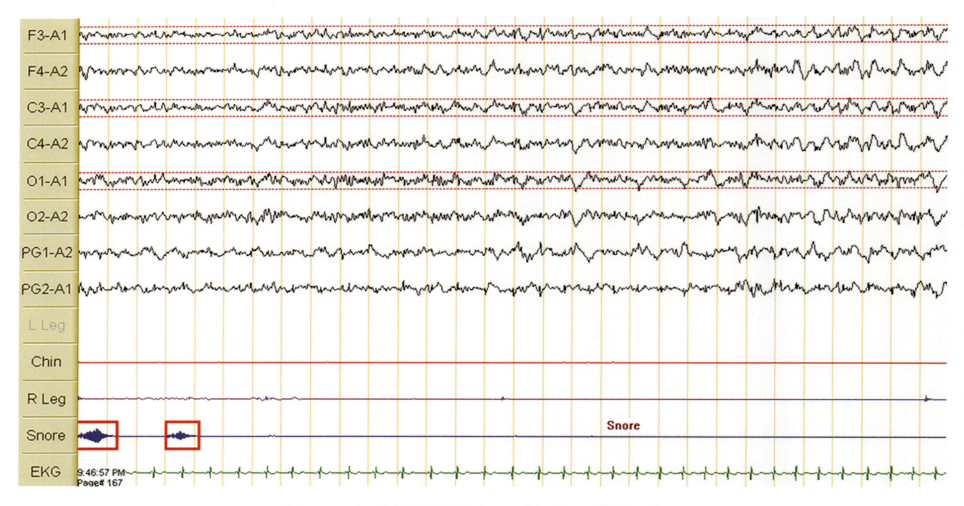

C: Same epoch with HFF 35 LFF 1 decreased dangling of EEG waveform.

Basic Concepts of Polysomnography Channels

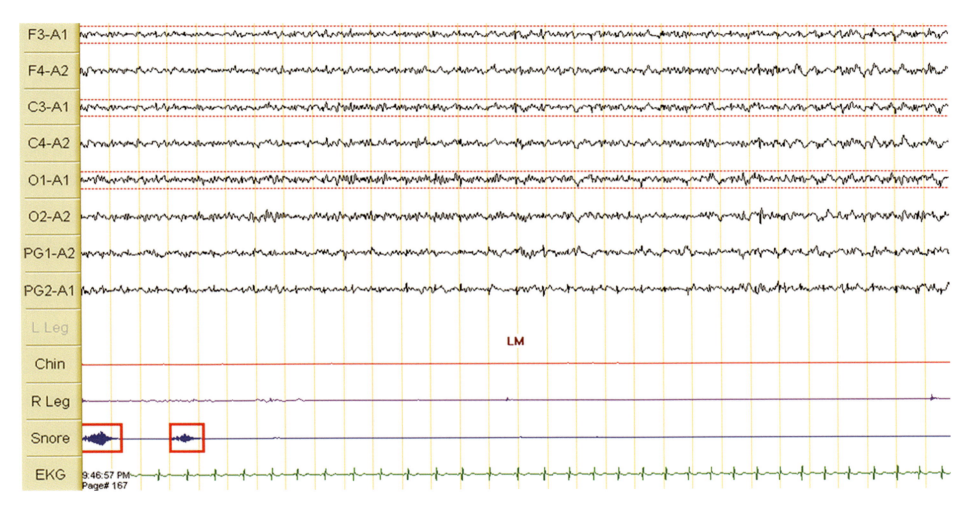

D: Same epoch HFF 35, LFF 3 theta waves disappeared.

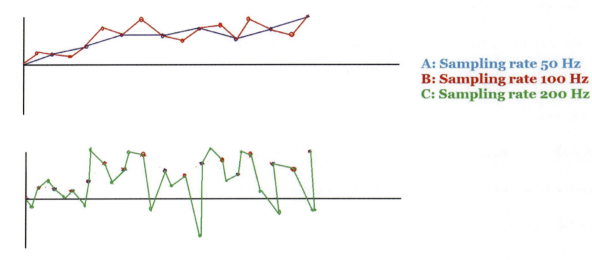

Figure 4.13 Waveforms depends upon the signal capturing rate: Low sampling rate can distort the signals while higher sampling rate improves the waveform.

to the change in visual clarity of waveforms owing to the resolution of the monitor. Computer monitors vary in the resolution (measured in pixels) and a monitor with lower resolution will not be able to show good quality signals. Hence, an optimum resolution monitor (at least 1600 x 1200 pixels) is necessary for good reporting.

Bit resolution refers to the number of available bits that will represent the analog signal into a digitalized form. For example, a system of 16 bits means than 2^{16} bits are available for representing a signal and that comes to 65536. With this information, we can calculate how many bits are available for depicting a unit voltage. Since most amplifiers allow signals between 5 volts on either side (positive and negative side, coming to a value of 10 V), one bit can depict a signal amplitude of 10/65536 = 0.000152 volts or 0.15 mV. Thus, any signal below this voltage will not change the amplitude or be depicted on the monitor.

4.4 PHYSIOLOGY AND RECORDING OF ELECTRICAL POTENTIALS

Electrical activity is recorded using surface electrodes that are placed over the area of interest. To improve the contact between the surface and the electrode (which reduces impedance), a conductive gel is applied after scrubbing the area to remove the dead skin. During the recording of an electrical activity, we actually measure the potential difference between two electrodes and the result is displayed as a waveform. In contrast to what has been mentioned earlier in the section of potentials, when electrical data are collected from the scalp (EEG), eye (EOG), muscles (EMG) and heart (ECG) during a sleep study, we do not get the activity from a single cell. Rather, the EEG depicts the sum of electrical activity occurring in all the cortical neurons that underlie the electrode (local potential, that is, the sum of all inhibitory and excitatory potentials). Similarly, we see the sum of electrical changes induced by the heart in an ECG and from muscles in an EMG (see Section 4.4.3). It must be remembered that an electrode records the electrical signals generated inside the body and, hence, during the study, we do not allow any external current to flow inside the body.

4.4.1 Electroencephalogram (EEG)

4.4.1.1 Source of Electrical Signals

The brain continues to discern stimuli from the external as well as the internal environment; hence, neurons keep working. Working, in other words, refers to the cross-talk between the neurons to pass on the processed information. Neurons are interconnected through synapses and information is transmitted from one neuron to another via neurotransmitters. In response to the action potential, neurotransmitters released from presynaptic neurons induce two kinds of potentials in the postsynaptic neurons: an excitatory postsynaptic potential (EPSP) and an inhibitory postsynaptic potential (IPSP) (Figure 4.2A-B). An excitatory neurotransmitter produces an EPSP while an inhibitory neurotransmitter produces an IPSP by opening different channels (Na^+ or Ca^{++} and Cl^- or K^+, respectively). These potentials are the major source of EEG signals owing to their slow activity. Pyramidal cells lie in the outermost layer of the cortex. Their dendrites are close to the outer side of the cerebral cortex. Hence, postsynaptic potentials of the pyramidal dendrites is the major source of EEG signals—non-synaptic activity, for example— action potentials of neurons are fast enough to be picked by the EEG; and hence, action potentials per se do not contribute to the EEG signals (Figure 4.14). However, when action potentials get synchronized across neurons as occurs during epilepsy and sleep-transient activity, they contribute to the EEG. In addition, intra-neuronal events, such as sub-threshold oscillations, movement of calcium across dendritic membranes and after-potentials and movement of ions across glial cells produce a dipole and contribute to the EEG signals.

Neuronal dendrites receive axons from subcortical structures, for example, reticular activating system and thalamus, in addition to axons from other areas of ipsilateral and contralateral cortex. Incoming synchronized stimuli from these fibers produce the synchronized postsynaptic potentials (excitatory or inhibitory) in the dendrites that are seen in the EEG as waveforms. Duration and

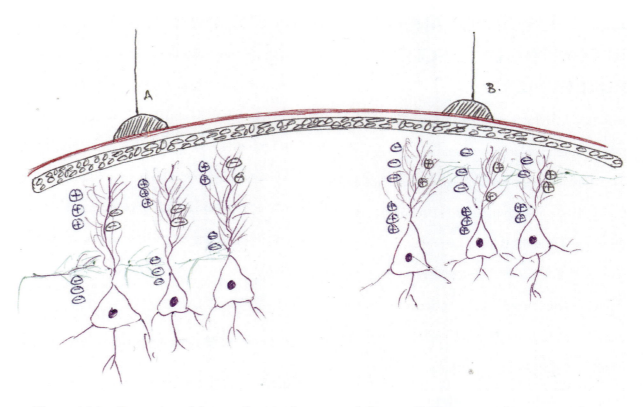

Figure 4.14 Generation of the waveform in electroencephalogram: Each neuron in the cortex can be in different phase of electrical polarity depending upon the incoming stimulus. Potential under any electrode is determined by sum of all positive and negative potentials in the area. For area A in this illustration, proximal to scalp surface, positive charges are more in number leading to overall positivity of area A. Opposite is true for area B, and hence, surface electrode will show negative polarity. Electrodes overlaying scalp record surface potential of the underlying cortical neurons. Final output in the form of waveform is the potential difference between two electrodes.

amplitude of these waveforms depend on the pattern of discharge of afferent fibers.

Depending upon the state of depolarization or hyperpolarization, the surface potential of the neuron changes. Because of this, electrical signals continue to appear in an EEG. However, their frequency and amplitude keep changing depending upon neuronal activity. A number of neurotransmitters and neuromodulators regulate the activity of an individual neuron. In general, the reticular activating system (see Chapter 1) innervates most of the areas of the brain. Thus, it helps in regulating the activity of a specific area (neuron and group of neurons). In addition, each neuron receives signals from various neighboring neurons as well as those present at some distance (neuronal circuits, for example, thalamocortical relay). Stimuli from these sources also influence the activity of a neuron, and in turn, the neuronal group. During the neuronal activity, stimuli from various sources increase and, thus, influence the electrical state of the neuron. During the active process, multiple neurons fire; however, some may be depolarized while others may be hyperpolarized. Different electrical states occurring simultaneously in a group of neurons result in a desynchronized low-amplitude activity, usually of the beta range. As a person closes eyes, information reaching the occipital cortex reduces, while synchronized activity continues in this area, and this generates the alpha waves. As a person goes into sleep, information reaching the brain from the external environment reduces, which results in reduced firing of neurons. With reduced stimuli from the environment, the activity of the cortico-subcortical circuits becomes more conspicuous. Thus, the activity becomes more synchronized and it changes to theta and with deeper sleep, to delta. As the frequency of activity goes down (from beta to alpha to theta to delta), the amplitude of waveforms increases. It must be remembered that activity in different areas of the brain could be specific to the state of wakefulness and stages of the sleep that are regulated by a complex process.

4.4.1.2 Electrodes—Active and Reference

EEG records the sum of the electrical state (depolarized or hyperpolarized) of the neurons that lie under it. While recording data from electroencephalography channels during polysomnography, we place six electrodes on the scalp, three on either side of the scalp, placed in frontal, central, and occipital positions. These electrodes are the active electrodes. They are referred to the opposite mastoid, that is, electrodes from the right side of the scalp are referred to the left mastoid (e.g., F2-M1; C2-M1 and O2-M1) and vice versa. Electrodes on the scalp are considered as 'active' electrodes and they gather information regarding electrical potential from the underlying cortex. Reference electrodes are chosen at a distant place to allow for a maximum potential difference between the two points, as electrical activity is minimal in the mastoid area, which results in a good waveform. Additionally, a reference electrode from the scalp is chosen as a reference (e.g., F2–C2). As the potential difference between both these areas may be small (as both are placed above electrically active cortex), this combination will not produce a good waveform. Moreover, because of the common mode rejection, certain important characteristic waveforms that are essential for the recognition of sleep stages may disappear (Figure 4.8).

However, bipolar electrode placement is important when we are looking for nocturnal seizures. During a seizure, one area of the brain becomes electrically different relative to the other areas. In such case, bipolar electrodes help in localizing that area, since electrodes are placed on the scalp, which also contains muscles, contraction of the scalp muscles produces electrical potential, which may appear in these electrodes, causing significant noise. Similarly, if any of these electrodes is placed over an artery, which carries the cardiac electrical potentials, ECG artifacts may appear in these leads (see the chapter on artifacts).

4.4.2 Electroocculogram (EOG)

This is used for recording eye movements. Two active electrodes are placed near the outer canthi of the eyes and they are referred to the opposite mastoid. Eyeballs have a difference in the electrical potential between the cornea and the retina (Figure 4.15). The cornea of the eye is positive relative to the retina. Thus, when the eyes move towards the left, the cornea of the left eye having positive potential, comes close to the electrode placed near the outer canthus of the left eye and this electrode will show a positive deflection. Since, the eyes have the conjugate movement, that is, both eyes move in one direction, the positively charged cornea of the right eye will move away from the electrode placed outside right canthus, and the negatively charged retina will come closer to it, this electrode will show negative deflection. Thus, in the electrooculogram, an out-of-phase deflection will appear. Similar changes will be seen when eyes are moved up or down, however, during vertical movement, deflections

Basic Concepts of Polysomnography Channels

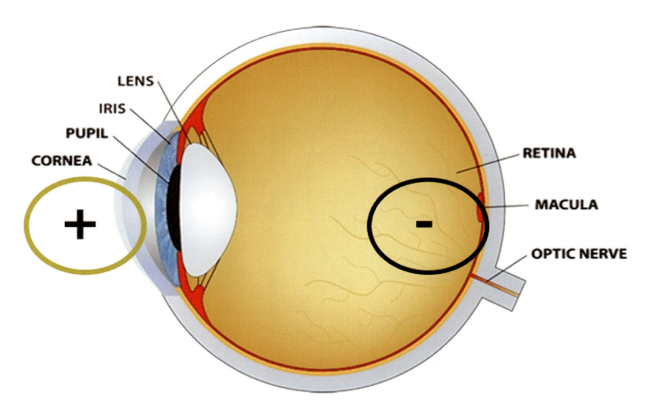

Figure 4.15 Difference of polarity between cornea and retina: Cornea is positively charged while retina is negatively charged.

may be seen in the frontal electrodes of the EEG, as eyes move towards or away from them (Figure 4.16A-B). These out-of-phase movements help in differentiating eye movements from delta waves and K complexes that may sometimes spill in the EOG leads due to their proximity to the frontal area of the brain (Figure 4.16C).

During wakefulness, when a person scans the environment, darting eye movements are seen. With the initiation of sleep, scanning reduces and slow eye movements replace darting eye movements. With increasing depth of NREM sleep, eye movements disappear completely. During REM sleep, fast saccadic eye movements are seen (Figure 4.17).

4.4.3 Electromyogram (EMG)

Usually, an EMG is recorded from two sites: the submental muscles and the anterior

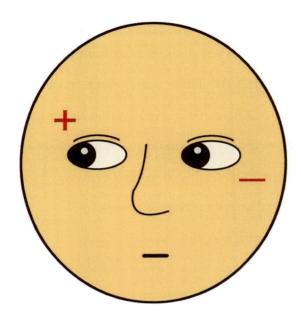

Figure 4.16A Eye movement produce changes in not only EOG electrodes but also frontal electrodes: A: Eye is a dipole, hence its movement produces a deflection in the EOG derivations; Right eye is showing positive deflection as positively charged cornea is moving towards it while left eye shows negative deflection as positively charged arena is moving away from it.

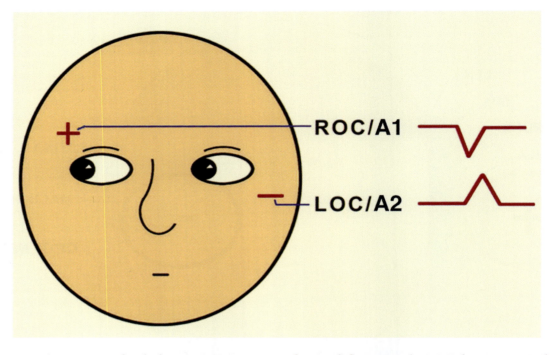

Figure 4.16B Eye is a dipole, hence its movement produces a deflection in the EOG derivations; Right eye is showing positive deflection as positively charged cornea is moving towards it while left eye shows negative deflection as positively charged arena is moving away from it.

Basic Concepts of Polysomnography Channels

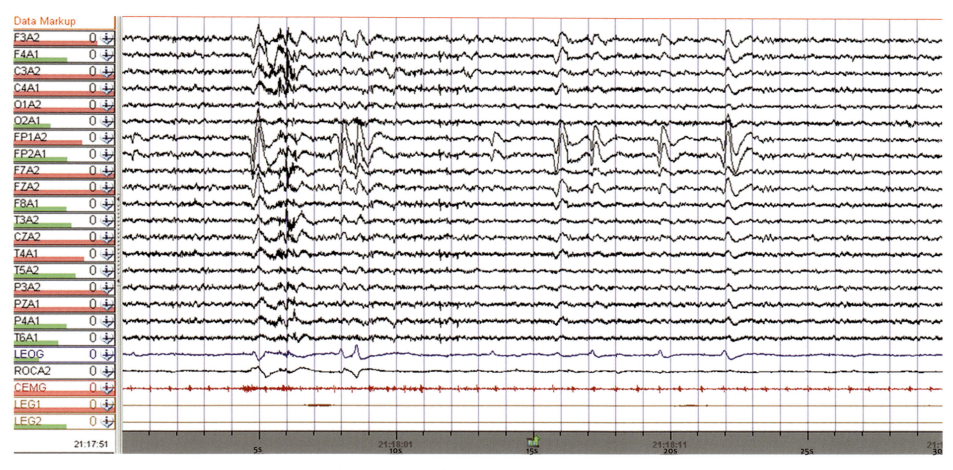

Figure 4.16C Waveform is also noticed in frontal derivations due to their proximity to the eye.

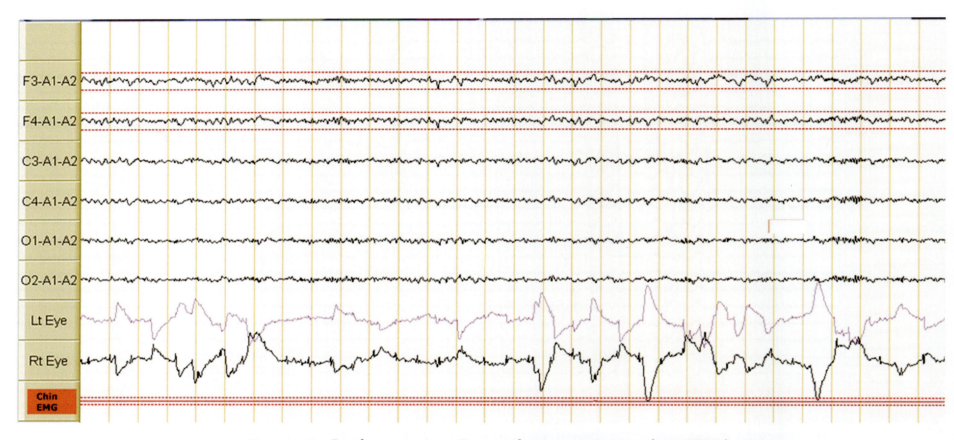

Figure 4.17 Rapid eye movements: Fast saccadic movements are seen during REM sleep.

Basic Concepts of Polysomnography Channels

tibialis muscles of both legs. However, in special cases where a movement disorder is suspected, an EMG may be recorded from the affected area, for example, masseter muscles during bruxism or arms in a suspected REM sleep behavior disorder. An EMG records the sum of electrical activity in the underlying muscles.

Even during the resting state, muscle maintains a basal tone. In other words, myocytes remain in the partially contracted state. As we have already discussed, contraction of a myocyte is brought by a change in the electrical potential. Thus, even at rest, some activity can be observed in the EMG channels (Figure 4.18A and B). During a muscle contraction, the firing frequency of the nerve fibers supplying the myocytes increases and more and more motor units (many myocytes innervated by single nerve fiber constitute one motor unit) become recruited. This results in the enhancement of the electrical potential, which can be picked by the EMG electrodes, and high amplitude signals appear in the EMG recording (Figure 4.19). However, recruitment is random, hence, the signal is also random and not synchronized. As sleep starts, basal tone reduces resulting in lowering of the amplitude of the EMG signal. During REM sleep, profound physiological atonia develops, leading to the diminution of signals in the EMG.

Signals from muscles are optimal if the distance between the muscles and the surface is minimal. However, the presence of subcutaneous fat increases the distances, hence, the amplitude of the signals decreases.

4.4.4 Electrocardiogram (ECG)

An ECG depicts the sum of the electrical activity of the heart during its pumping process. The heart has its own conduction system that is present below the endocardium. The conduction system of the heart is specialized where impulses generated in the sino-atrial (SA) node traverse down to the atrioventricular (AV) node and then to the ventricles. Upon activation of the conduction system, changes akin to the skeletal muscles occur that result in contraction and relaxation.

The heart has four chambers, two atria, and two ventricles. They do not contract together. First, the conduction system depolarizes the atria (when the ventricles remain at resting membrane potential), and after a few seconds, a depolarization wave proceeds to the ventricles

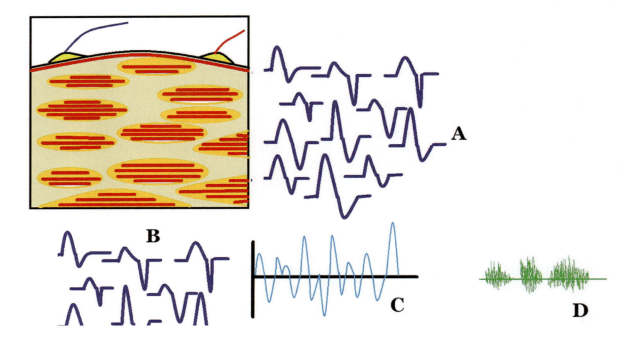

Figure 4.18A Generation of muscle potential and appearance of EMG: A: Each motor unit generates electrical potential; however, it contributes to the measured potential in EMG depending upon proximity of the motor unit and the measuring electrode. B: Furthermore, different motor units get recruited at different frequencies and this also changes the waveform. C: All the potentials generated by different motor units get summed up a given time to generate an output. D: This is why signal during a muscle contraction appears constellation of multiple small potential changes whose amplitude varies temporally.

(by that time atria reaches the resting membrane potential). When the atria depolarize, their outer surface becomes positive compared to the ventricles. Hence, a dipole is created and current flows from the right side (because ventricles are on the left side as compared to atria) and also from the backside of the chest to front (because atria are close to back while ventricles are closer to the anterior chest wall). Localized change in the membrane potential of the heart with reference to the other part creates a dipole (which keeps changing temporally) that can be picked up by surface electrodes during an ECG. The direction of waveforms in the ECG depends upon the lead used. For example, bipolar lead attached to the right arm and left arm, with the negative pole of the channel to the right side and the positive towards the left side. Since the current from the heart is flowing

Basic Concepts of Polysomnography Channels

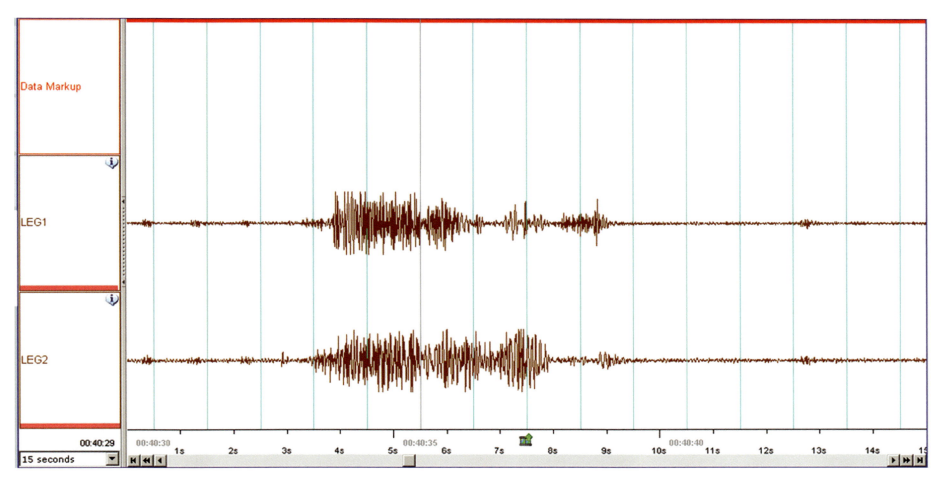

Figure 4.18B Real time EMG data.

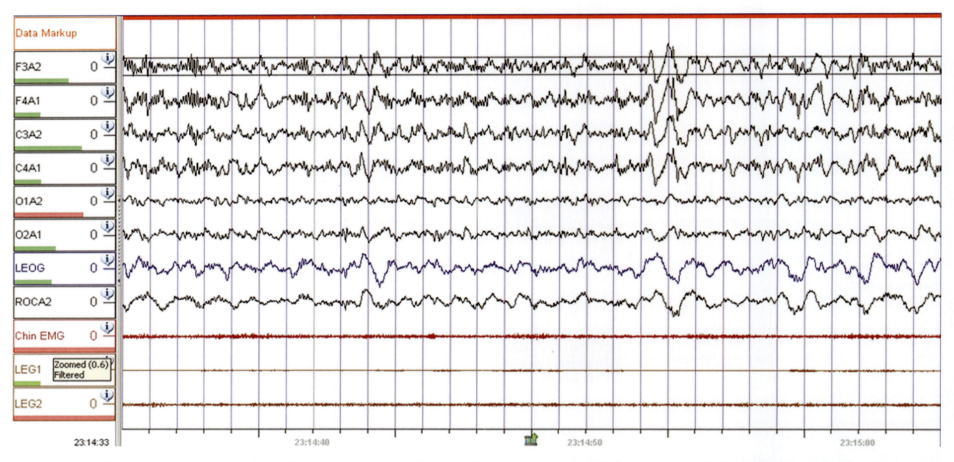

Figure 4.19 Changes in muscle tone are dependent upon state of wakefulness and sleep, sleep stages and movement. A: As the sleep ensues, muscle tone reduces, with complete atonia during REM and increment of tone with movement. This epoch depicts chin tone during N3.

Basic Concepts of Polysomnography Channels

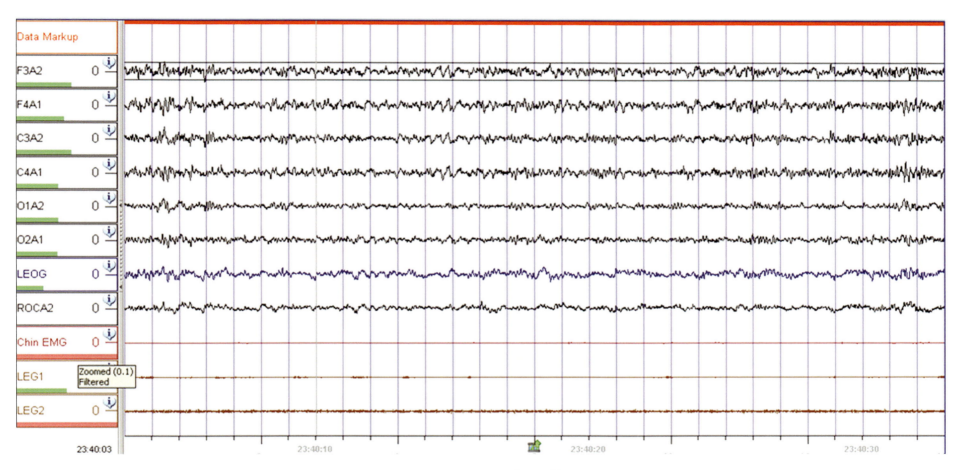

Figure 4.19B Atonia during REM.

from right to left, it will create a positive deflection in this lead and will be observed as the first half of the P wave in the ECG. The second half of the P wave reflects that action potentials in the atrial muscles are coming back to the resting membrane potential. In this manner, because of the changing dipoles in the heart, QRS complex and T waves are generated.

4.5 RESPIRATORY DATA

4.5.1 Airflow Measures

Airflow is usually recorded using two different modalities—thermocouple or thermistor and pressure transducer.

Thermocouple or thermistors are made up of a combination of two metals, which expand, or contract as the temperature changes (Figure 4.20). In the nasal/oral airway, inhaled air is cooler than the

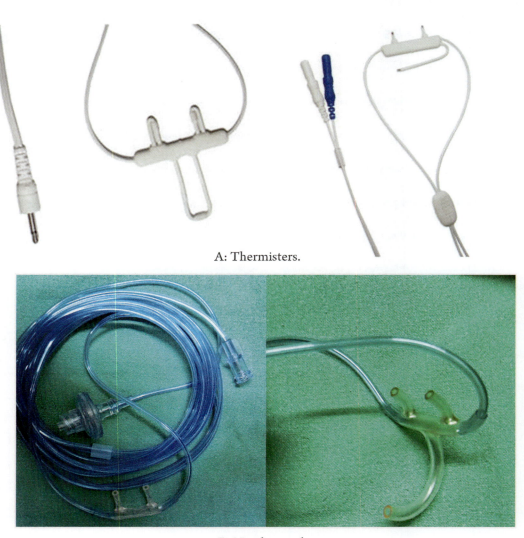

Figure 4.20 Thermistors and nasal cannula from various manufacturers.

A: Thermisters.

B: Nasal cannula.

exhaled air. With that temperature change, these metals also change their property that generates the signals. Since they can detect a small change in the temperature, they are sensitive to detect apnea, where, by definition, air does not flow through the oro-nasal passage. However, during hypopnea and during airflow limitation as seen during the upper airway resistance syndrome, some air continues to flow causing temperature changes and generating a signal that usually has an equal amplitude to that of a normal breath. In short, thermistors and thermocouples are not able to detect flow limitation and hypopneas (Figure 4.21).

On the other hand, a pressure transducer detects the changes in the pressure of air column, which is transmitted to a piezoelectric detector, which in turn generates an electrical signal. Piezoelectric sensors are based on the piezoelectric principle. Certain materials (e.g., quartz) have the ability to produce an electrical charge when pressure is applied to them. The electrical signal produced by piezoelectric sensors is proportional to the degree of pressure applied; hence, these sensors provide a waveform that is concordant to the depth of respiration. Because of this quality, pressure transducers are optimal for recording airflow limitation.

4.5.1.1 Concepts of Aliasing, Filter Setting, Sampling Rate, and Polarity

These are applicable to the respiratory parameters as well. Hence, we need to define a sampling rate, and LFF and HFF. Another important decision is to choose the signal recording from AC or DC input.

In DC inputs, constant voltage input will produce constant voltage out, so the input signals are proportional to the output waveform. On the other hand, in AC channels, constant voltage input will be filtered. This is important as during flow limitations (hypopnea/apnea), airflow becomes constant for some time resulting in a constant voltage output from the nasal transducer. In AC channels, it will be filtered and will not produce any waveform. Hence, in machines with AC channels, a low frequency filter is set to 0.01 Hz that results in a time constant of approximately 5 seconds. With this setting, waveforms in AC will appear similar to DC. Nowadays, most of the machines come with DC inputs for respiratory channels. For the high-frequency filter, a decision has to be made whether recording of snoring is required embedded in the respiratory flow waveform or not. Snoring is generated by partial resistance in the upper airway and this resistance can produce minor changes in the flow (Figure 4.22A-B). If the HFF is set to 10–15 Hz,

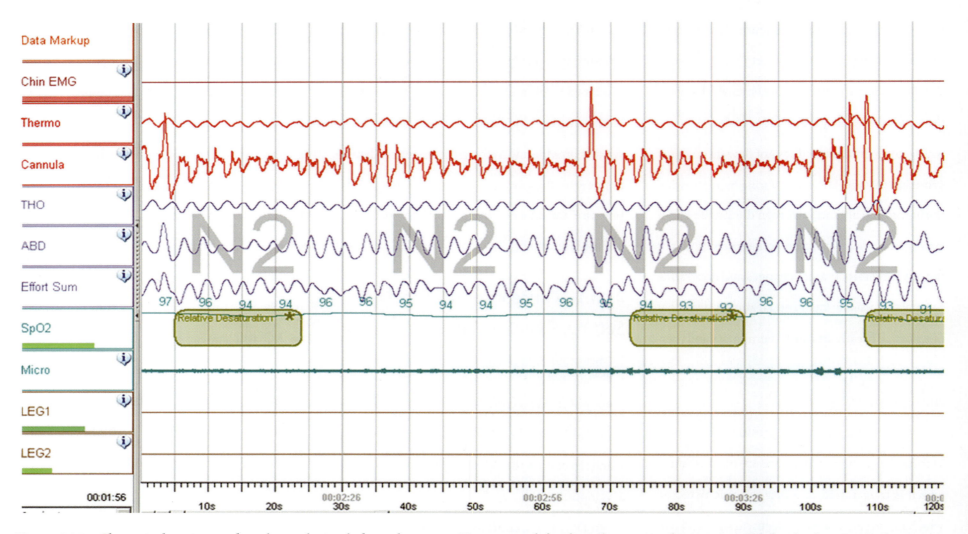

Figure 4.21 Change in thermistor and nasal cannula signals during hypopnea: Hypopnea is defined as at least 30% reduction in amplitude of airflow. During hypopnea, thermistor signals change minimally while signals in pressure transducer to a large extent.

snoring signals can be removed. The setting of HFF is required in DC systems as well.

The respiratory waveform should be placed in the correct polarity with the rounded up tip and the sharp part at the bottom to get a clear idea about flow (Figure 4.23A-B). Sampling rate has to be adjusted depending on the decision to record snore signals in the airflow waveform or not. If filtered signal (without snoring) is preferred, the sampling rate may be 10 Hz, however, for recording snoring, it needs to be increased to 200 Hz.

4.5.2 Chest and Abdominal Movements

With respiration, the volume of the chest and abdomen changes, and measurement of the change of volume is known as plethysmography. Recording of the chest and abdominal movements help to differentiate obstructive sleep apnea from central sleep apnea. It is important since the pathophysiology and management of both are different.

For measuring the chest and abdominal movements, belts are tied around the chest and around the abdomen. During inspiration, chest and abdomen both expand and during expiration, both deflate. Belts tied around the chest and abdomen also sense a strain during expansion (inspiration) and come back to a normal state during exhalation. This is recorded through sensors attached to the belts.

These elastic belts are made up of nylon and they are connected to a small sensor from both sides. The simplest type of sensor is a piezoelectric strain gauge that emits current when a change in the strain occurs in the belts during the respiratory movement. However, these strain gauge sensors have important limitations. For example, sometimes the position of the body may limit the movement of the belt, and thus, because of inappropriate pressure transferred to the sensor, the movement may be missed or may be exaggerated. Moreover, the output signal of the piezo technology is not linear; therefore, it cannot be used to assess hypopnea. Additionally, the piezo technology can produce a false paradoxical breathing signal when tension is applied to the belt during patient movements.

To overcome this limitation, another technology was adopted that uses the inductance. Faraday's law states that when a current is passed through a loop of wire, it creates a magnetic field around it. When the girth of this loop is changed, it induces an opposing current that is directly proportional to the change in girth (Lenz's law). Thus, in respiratory inductance plethysmography (RIP), small alternate

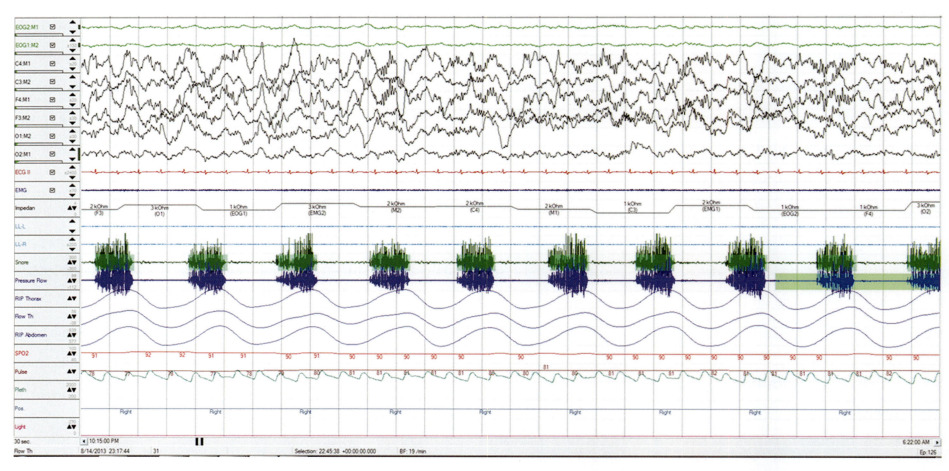

Figure 4.22A Snoring embedded in respiratory waveform: Snoring signals from the airflow waveform can be removed by appropriate filter setting. A: Setting high frequency filter to 50 Hz derivations to appearance of snoring. In this illustration, respiratory waveform is flat as LFF has been mistakenly set as 1 Hz.

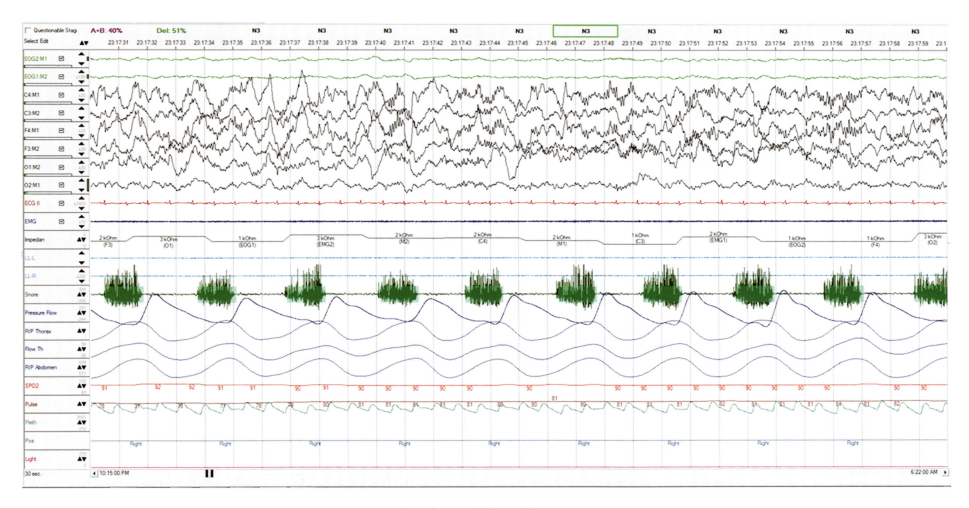

Figure 4.22B Setting HFF to 1 Hz removes snoring.

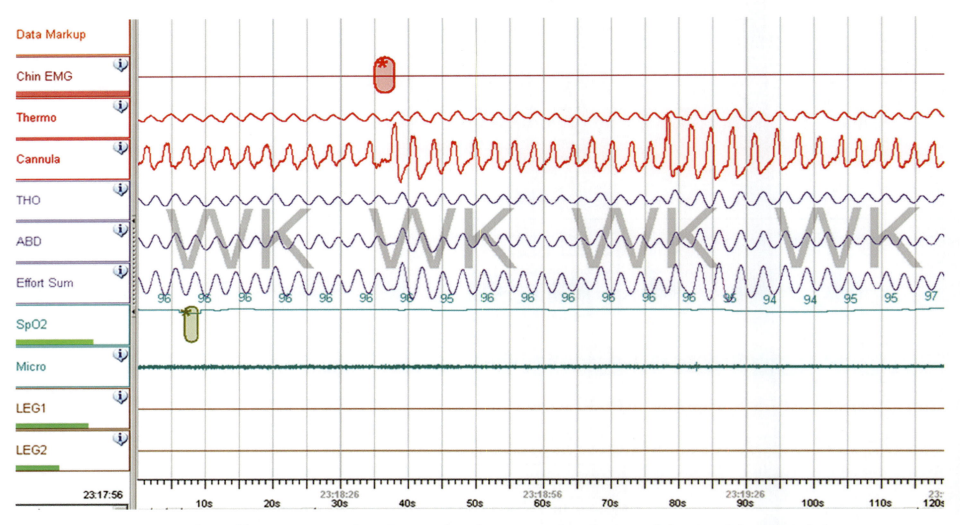

Figure 4.23A Polarity of the respiratory waveform. Correct polarity for respiratory flow wave is rounded up tip and sharp part at the bottom.

Basic Concepts of Polysomnography Channels

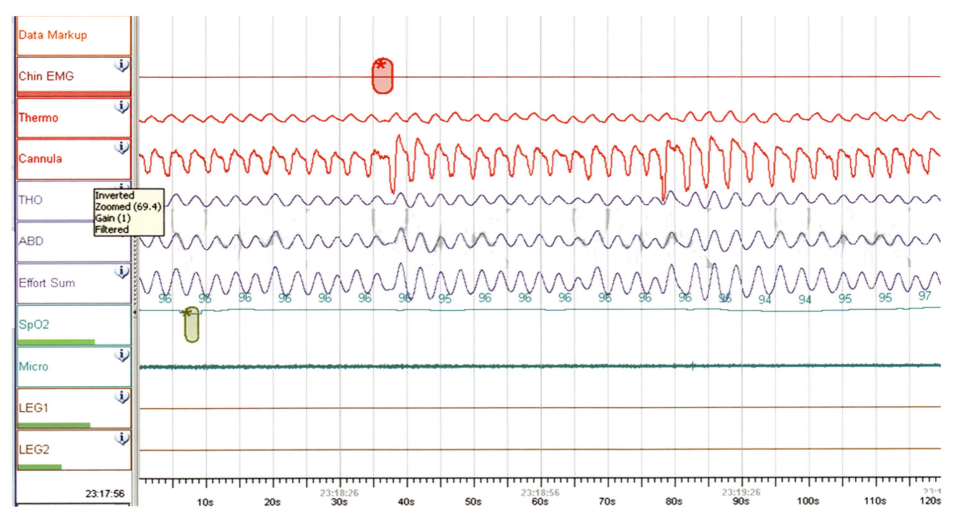

Figure 4.23B Incorrect polarity of respiratory flow waveform. This may be rectified by reversing the polarity through channel properties.

current is passed through a wire that is woven in a belt in a zig-zag manner. These belts are tied around the chest and abdomen. The magnetic field is generated around the wire in resting position changes during respiratory movements because of the expansion of girth of the chest and abdomen (Figure 4.24). This induces an opposing current that can simply be measured as a change in the frequency of the applied current. These signals can be seen as a waveform. RIP sensors are superior over strain gauges because their signals are unaffected by the trapping of belts. The output signal of the RIP technology is linear and, hence, more accurate than the piezo technology. If calibrated appropriately, RIP provides accurate information regarding paradoxical breathing and flow volume loops.

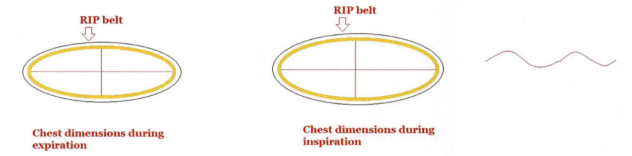

Figure 4.24 Magnetic field created by RIP belts around the chest: Expansion of chest alters the magnetic field around the chest and thus the output signal.

4.5.2.1 Concepts of Aliasing, Filter Setting, and Sampling Rate

To overcome aliasing, the sampling rate is set at 10 Hz. Just like respiratory flow signals, when these belts are attached to an AC channel, LFF is set to 0.1 Hz or lower while HFF is set to 35 Hz.

4.5.3 Oxygen Saturation

In the body, oxygen is carried to the peripheral tissues through hemoglobin. Oxygen of the inhaled air diffuses through the alveoli of the lungs to the capillaries. These capillaries carry the deoxygenated blood (hemoglobin in red blood cells (RBC) without oxygen) that is propelled by the right ventricle of the heart through the pulmonary artery. Because the concentration of oxygen is higher in the inhaled air, it diffuses through the alveolar wall and capillary wall into the RBC, where it binds with the hemoglobin (Figure 4.25). One molecule of hemoglobin can carry eight molecules of oxygen. Then blood enters the left side of the heart through the pulmonary veins from where it is pumped to the whole body through the aorta. The aorta divides into small arteries, arteries into arterioles, and arterioles in capillaries. Capillaries form the smallest part of the vascular system and are present close to the cells in the peripheral tissues. This is how oxygen (and other nutrients) is delivered (through diffusion) to the peripheral tissues. Areas of the body like finger, nose, ear lobe, and forehead have rich vascularity; hence, they are the best sites to measure oxygen concentration.

Pulse oximetry is based on the principle of differential absorption of infrared (940 nm) and red (660 nm) lights by oxyhemoglobin and deoxyhemoglobin. Oxyhemoglobin absorbs a greater amount of infrared light and less of red lights (that is why it appears red), while the deoxyhemoglobin possesses the opposite quality. Light-emitting diodes in pulse oximeter emit lights of these two wavelengths that are made to traverse through the tissue (finger, nose, ear lobe). These are present on one side of the probe. On the other side of the probe, the amount of lights that has passed is sensed by a photodiode (Figure 4.26). Relative amounts of red and infrared lights are measured using a mathematical formula that allows calculation of the concentration of oxygen bound to hemoglobin.

Considering this principle, it is simple to decipher that any condition that reduces the flow of blood to peripheral tissues (such as congestive heart failure, any kind of pressure on the artery supplying the area where probe is applied), anything that interferes with transmission of light

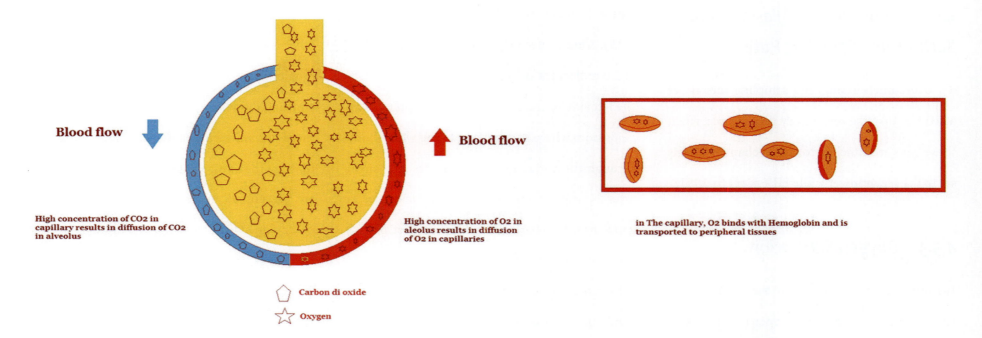

Figure 4.25 Oxygen exchange in lungs: Because of the concentration gradient, Atmospheric oxygen diffuses through the alveolar and capillary to enter the RBC. It has high affinity for hemoglobin so it converts the deoxygenated hemoglobin to oxygenated hemoglobin. Reverse happens in the peripheral tissues.

from one arm to other (nail polish, if the probe is applied to the finger), reduced number of RBCs (severe anemia), change in the property of hemoglobin (binding of carbon mono oxide; hemoglobinopathies), poor probe placement, and excessive movement during recording can provide falsely low or high values of saturation.

For pulse oximetry used during a sleep study, a fast sampling rate oximeter is recommended (shorter interval, for example, 3 seconds or less) to improve sensitivity as patients with OSA usually have intermittent hypoxemia. Overnight pulse oximetry is an important parameter for the evaluation of respiratory disturbances during sleep, particularly for scoring hypopneas for which desaturation is one of the criteria. In addition, it indicates the severity of SDB and the need to supplement the PAP device with oxygen therapy.

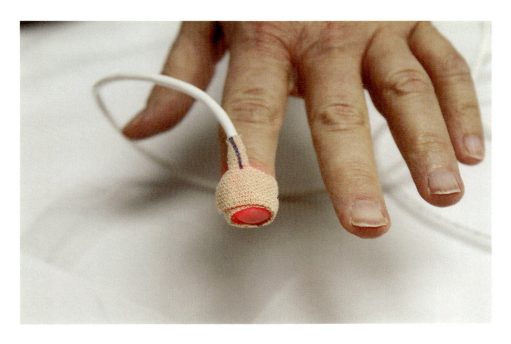

Figure 4.26 Pulse oximeter: Light emerges from one side of the oximeter and reaches the other after traversing the tissues. (Photo by Quinn Dombrowski. Creative Commons Attribution-Share Alike 2.0 Generic license).

4.5.4 Record of Snoring

As has been already discussed, snoring may be recorded through the pressure transducer. However, most modern machines have a microphone or piezoelectric sensor that is placed on the larynx. This microphone sends the signals to the software where these are shown in the form of waveforms. The amplitude and duration of these waveforms vary according to the intensity and duration of snoring (or any other sound produced in close proximity to the probe). Thus, it records not only the snoring but also the teeth grinding and vocalization during sleep.

In addition to the sensor placed on the larynx, some laboratories also record all kinds of audio signals through an additional microphone that is mounted close to the head end of the bed. This not only helps in communicating with the patient during the study, if a need arises but also supplements visual information recorded during sleep. Signals from the microphone are synchronized with video recording and other data, hence, they play an important role in cases of sleep seizures and parasomnias. Snoring can also be recorded via nasal pressure cannula, where the snoring signal can be seen superimposed on the airflow waveform signal when the signal from airflow sensor cannula is unfiltered.

4.6 BODY POSITION

Body position monitoring is important during PSG due to the sleep-dependent nature of SDB. Body position monitoring allows the technician to capture changes in SDB breathing in different positions (supine, lateral and prone) and to accurately gauge the true severity and appropriate PAP titration. Body position sensor has a gyroscope that determines the position of the body. It sends different signals depending upon the body position: right, left, supine, prone, and upside. These different signals are then converted into a waveform and depicted in the polysomnography. The body position sensor reports the position of the sensor rather than the position of the human body; therefore, it is essential to assure that the sensor is oriented correctly to the patient.

4.7 VIDEO DATA

Most of the sleep laboratories that are dealing with cases of suspected sleep seizures and parasomnias also record video data that are synchronized with data from other channels. Since a dark environment is required during sleep, infrared

Basic Concepts of Polysomnography Channels

light is emitted from the camera, and its sensors record the reflected infrared signals. The focus and direction of the camera may be manipulated from the monitoring room with the help of special software. With these manipulations, a sleep technician can record even the smallest movement during any activity occurring in sleep.

4.8 CARBON DIOXIDE MONITORING (PCO_2)

In pediatrics, end-tidal PCO_2 ($ETCO_2$) is considered a standard practice during PSG monitoring. On the other hand, in adult patients, the AASM recommends the use of arterial PCO_2, $ETCO_2$, or transcutaneous PCO_2 for detection of hypoventilation during a diagnostic sleep study, and the use of arterial PCO_2 or transcutaneous PCO_2 for detection of hypoventilation during PAP titration. Both $ETCO_2$ and transcutaneous PCO_2 are noninvasive validated indirect methods to predict arterial PCO_2 ($ETCO_2$ reflects exhaled CO_2 at an end-tidal sample of exhaled gas. Infrared spectroscopy is the technique usually used to assess $ETCO_2$ in the exhaled gas. When the patient is being ventilated using a closed circuit (e.g., endotracheal tube), $ETCO_2$ can be measured directly with good accuracy. However, during a sleep study, $ETCO_2$ is measured using a nasal cannula in a spontaneously breathing patient; therefore, side-stream sampling often occurs. When a patient is on a noninvasive PAP device, the increased flow within the open circuit results in dilution of the exhaled gas. Therefore, the numerical value or the displayed waveform of $ETCO_2$ may not accurately reflect arterial PCO_2. Mouth breathing may influence the displayed signal too. Transcutaneous PCO_2 is obtained through the skin where the electrode warms the skin surface, increasing local capillary perfusion, and measures the CO_2 gas as it diffuses from the dermis across a gas-permeable membrane. Currently available devices require less heating (42°C for adults and 41°C for neonates); therefore, resulting in less discomfort and less potential for skin damage. Nevertheless, the sensor may need to be repositioned at least once during a sleep study. The absolute value of transcutaneous PCO_2 is affected by skin thickness and capillary density. Therefore, it is important to place the electrode at a site of high capillary density and thin skin. This presents no problem in the newborn, in which these conditions are usually present. However, in adults, there is greater variation from site to site. The suggested locations for best transcutaneous measurements are the forehead, forearm, chest, or abdomen. Transcutaneous PCO_2 provides a good alternative to $ETCO_2$ during PAP titration to assess the response to treatment.

FURTHER READING

1. Patil, S. P., (2010). What every clinician should know about polysomnography. *Respiratory Care* 55, 1179–1195.
2. Olejniczek, P. (2006). Neurophysiologic basis of EEG. *J Clin Neurophysiol.* 23, 186–189.
3. Bucci, P., & Galderisi, S. (2011). Physiologic basis of EEG signals. In: *Standard Electroencephalography in Clinical Psychiatry: A Practical Handbook.* Boutros, N., Galderisi, S., Pogarell, O., Riggio, S. (Eds.). Wiley–Blackwell, Oxford Chichester, pp. 7–12.
4. Hall, J. E., & Guyton, A. C., (2011). Guyton and Hall textbook of medical physiology. Philadelphia, PA, Saunders Elsevier.
5. Chan, E. D., Chan, M. M., & Chan, M. M., (2013). Pulse oximetry: understanding its basic principles facilitates appreciation of its limitations. *Resp Med.* 107, 789–799.
6. Kirk, V. G., Batuyong, E. D., & Bohn, S. G., (2006). Transcutaneous carbon dioxide monitoring and capnography during pediatric polysomnography. *Sleep.* 12, 1601–1608.

REVIEW QUESTIONS

1. Resting membrane potential differs from action potential as:
   A. it is not dependent upon the transmission across synapse
   B. it has smaller time duration than action potential
   C. not seen in muscles
   D. mainly dependent upon the concentration of calcium ions
2. At the termination of action potential following pump bring the cell to pre-excitatory stage:
   A. Na-Cl pump
   B. Ca^{+2} pump
   C. Na^+-K^+ pump
   D. Cl^- pump
3. EEG waves depict:
   A. electrical activity of single neuron that underlies electrode
   B. electrical activity of a small group of neurons that underlie electrodes
   C. cumulative electrical activity of all the cortical neurons irrespective of electrode placement
   D. cumulative electrical activity of all the subcortical neurons that underlie electrode
4. During a muscle contraction:
   A. all the motor units contract together
   B. all the motor units contract sequentially one after another
   C. all the motor units contract in two divided parts
   D. all the motor units contract so as they reach a peak followed by relaxation
5. Amplifiers works to:

A. amplify the frequency of electrical signals

B. amplify the resistance of electrical signals

C. amplify the voltage of electrical signals

D. amplify the duration of electrical signals

6. Differential discrimination refers to:

    A. accepting the identical potentials between two electrodes

    B. rejecting the identical potentials between two electrodes

    C. accepting non-identical potentials between two places

    D. rejecting non-identical potentials between two places

7. During a digitalized recording, appearance of waveform depends upon all EXCEPT:

    A. sampling rate

    B. monitor aliasing

    C. bit resolution

    D. RAM of system

8. Ventricular contraction in EKG is depicted by:

    A. QRS complex

    B. P wave

    C. T wave

    D. U wave

9. Ideally, for higher amplitude, during EEG acquisition reference electrodes are placed at:

    A. area that is electrically active

    B. area that is electrically neutral

    C. area with frequent change of potentials

    D. area having equal potential to the active electrode

10. EOG produces waveforms because of:

    A. negative charge in cornea and positive charge in retina

    B. negative charge in cornea and positive charge in eye muscles

    C. positive charge in cornea and negative charge in retina

    D. positive charge in cornea and negative charge in eye muscles

11. Best measure for the hypopnea is:

    A. pressure transducer

    B. PVDF belt

    C. strain gauge

    D. thermistor

12. Pulse oximetry is important for:

    A. scoring hypopnea

    B. scoring obstructive apnea

    C. scoring hypoventilation

    D. scoring central apnea

ANSWER KEY

1. A   2. C   3. B   4. D   5. C   6. B

7. D   8. A   9. B   10. C   11. D   12. A

5. PREPARATIONS FOR THE SLEEP STUDY

LEARNING OBJECTIVES

After reading this chapter, reader must be able to:

1. Understand the optimal environment and setting of a sleep laboratory.
2. Able to check the equipment for electrical safety.
3. Prepare the patient for optimal recording of data during polysomnography.
4. Provide instructions to the patient for optimal recording.

CONTENTS

5.1 Sleep Laboratory 116

5.2 Preparing the Machines 116

5.3 Preparing the Patient 117

Review Questions 118

Answer Key 119

A good recording requires that the room, equipment, and patient are well-prepared before the process of gathering data is initiated. These issues are important as they may influence the quality of data and, thus, your conclusion and as well as the safety of the patient. It is important that you choose the correct machine (Chapters 3 and 9), and have correct montage (Chapter 10) for recording the data. Adequate measures to prevent infection should be taken care of before subjecting a patient to the sleep study (Chapter 24). This chapter does not discuss topics that have already been covered in other chapters.

5.1 SLEEP LABORATORY

The room in which a patient will stay for the recording plays an important role in the quality of the study. The sleep laboratory should be situated away from a crowded and noisy area and its décor should be attractive and homely. It must have adequate ventilation and provision for daylight. It should include a comfortable bed television, and an attached bathroom. The room size should be adequate enough to alley anxiety in claustrophobic subjects. A couch must be provided for a family member to stay with the patient. The room must be air-conditioned and sound proof, and should have thick curtains to create a dark environment for daytime recordings. Air-conditioning is necessary for multiple reasons—it prevents dust from entering into the room, which can interfere with signals, and it also helps to maintain a comfortable temperature for a good sleep. Many patients with OSA sweat a lot because their work of breathing is increased. This is one of the major reasons for falling off the electrodes and artifacts in the recording. Having an adequate temperature of around 22°C minimizes sweating. A blanket may be provided in case the patient feels cold.

The laboratory must provide means for the patient to communicate with the sleep technician anytime during the recording, without reaching for the phone. For this, a microphone and a speaker may be placed near the head-end of the bed.

5.2 PREPARING THE MACHINES

Polysomnography machines require electrical power for their working and so they are connected to the electrical supply that usually has 240 V current. These machines also have capacitors and transformers, in addition to the integrated circuits. Normally, these parts are placed close to each other to enable the electric flow between them. This results in the generation of an electromagnetic field, which can generate "stray capacitance" and "stray inductance." This, in turn, can generate a small amount of current in the system that can flow through the leads, especially the leads that attach directly to the body (e.g., EEG, EKG, and EMG) and may be delivered to the patient's body. Another source of aberrant current is any short-circuit inside the polysomnography recorder that connects the patient directly to the power supply.

To avoid both these unwanted scenarios, almost all manufacturers take optimal precautions of "grounding" the machine. A wire is attached to the chassis of the motherboard that "sucks" unwanted current and delivers it to the "ground (earth)" through a large round pin in the socket. In absence of this ground, the current will flow towards the patient. Which path the aberrant current will prefer—"earth" or the "patient's body" depends upon the relative resistance of each. If the resistance is lesser in the "earth," the current will prefer this path.

Hence, it is of utmost importance that the "earth/ground" wire of the plug is working properly and has minimal resistance. This should be regularly checked by an electrical engineer.

5.3 PREPARING THE PATIENT

Before conducting a sleep study, it is important to prepare the patient for it. If the patient is not explained the procedure in advance, the equipment and the web of wires could prevent his/her sleep, leading to what is known as the "first night effect." Hence, it is advisable to share a printed flyer with the patient explaining the reason and procedure of the study. The flyer must also depict the "Dos and Don'ts before the sleep study" to allow the patient to be better prepared for the procedure. It is also advisable to explain the entire procedure to the patient using appropriate visuals to allay all fears and anxiety. A change of place is another factor contributing to the "first night effect." A change in environment prevents a patient from going to sleep. To minimize this, the patient should be advised to report at least 8–10 hours ahead of the bedtime. Once in the sleep laboratory, the patient may spend time in some leisure activities in order to get acclimatized to the environment.

Some general precautions should be taken while a patient is in the sleep laboratory. These are listed in Table 5.1. Further, a patient also needs to be prepared for effective monitoring, for example,

Table 5.1 Instructions to the Patients Before Sleep Study

1. Keep your body clean. For the proper application of electrodes on your body, the skin should be clean, dry and free of dirt and any grease. It is better that you take bath two hours before the start of polysomnography and clean your hair with shampoo. If it does not interfere with your religious practices/beliefs, men are requested to shave the beard. After this, please do not apply any kind of oil or cream on your skin/hair.
2. Cotton cloths are preferred during the test. Clothes made up of synthetic material may have electrostatic discharges, which may interfere with the signal acquisition.
3. You are advised to wear pajama and t-shirt. While choosing your clothes, please consider that sleep technician may enter the room in the night to replace the fallen electrode.
4. Although the blanket/bedsheet/pillow is available in the sleep laboratory, however, if you prefer, you may bring your own.
5. Please do not bring any valuable item with you.
6. Women should avoid wearing jewelry during the test.
7. Avoid taking nap in the day before your sleep study is planned.
8. Take your medications as you are taking on other days. However, please discuss them before with your sleep physician before you come for the study.
9. It is advisable that you do not consume any substance of abuse before the test. Please consult your sleep physician in this regard.
10. If you prefer to sleep with any particular object, for example, any toy (especially children), you may bring it along with.

an abrasive gel needs to be applied to remove the dead layers of the skin from the body. This is done to reduce the resistance between the electrode and the skin such that the EEG, EMG and EKG electrodes are able to accurately capture the electrical signals from the body. However, the abrasive gel should not be applied to an inflamed area or any area where a breach in the skin is visible. Similarly, the patient's history should be studied in advance to identify any history of allergy to the gel, and if found, the gel should be avoided.

REVIEW QUESTIONS

1. Following are the possible reasons that can deliver an electrical shock to patient during polysomnography, EXCEPT:
   A. stray inductance

B. double grounding of the machine

C. short circuit in the machine

D. low resistance in machine grounding

2. Reverse first night effect refers to:

   A. poorer subjective sleep in sleep laboratory

   B. poorer subjective sleep at home

   C. better subjective sleep in sleep laboratory

   D. better subjective sleep at home

3. Skin of the patient has to be prepared before application of:

   A. electrodes so as to reduce the inductance

   B. electrodes so as to increase the inductance

   C. oximeter so as to reduce the inductance

   D. oximeter so as to increase the inductance

4. Following may interfere with oximetry data, when placed on finger, EXCEPT:

   A. poor blood supply to finger

   B. methemoglobin

   C. nail paint

   D. adequate oxygen saturation in blood

5. Acclimatization refers to:

   A. adapting to a familiar environment

   B. adapting to an unfamiliar environment

   C. adapting to the cold places

   D. adapting to warm places

ANSWER KEY

1. D   2. C   3. A   4. D   5. B

6

PLACEMENT OF LEADS FOR THE SLEEP STUDY

LEARNING OBJECTIVES

After reading this chapter, the reader should be able to:

1. Place sensors at correct position during a sleep study.
2. Get the optimal output from the sensors.
3. Discuss why sensors are placed at that very position as suggested in manual from American Academy of Sleep Medicine.

CONTENTS

6.1	Electroencephalogram (EEG)	122
6.2	Electrooculogram (EOG)	129
6.3	Electromyogram (EMG)	131
6.4	Electrocardiogram (ECG)	132
6.5	Placement of Measures of Respiration	137
6.6	Body Position Sensor	140
Further Reading		142
Review Questions		143
Answer Key		144

It is important to place all the recording channels in proper position to obtain optimal data. If the sources of the signals are not placed at defined places, the recording will be of suboptimal quality and the data captured would not be accurate. This would, in turn, compromise the comparability of data obtained from two different persons. To counteract these issues, the AAMS has recommended appropriate places and methods of hooking up the patients. In this chapter, we will guide you on how to place different sensors during a sleep study.

6.1 ELECTROENCEPHALOGRAM (EEG)

For recording an EEG, electrodes are placed bilaterally in the frontal, central and occipital positions. These are referenced to the opposite mastoid (not the ear).

These electrodes should be selected based on the understanding of the neuro-biological mechanisms of different stages of sleep. Their placement will help to collect the necessary information during the mpolysomnogram.

1. Frontal electrodes best pick the delta activity (exhibit N3 sleep)
2. Central electrodes best pick the Vertex waves, K complexes, spindles and sawtooth waves (determine N1, N2 and REM stage)
3. Occipital electrodes best pick the alpha activity that can be used to differentiate wakefulness and sleep.

Bilateral placement of electrodes (primary and back-up) helps to determine various sleep stages, even when one or more electrodes on one side of the scalp develop any artifact or fall during the sleep study (Figure 6.1). These electrodes are referenced to the opposite side mastoid.

The recommended derivations are shown in Box 1.

However, if for some reasons, these derivations cannot be placed, other acceptable derivations are shown in Box 2.

Box 1 EEG Derivations for the Polysomnography Recommended by AASM*

Primary	Backup
F4-M1	F3-M2
C4-M1	C3-M2
O2-M1	O1-M2

*American Academy of Sleep Medicine, 2016

Box 2 EEG Derivations for the Polysomnography Acceptable to AASM*

Primary	Backup
Fz-Cz	Fpz-C3
Cz-Oz	C3-O1
C4-M1	C3-M2

It is important to note that the shape and size of the head varies from one person to the other and so proper measurements should be taken to place the leads. Further, as mentioned above, different waves that are used to classify sleep stages appear at different areas of the brain. Hence, any misplacement

Placement of Leads for the Sleep Study

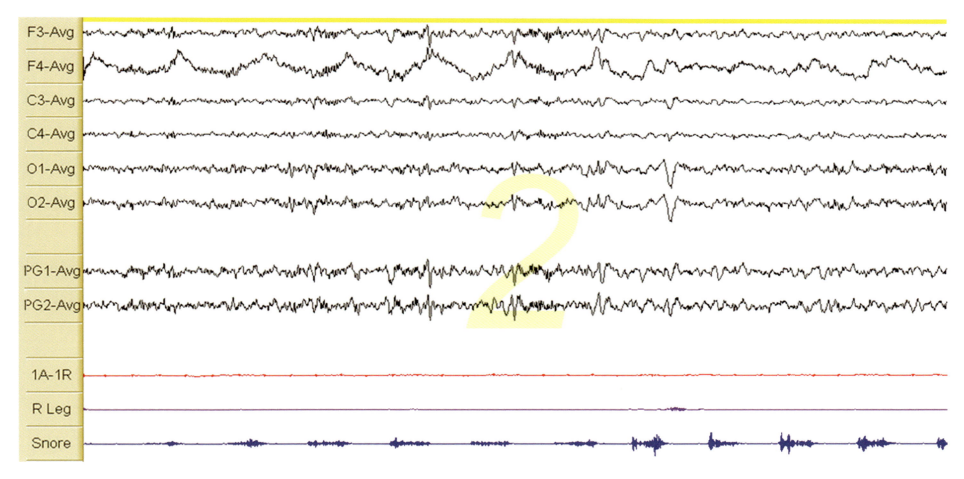

Figure 6.1 Malfunction of one electrode: F4-Avg does not show waves clearly due to poor contact between F4 and skin, but waveforms from frontal area can be recognized in F3.

of these leads may obscure the waves and affect reporting. These cephalic electrodes should be referenced to the contra-lateral mastoid. Mastoid is an area of electrical silence and so it does not contaminate the electrical signals. Rather, it increases the distance between the leads and amplifies the electrical potential (see Chapter 4).

To overcome the problem, EEG electrodes should be placed according to the International 10–20 system. The figures 10 and 20 denote the distance between the electrodes that area measured in percentages (10% or 20%) of the length and breadth of the fixed anatomical points of the scalp.

6.1.1 Placing the EEG Electrodes According to 10–20 Systems

The material required are:

1. Washable non-irritant marker;
2. Non-elastic measuring tape;
3. Abrasive material to clean the area;
4. Electrodes; and
5. Conductive gel.

First, identify the nasion on the front of the face. Nasion is the depressed area on the top of the nose and between the eyes where it meets the frontal bone (Figure 6.2). Second, identify the inion at the backside of the lowermost part of skull. Run your finger from the back of head towards the neck in the median plane. You will feel a protuberance here. This is the inion (Figure 6.3). Using a non-elastic measuring tape, measure the distance between these two points. Suppose this distance is 34 cm. Now, calculate the following:

10% = 3.4 cm

20% = 6.8 cm

50% = 17 cm

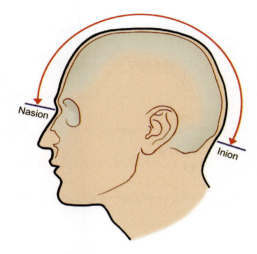

Figure 6.2 Nasion (the depressed area where the nose meets forehead between eyes).

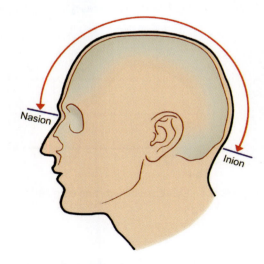

Figure 6.3 Inion (the protruded area on the lower part of back of skull (occipital bone) where it meets neck).

Placement of Leads for the Sleep Study

Place a mark 10% above the nasion—Location of FPz

Another mark 20% above the location of the FPz (30% from nasion): Fz

Another mark at 50% from the nasion—Cz

Another mark 10% above the inion—Location of Oz

Another mark 30% above the inion—Location of Pz (Figure 6.4)

Note: In case you are conducting a sleep study which does not require an extended EEG montage (for seizure or parasomnia), only Fz, Cz and Oz markings are important.

In step 2, identify the pre-auricular points, which are just anterior to the tragus of the ear on both sides (Figure 6.5). Measure the distance between the pre-auricular points on the two sides, but make sure that the measuring tape passes

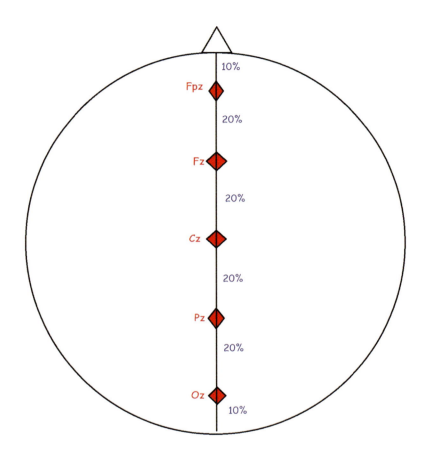

Figure 6.4 Markings in the anteroposterior plane according to 10–20 system: If you are doing sleep study where extended EEG montage (for seizure or parasomnia) is not required, only Fz, Cz and Oz markings are important.

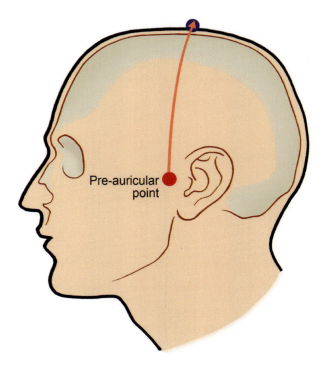

Figure 6.5 Pre-auricular point (it lies in front of tragus of ear).

through Cz. Suppose, this distance is 28 cm. Now calculate the following:

10% = 2.8 cm

20% = 5.2 cm

Place a mark 10% above pre-auricular points: Location for T3 (left) and T4 (Right)

Another mark 30% above the Pre-auricular points—Location for Central electrodes (C3: left; C4: Right) (Figure 6.6).

Note: If you are conducting a sleep study which does not require an extended EEG montage (for seizure or parasomnia), only C3 and C4 markings are important.

In step 3, measure the distance between FPz and Oz by encircling the tape around the head. This must pass through the site for T3 and T4 on the respective sides. Suppose, this is 48 cm.

Now calculate the following:

5% = 2.4 cm

10% = 4.8 cm

Placement of Leads for the Sleep Study

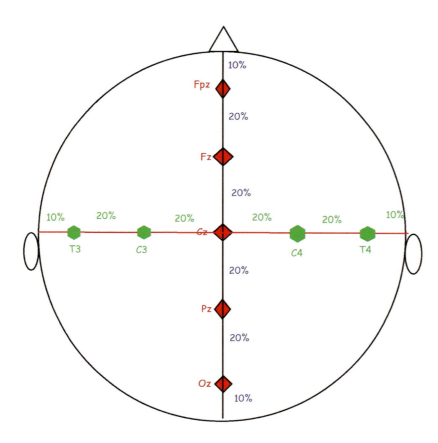

Figure 6.6 Marking in the coronal plane: If you are doing sleep study where extended EEG montage (for seizure or parasomnia) is not required, only C3 and C4 markings are important.

Place a mark at 5% distance on either side of the nasion (Location for Fp1 on the left and Fp2 on the right) and inion (Location for O1 on left and O2 in right side)

Place a mark at 15% distance on either side of the nasion (Location for F7 on the left and F8 on the right)

Midway between T3 and O1 is the marking for T5 and similarly on the right side between O2 and T4 is the marking for T6 (Figure 6.7A-B).

Note: If you are conducting a sleep study which does not require an extended EEG montage (for seizure or parasomnia), only O1 and O2 markings are important.

In step 4, measure the distance between F7-Fz and Fz-F8. Place a mark halfway (50%) between F7-Fz and Fz-F8. This will provide you the location of F3 (left side) and F4 (Right side). Similarly, midway between T5 and Pz, gives you

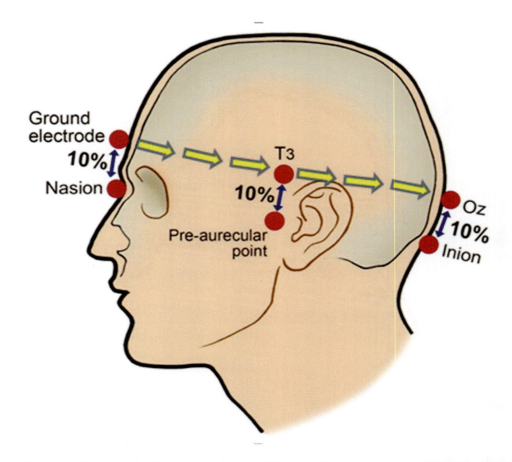

Figure 6.7A Markings on the head circumference: If you are doing sleep study where extended EEG montage (for seizure or parasomnia) is not required, only O1 and O2 markings are important.

the place for the P3, and similarly on the right side, midway between T6 and Pz provides the place for P4 (Figure 6.8).

Note: If you are conducting a sleep study which does not require extended EEG montage (for seizure or parasomnia), only F3 and F4 markings are important.

In step 5, find the mastoids at the back of the ears. This is the hard bone just behind the pinna. This is the site for reference electrodes (M1 on the left and M2 on the right side)

For regular polysomnography, you require the following electrodes:

Frontal: F3 and F4

Central: C3 and C4

Occipital: O1 and O2

Mastoid: M1 and M2

Clean the area with an abrasive gel, as discussed in Chapter 5. Take adequate amount of

Placement of Leads for the Sleep Study

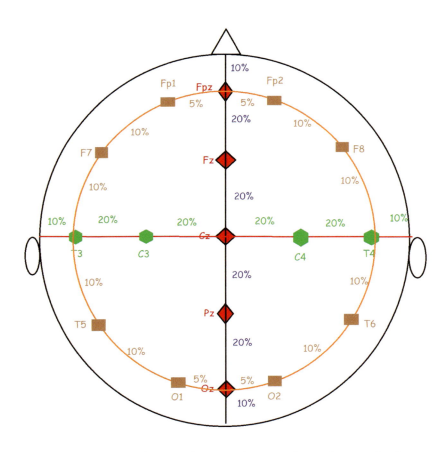

Figure 6.7B Markings on the head circumference: If you are doing sleep study where extended EEG montage (for seizure or parasomnia) is not required, only O1 and O2 markings are important.

conductive paste in the gold cup. Place the electrode and fix it.

6.2 ELECTROOCULOGRAM (EOG)

Two EOG electrodes are placed, one cm below the outer canthus of the left eye (EO1) and another (EO2) 1 cm above the outer canthus of the right eye (Figure 6.9A-B). These electrodes are referenced to the opposite side of the mastoid. Before placing the electrodes, clean the area adequately with an abrasive gel to improve conductance.

Placing one electrode above and below has special value. Cornea is positively charged compared to the retina and eyes show conjugate movement, that is, both the eyes move together

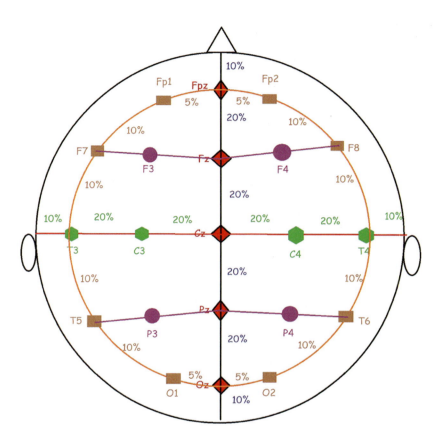

Figure 6.8 Markings for frontal and parietal electrodes: If you are doing sleep study where extended EEG montage (for seizure or parasomnia) is not required, only F3 and F4 markings are important.

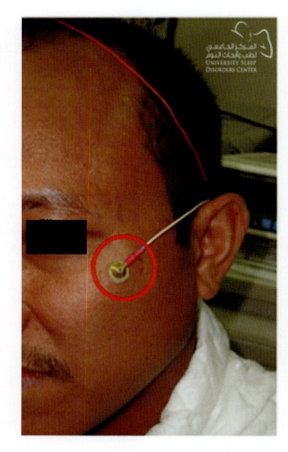

Figure 6.9 A Placement of EOG electrodes: One EOG electrode is placed one cm above and other one cm below to the outer canthi of eyes on opposite sides. Placement of EOG electrodes: One EOG electrode is placed one cm above and other one cm below to the outer canthi of eyes on opposite sides.

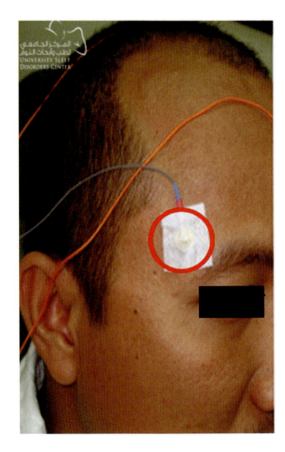

Figure 6.9 B Placement of EOG electrodes: One EOG electrode is placed one cm above and other one cm below to the outer canthi of eyes on opposite sides. Placement of EOG electrodes: One EOG electrode is placed one cm above and other one cm below to the outer canthi of eyes on opposite sides.

in one direction whether is up, down, right or left unless the patient has a weak extra-ocular muscle(s) or an artificial eye. Placing the EOG electrodes above and below to outer canthi of both eyes helps in detecting signals of opposite polarity in both leads, which makes them easily recognizable during sleep (Figures 6.10–6.13).

However, if, for some reasons these derivations cannot be placed, EOG electrodes on both sides can be placed 1 cm below and 1 cm lateral to the outer canthi of the eyes. These are then referenced to Fpz instead of the opposite side of the mastoid (Figure 6.14).

6.3 ELECTROMYOGRAM (EMG)

Electromyogram is placed at two sites in the body. First is to measure the tone of the mentalis-submentalis muscle, which helps in scoring of the sleep stage and another one to measure the periodic limb movements during sleep (PLMS).

To measure the tone of mentalis-submentalis muscles, electrodes are placed close to the chin. One electrode is placed in the midline, 1 cm above the lower border of the mandible and the other two are placed 2 cm below the lower edge of the mandible and two centimeters lateral to the midline on either side (Figure 6.15).

Placement of these electrodes is difficult in bearded persons and they often fall off.

For PLMS, electrodes are placed on the middle one-third of the anterior tibialis muscle. This may be identified by asking the patient to extend his/her lower limb and to dorsiflex the foot. As the person makes this movement, you will feel some muscle moving just lateral to the shin. This

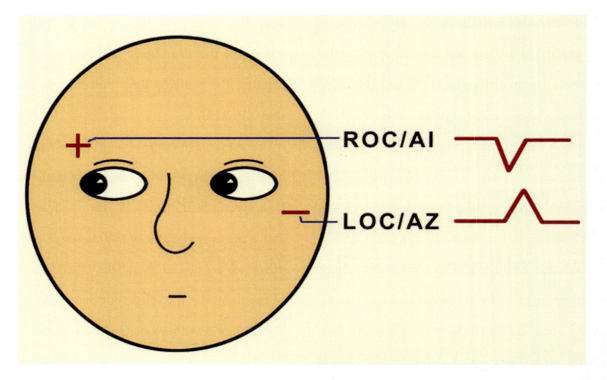

Figure 6.10 Waves during eye movement: Placement of electrodes above and below the outer canthi of both eyes results in signals of opposite phase during any movement of eyes—vertical or horizontal.

is the anterior tibialis muscle. Now divide it into three parts and place the electrode at the borders of the middle one-third of the muscle. These electrodes should be placed on both legs and separate channels (Right leg and left leg) should be used for both legs (Figure 6.16).

6.4 ELECTROCARDIOGRAM (ECG)

For the ECG, usually two leads are used—I and II. Opposed to the traditional ECG recording, where these electrodes are placed on the limbs, for sleep study, electrodes are usually placed on the chest. Place one electrode at the junction of the chest with the right shoulder, another at a similar location on the left side, and the third below the apex of the heart. This point should align the right shoulder and left hip (Figure 6.17).

Placement of Leads for the Sleep Study

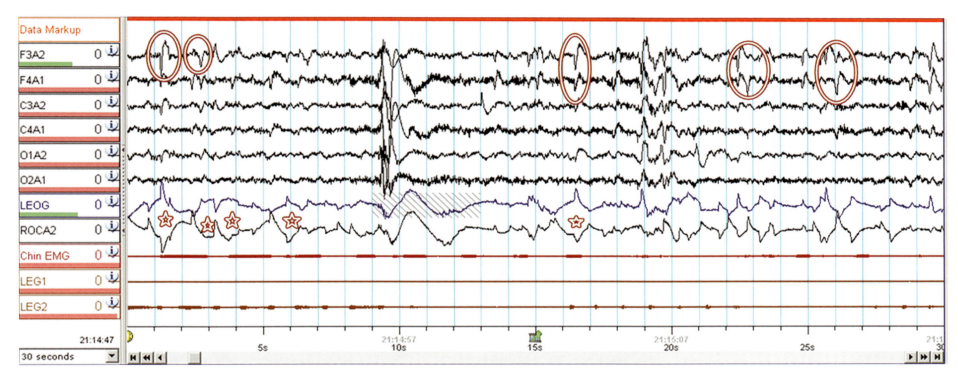

Figure 6.11 Eye movements during wake state-opposite phase: Eye movements during wakefulness. Waveform around each star represents one eye movement. Appreciate that when signals of LEOG lead are moving up, signals in other REOG/ROC lead are moving in opposite direction. Thus, eye movements can be easily recognized. In addition to the EEG signals, deflections in the frontal channels (in the circle depicting blinking) differentiate eye movements that occur during awake state from those seen during REM sleep.

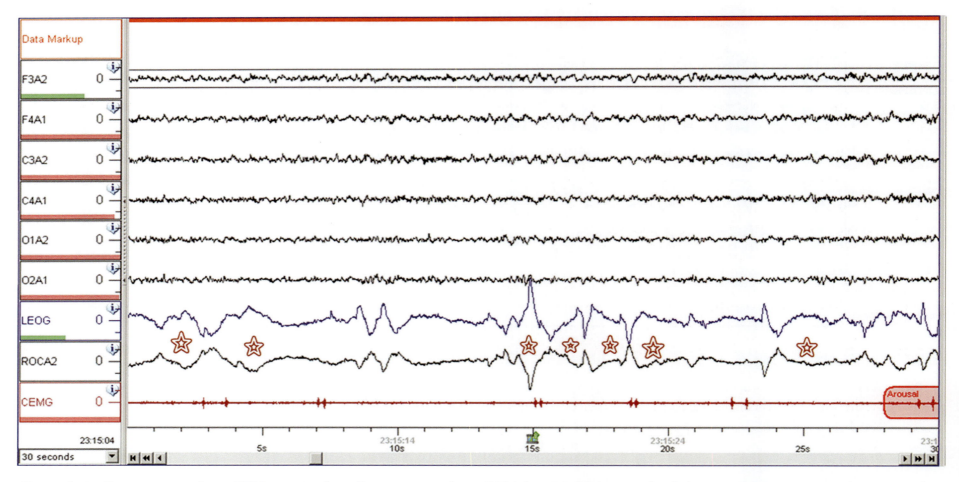

Figure 6.12 Eye movements during REM-opposite phase: Eye movements during REM sleep. Waveform around each star represents one movement. Appreciate that when signals of one EOG lead are moving up, signals in other EOG lead are moving in opposite direction. Thus, eye movements can be easily recognized. Please see that waveforms in EEG are different and blinking artifacts are absent.

Placement of Leads for the Sleep Study

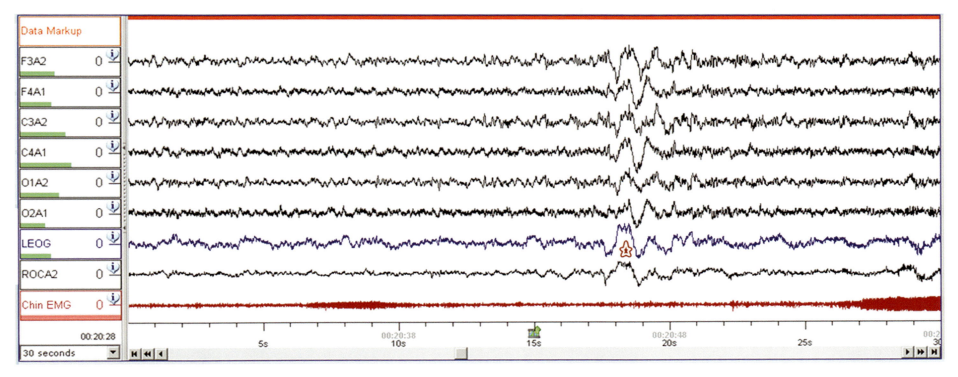

Figure 6.13 Sometimes K complexes and delta waves may appear in eye derivations. These waves may be differentiated from true eye movements as true eye movements are out of phase deflections while these are "in-phase" deflections (marked with star).

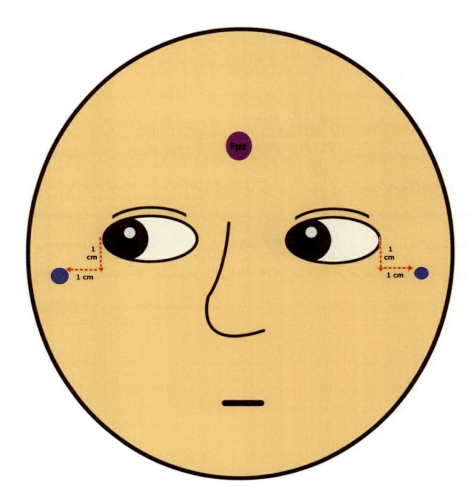

Figure 6.14 Acceptable EOG derivations: These derivations are referenced to the Fpz instead of opposite mastoid. Since they are placed lying at same level, vertical movements will appear "in-phase" while horizontal movements will appear "out of phase."

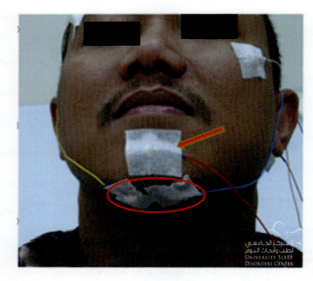

Figure 6.15 Placement of chin EMG electrodes: One side of the chin electrode works as back-up of the other one.

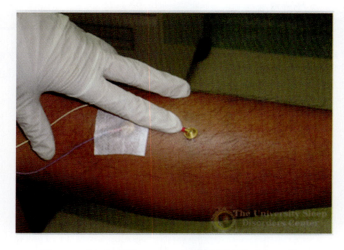

Figure 6.16 Placement of leg EMG electrodes.

Placement of Leads for the Sleep Study

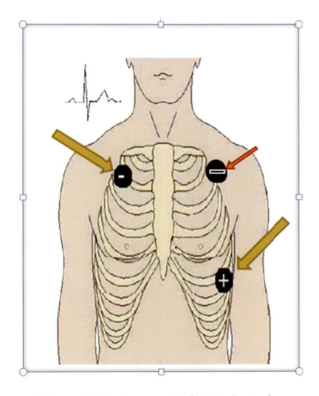

Figure 6.17 Placement of ECG electrodes.

6.5 PLACEMENT OF MEASURES OF RESPIRATION

6.5.1 Nasal Airflow

Nasal airflow is measured using a thermistor and a pressure transducer. Pressure transducer is usually attached to a cannula that is kept close to the nostrils (Figure 6.18 and 6.19A). Cannula may have one end that is to be fixed on the sensor. In the laboratories, which are recording the end tidal CO_2 as well, this cannula has two ends. Other end of the cannula is attached to the capnograph. The cannula must be properly fixed on the face with tape so that it does not get misplaced during the study. Some companies supply cannula that has a provision for the placement of thermistor in itself (Figure 6.19B). This kind of cannula reduces the crowding of instruments in nostrils. If you do not have this kind of cannula, you may place the thermistor along with the cannula.

6.5.2 Respiratory Effort

Respiratory effort is measured by either an elastic belt or an RIP belt which is placed over the chest and abdomen. These belts should not be too loose or too tight. In the former case, they will fail to record the movement, while in the latter case, they will produce discomfort. A rule of thumb is to lengthen the belt to the two-third of the circumference of the chest and abdomen, and then using its elastic property, tie it around the chest and abdomen. The belly should not be twisted as it will not only produce discomfort, but in case of an RIP belt, also interfere with proper signal generation.

On the chest, it should be placed at the level of nipples. Fixing the chest belt may be a bit

138 Clinical Atlas of Polysomnography

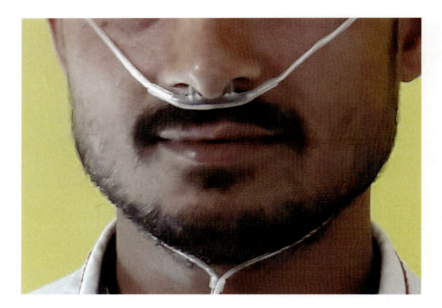

Figure 6.18 Placement of nasal cannula: Nasal cannula is placed in nose and can be fixed on the cheeks using a surgical tape.

Figure 6.19A Placement of thermistor: Thermistor is placed in the nose.

Placement of Leads for the Sleep Study

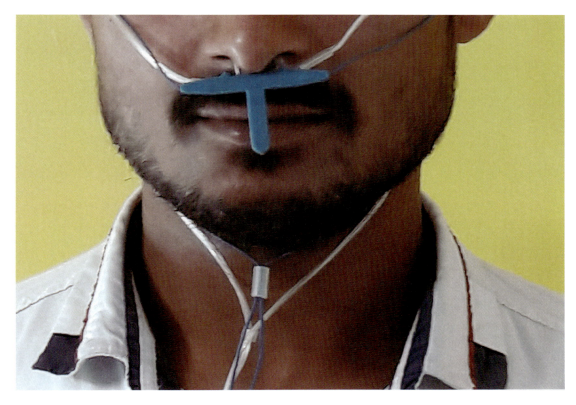

Figure 6.19B Thermistor should be placed along with the nasal cannula.

tricky in case of females, where it should be fixed just below the breasts. Placing it over the breasts increases the chances of the belt sliding, which results in improper signals (Figures 6.20 and 6.21).

Over the abdomen, the belt should be placed at the level of the naval. You must remember that abdominal girth reduces as a person assumes supine position from standing or sitting. So you may need to change the girth of the belt accordingly. In case of obese persons, it may be fixed just below the rib cage as it may slide if placed at the level of the naval.

6.5.3 Oxymeter

Two kinds of oximeter probes are usually supplied along with the machine—ones that have free ends that get fixed with the help of an adhesive tape (Figure 6.22). This tape get adhered

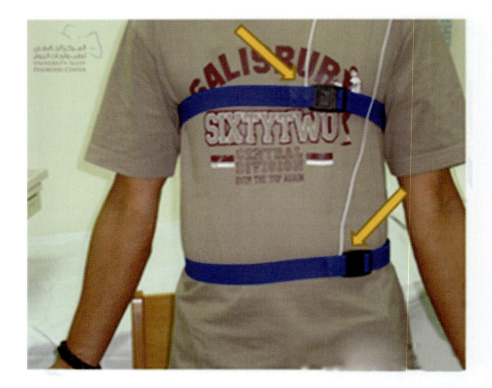

Figure 6.20 Placement of RIP belts: RIP belt for the chest is placed at the level of nipples while that of abdomen is placed at the level of navel.

to the finger, thus, it prevents it from falling off. However, many companies supply a two-pronged sensor that may fall off during a study. To prevent misplacement, such probes are usually fixed on the finger with the help of an adhesive tape. Adequate precautions should be taken before placing them, as discussed in Chapter 4.

6.5.4 Snore Microphone

It should be placed on either side of the midline close to the Adam's apple. This should be firmly fixed at that place with surgical tape so that it does not fall off during sleep.

6.6 BODY POSITION SENSOR

Body position sensor is usually placed over the chest and it gets fixed to the respiratory chest

Placement of Leads for the Sleep Study

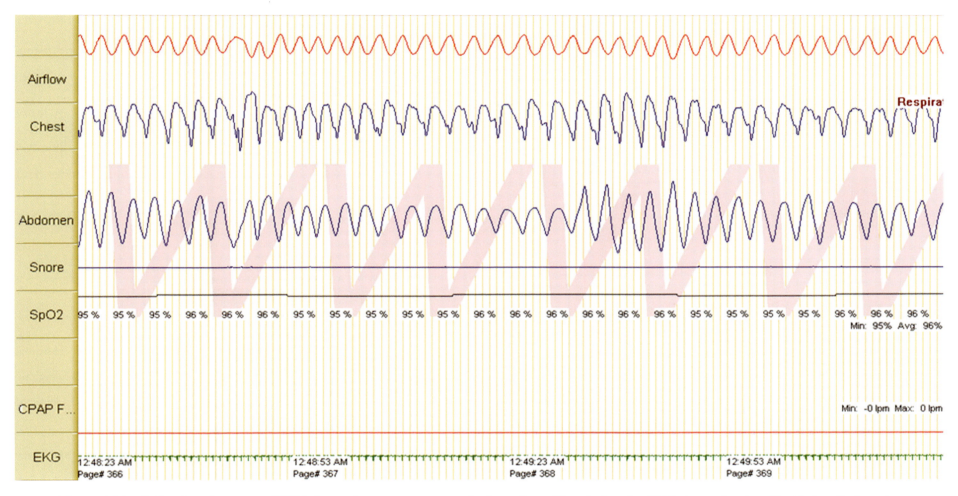

Figure 6.21 In females, placement of RIP belt over breasts generate notched waveform. Thus, in females, RIP belt may be placed just below the breasts.

belt. Most of the sensors have a diagram over them that guides you regarding their placement. Please ensure that you place it correctly to obtain correct data (Figure 6.23). A loose sensor may give you spurious data that may influence the interpretation and management plan.

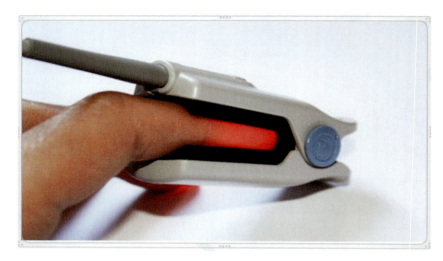

Figure 6.22 Placement of oximeter.

FURTHER READING

1. Berry, R. B., Brooks, R., Gamaldo, C. E., Harding, S. M., Lloyd, R. M., Marcus, C. L., & Vaughan, B. V., (2017). For the American Academy of Sleep Medicine. The AASM manual for scoring of sleep and associated events: rules, terminology and technical specifications. Version 2.4. www.aasmnet.org. Darian, Illinois: *American Academy of Sleep Medicine.*

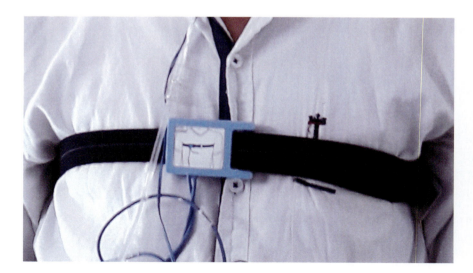

Figure 6.23 Placement of body position sensor: Body position sensor should be placed in midline and is usually mounted on the chest belt.

REVIEW QUESTIONS

1. Back up EEG electrodes are placed at:
   A. Same positions on the opposite hemisphere
   B. Different position on the same hemisphere
   C. Anywhere on both the hemispheres
   D. On the frontal prominences

2. Central electrodes capture all of the following EXCEPT:
   A. K-complexes
   B. Vertex waves
   C. Sawtooth waves
   D. Delta waves

3. EOG electrodes are placed at:
   A. 1 cm outside the opposite outer canthi
   B. 1 cm above and below the opposite outer canthi
   C. 1 cm lateral and then above and below the opposite outer canthi
   D. 1 cm below the outer canthi

4. Chin EMG picks the signals from:
   A. Hypoglossal muscle
   B. Submentalis muscles
   C. Platysma
   D. Sternohyoid muscles

5. Ideal impedance for the electrodes is:
   A. 16–20 K Ohm
   B. 11–15 K Ohm
   C. 6–10 K Ohm
   D. 0–5 K Ohm

6. Preferred lead for the EKG during sleep study is:
   A. III
   B. II
   C. I
   D. IV

7. In patient with predominant oral breathing, for measuring airflow:
   A. Cannula may be used that has additional port for oral airflow
   B. Cannula cannot be used to measure airflow
   C. Thermistor is better choice than cannula to measure airflow
   D. Oro-nasal mask with PAP at 4 cm H_2O can be used

8. Following must be taken care of while placing RIP belts:
   A. Belt length should be 1/3 of the girth of chest/abdomen where it is to be placed
   B. Belt length should be 2/3 of the girth of chest/abdomen where it is to be placed
   C. Belt length should be kept to the minimum

D. Belt length should be kept at the maximum

9. Body position sensor should be placed:
   A. Forehead
   B. Abdominal belt
   C. Chest belt in midline of chest
   D. Arm that is on upper side while the patient sleeps

10. Snore microphone should be placed:
    A. Junction of the upper 1/3 and lower 1/3 of neck in midline
    B. On the Adams apple in midline
    C. Close to the Adams apple on either side
    D. Close to the nose, just lateral to the nares

ANSWER KEY

1. A   2. D   3. C   4. B   5. D   6. C
7. A   8. B   9. C   10. C

7

STARTING AND CLOSING THE STUDY

LEARNING OBJECTIVES

After reading this chapter, the reader should be able to start and close the recording in Level 1 polysomnography machines from various manufacturers.

CONTENTS

Steps at Starting and Closing the Sleep Study 146
Review Questions 178
Answer Key 178

In this chapter, we will discuss, step-by-step, how to start and close a sleep study in Level 1 polysomnography machines from three manufacturers—Philips Respironics, Somnomedics, and Cadwell. This does not involve any technical information, hence, only illustrations are used for the understanding (Figures 7.1–7.31).

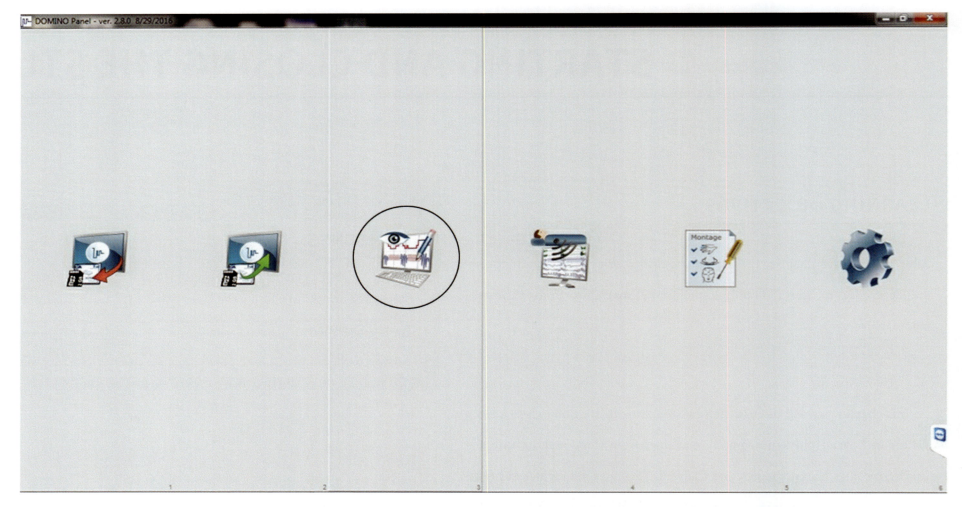

Figure 7.1 Handling Somnoscreen Software: On opening the software this page appears on the screen. Click on "New Study."

Starting and Closing the Study

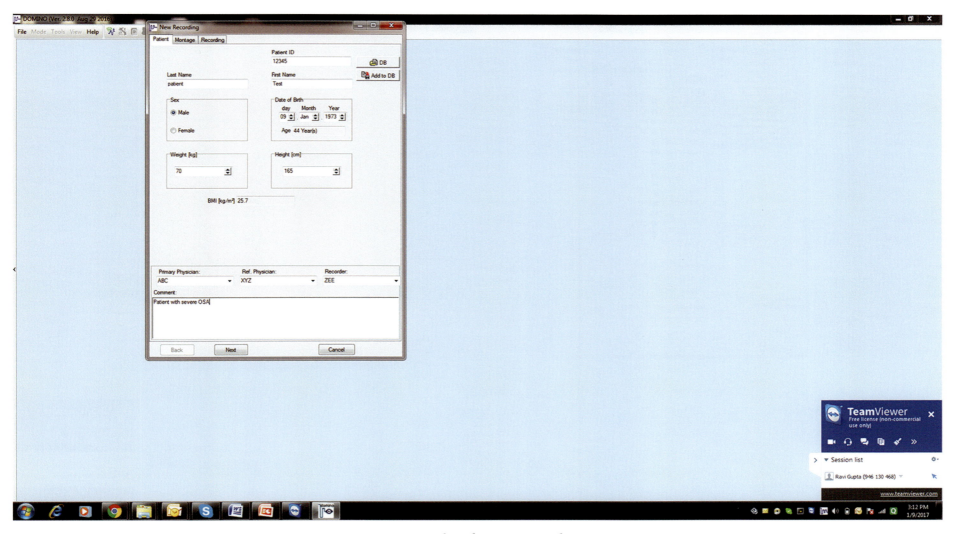

Figure 7.2 Enter the information regarding patient.

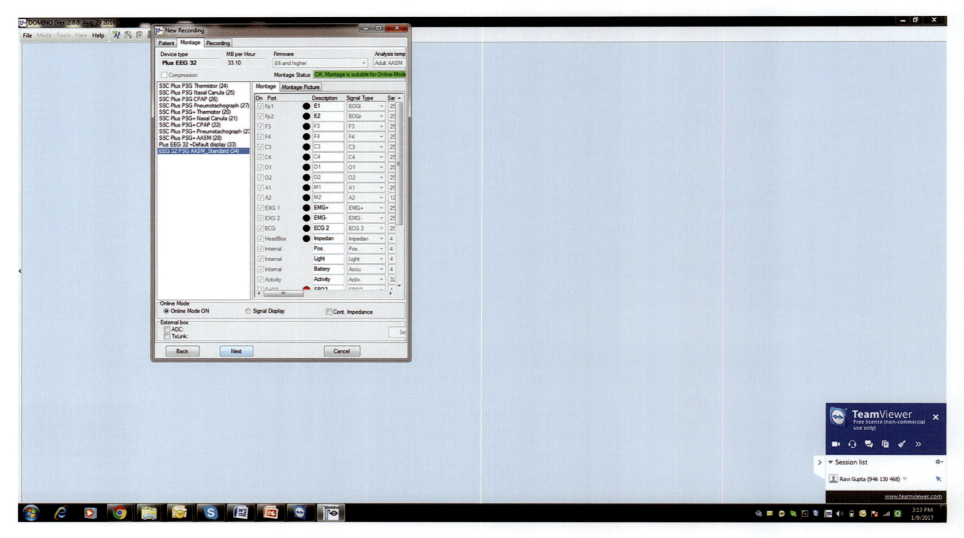

Figure 7.3 Click on Montage and choose the required montage.

Starting and Closing the Study

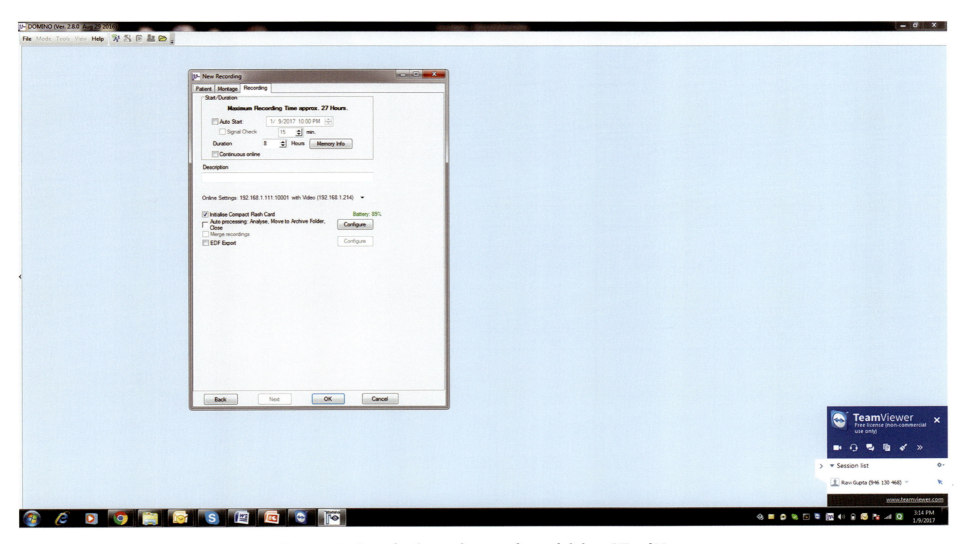

Figure 7.4 Enter details regarding recording and click on OK and Next.

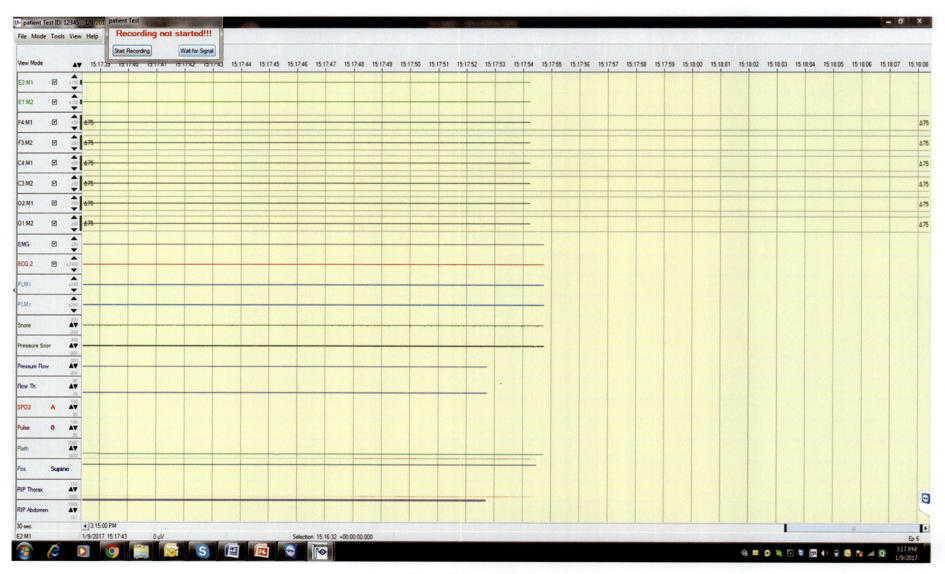

Figure 7.5 Click on Start Recording.

Starting and Closing the Study

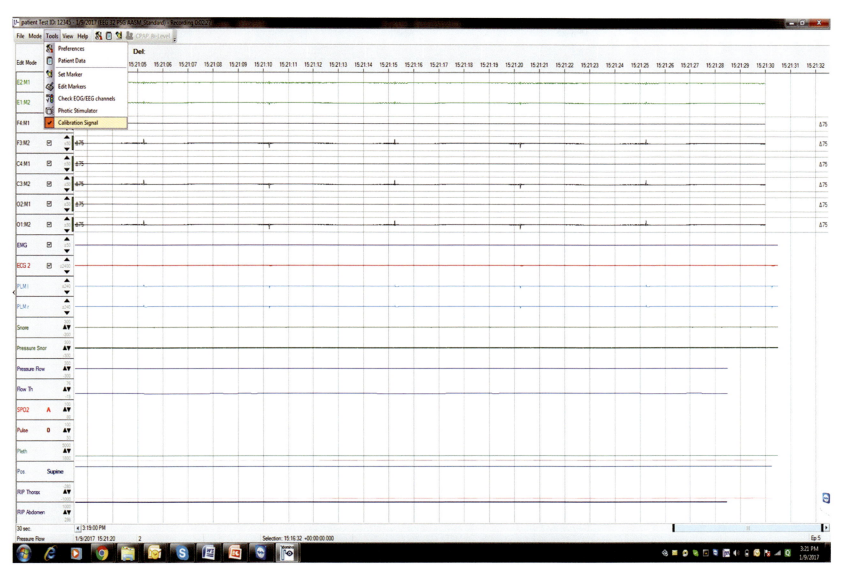

Figure 7.6 Start Calibration (Chapter 8). Click again to end calibration.

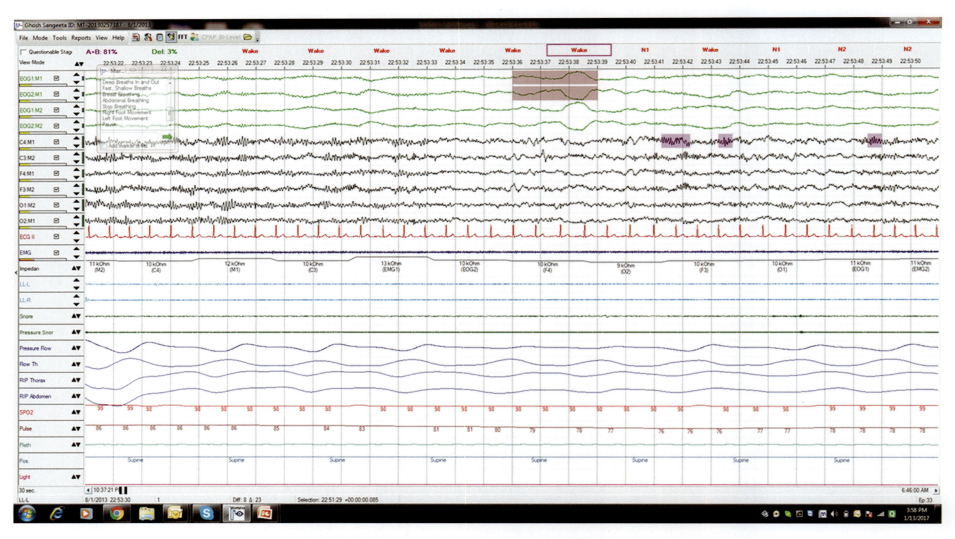

Figure 7.7 At the end of calibration, these signals will appear on the page. This is real-time data.

Starting and Closing the Study

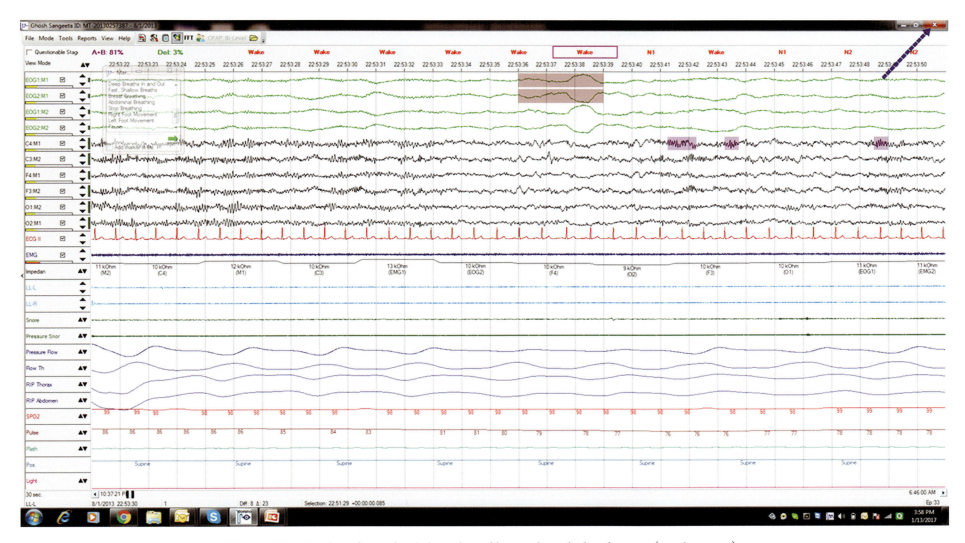

Figure 7.8 To close the study, click on the red box in the right-hand corner (see the arrow).

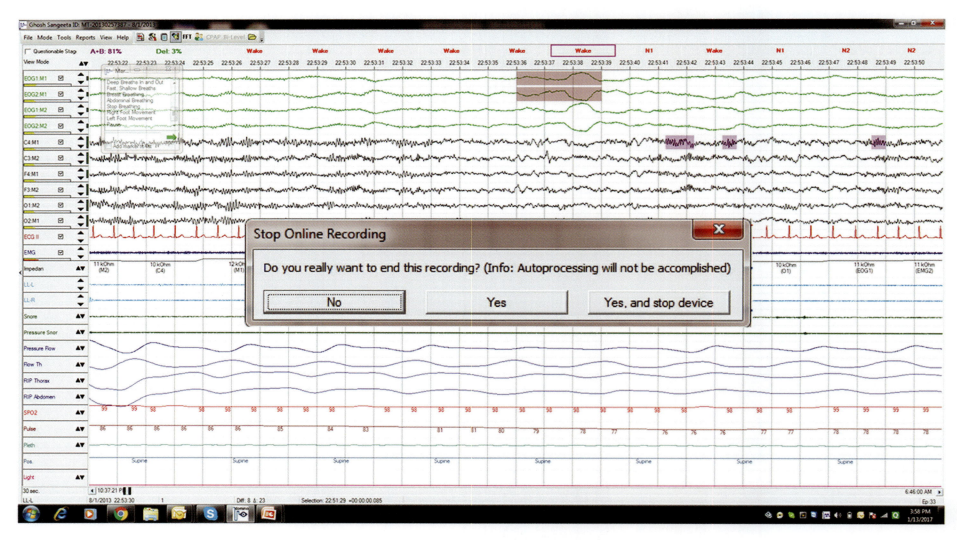

Figure 7.9 This dialogue box will appear. Click on Yes and Stop Device.

Figure 7.10 Handling Philips Alice 5 Software: Click on the G3 icon on desktop.

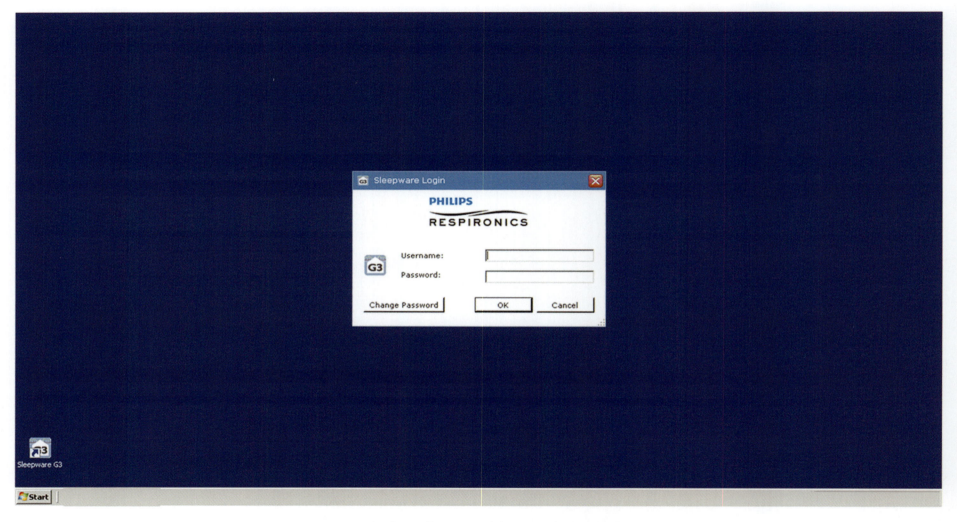

Figure 7.11 This box will appear. Enter your information. Click OK.

Starting and Closing the Study

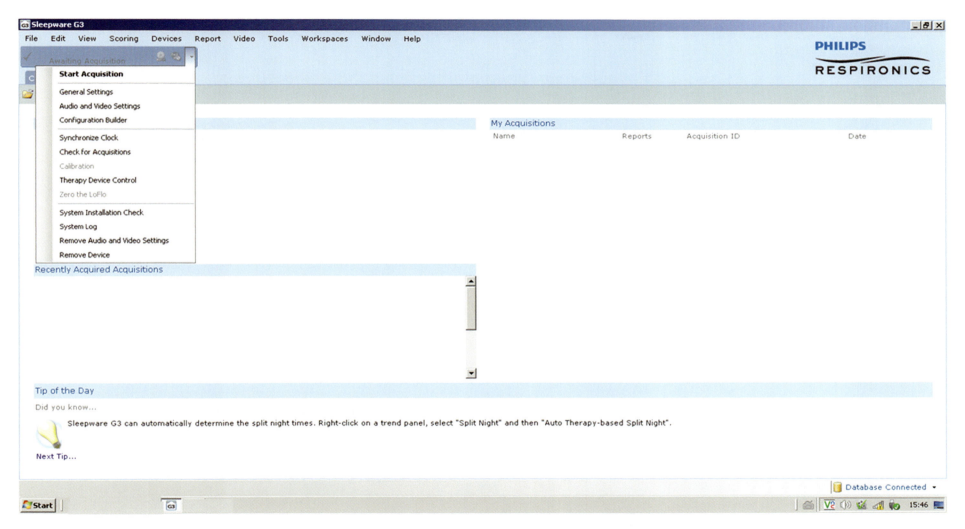

Figure 7.12 This page will appear. Click on Start Acquisition in the drop-down menu.

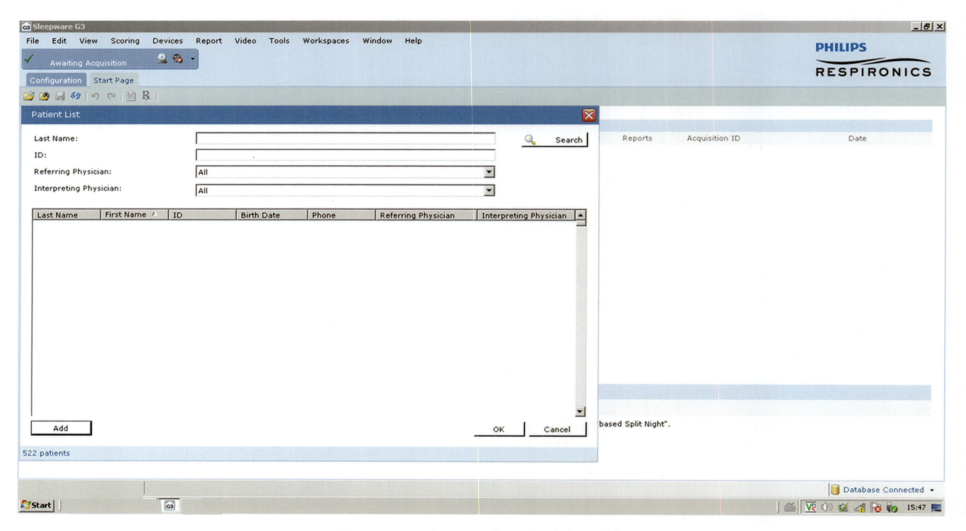

Figure 7.13 This page will appear. Click on Add.

Starting and Closing the Study

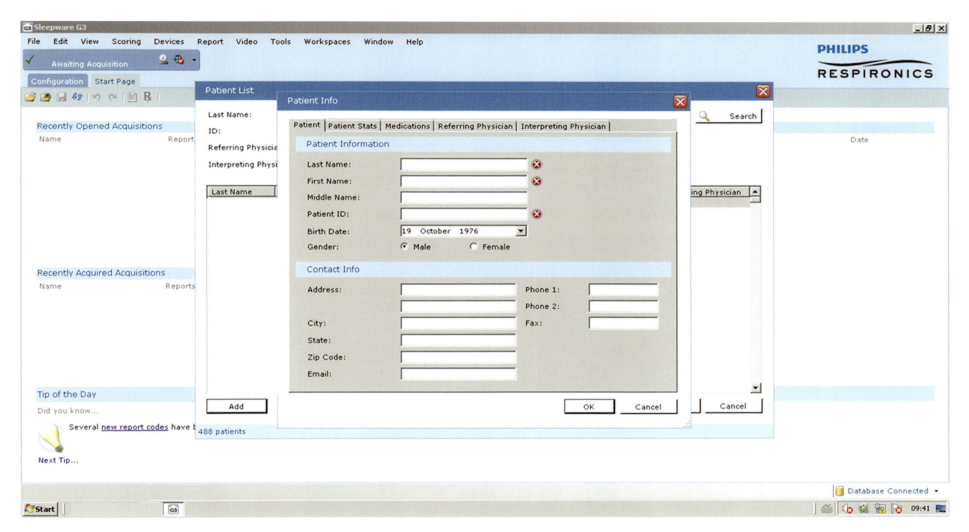

Figure 7.14 Fill in the patient's information.

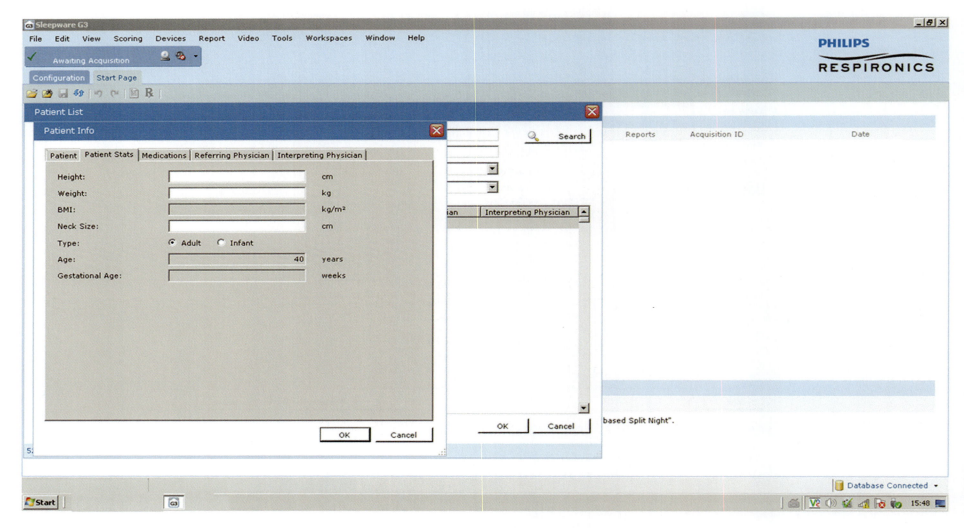

Figure 7.15 Fill the anthropometric data.

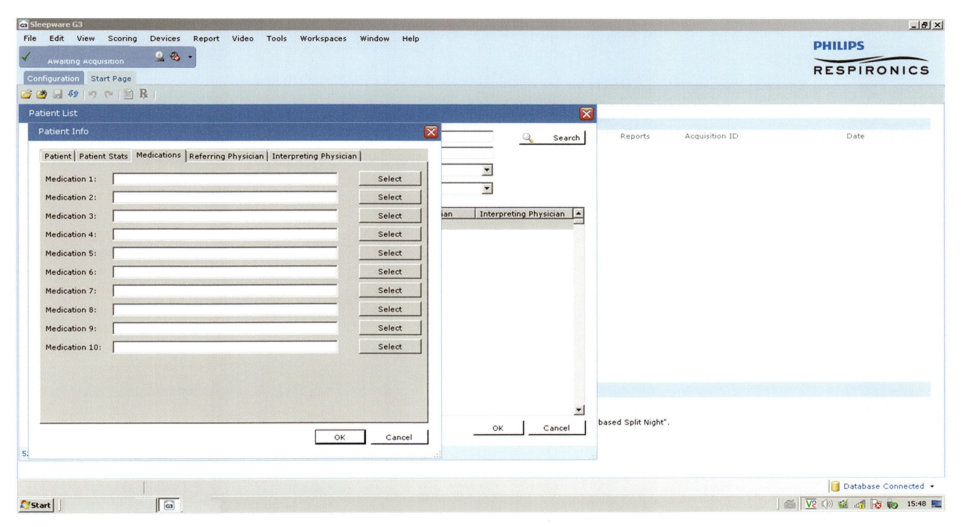

Figure 7.16 Fill in the details of Medications.

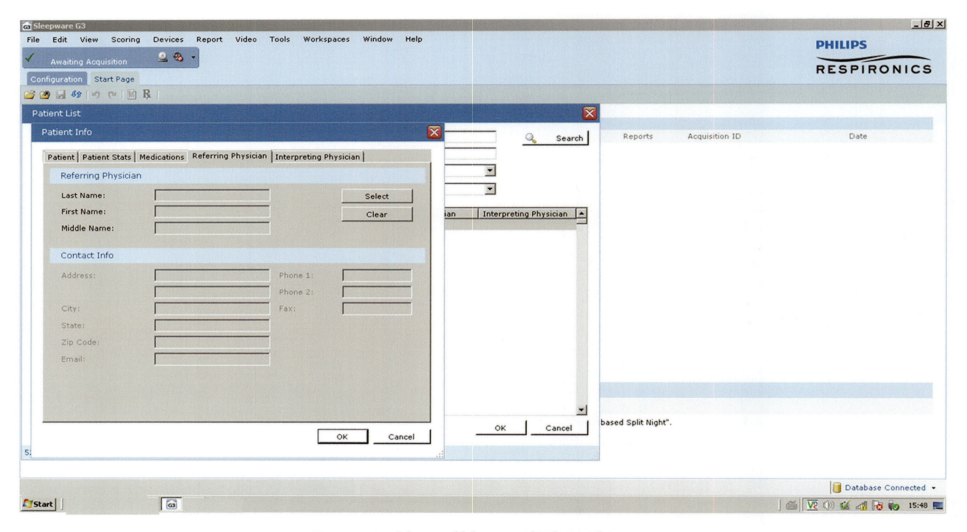

Figure 7.17 Select or add the name of Referring Physician.

Starting and Closing the Study

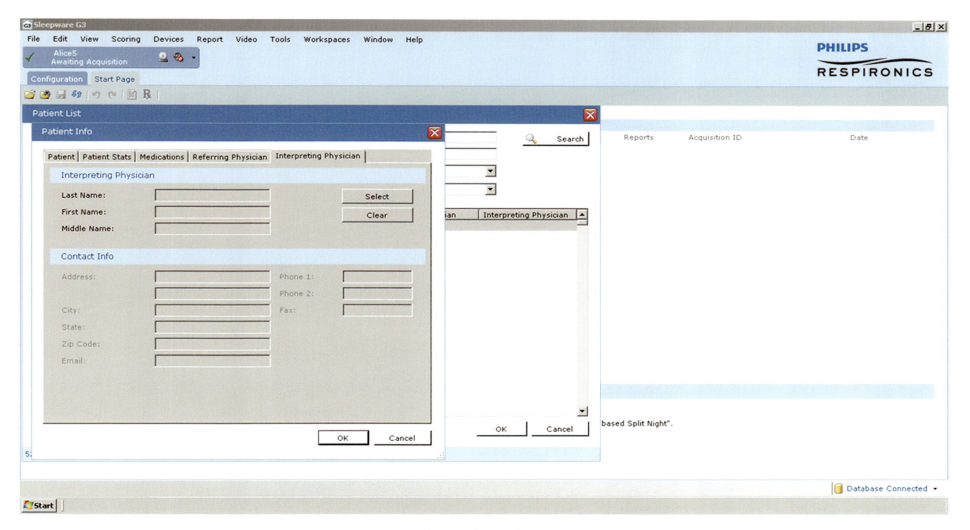

Figure 7.18 Fill in the details of Interpreting Physician.

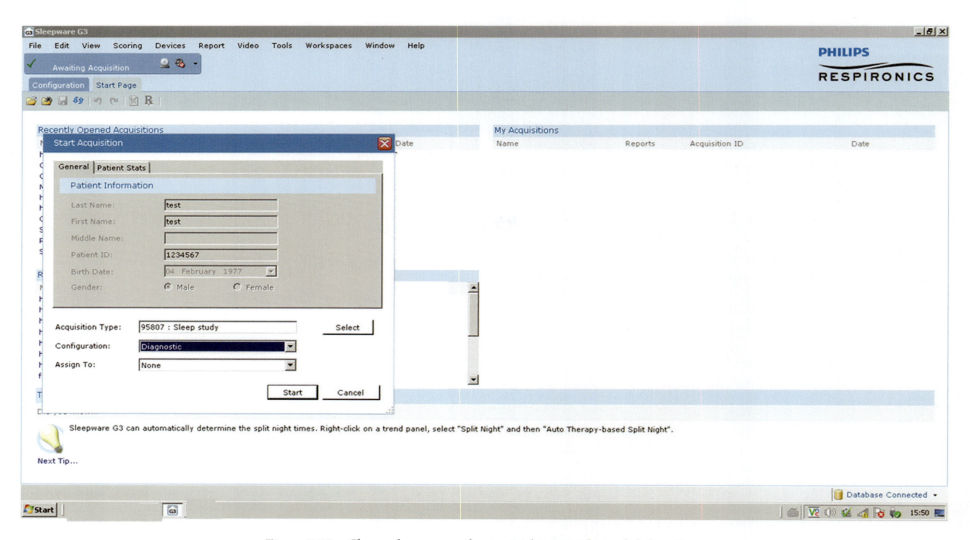

Figure 7.19 Choose the montage that you wish to record in and click on Start.

Starting and Closing the Study

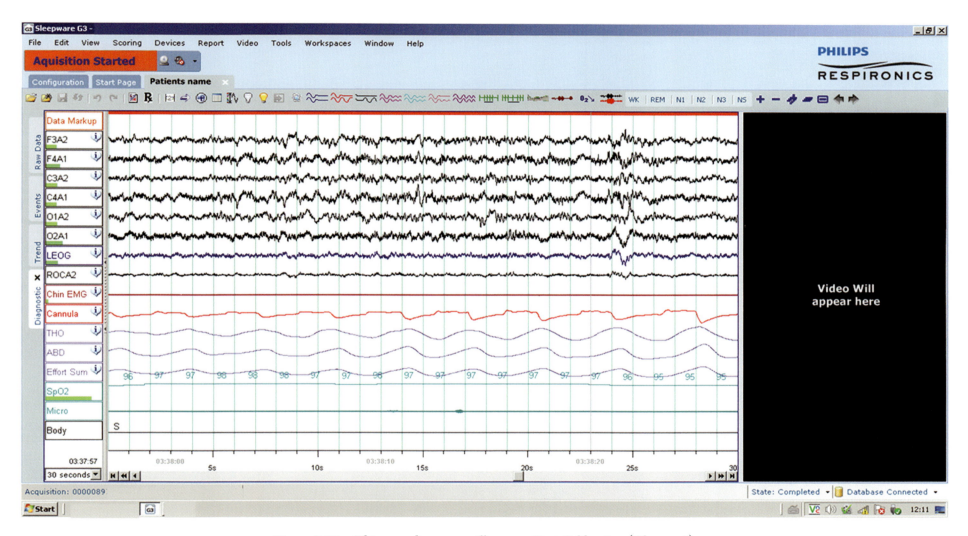

Figure 7.20 This recording page will appear. Start Calibration (Chapter 8).

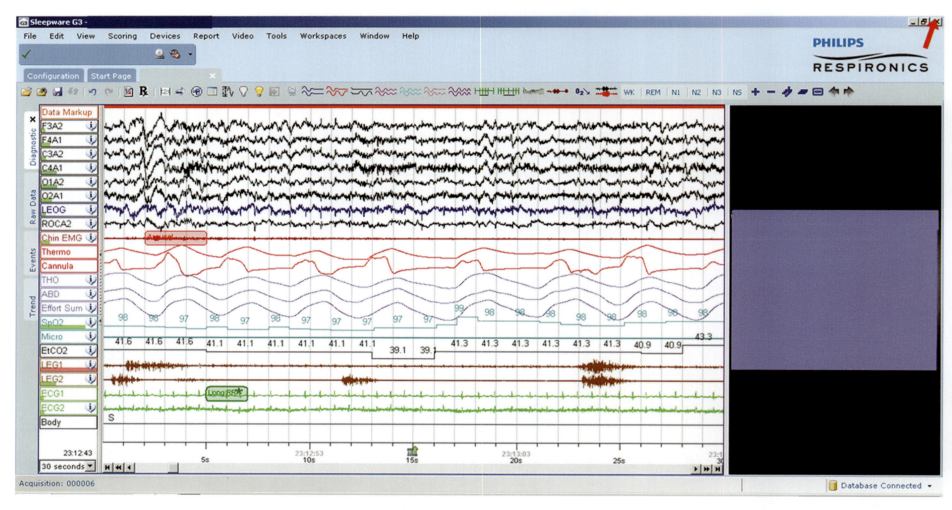

Figure 7.21 Click on this arrow in top right-hand corner to stop the study.

Starting and Closing the Study

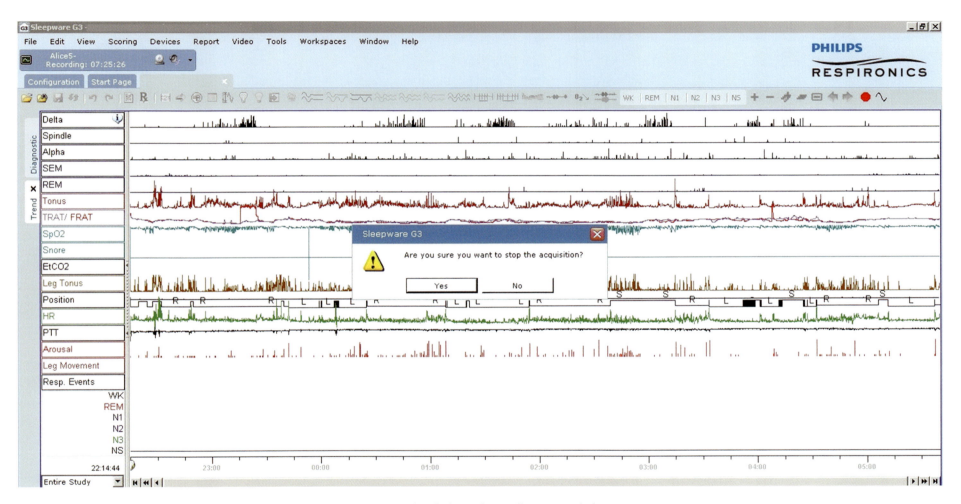

Figure 7.22 This dialogue box will appear. Click on Yes.

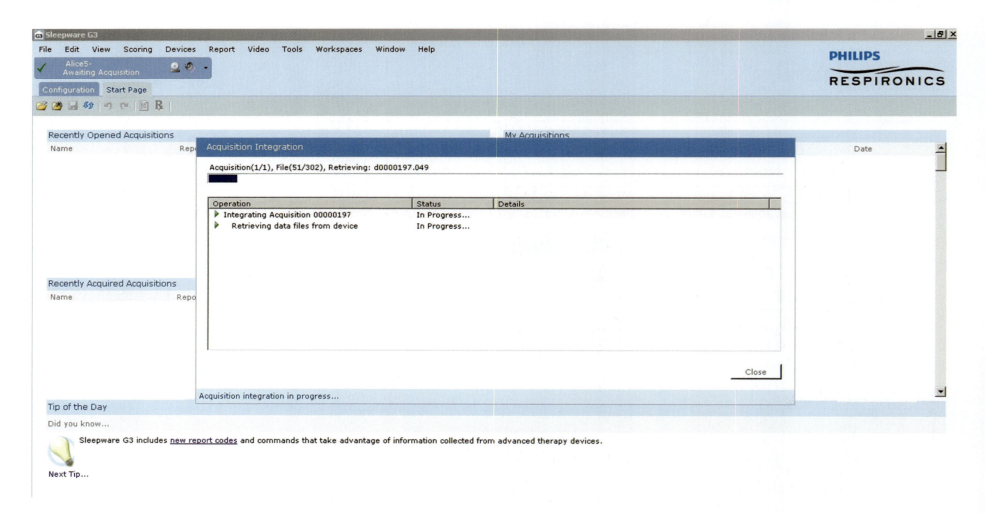

Figure 7.23 This box will appear. Let the file transfer be completed.

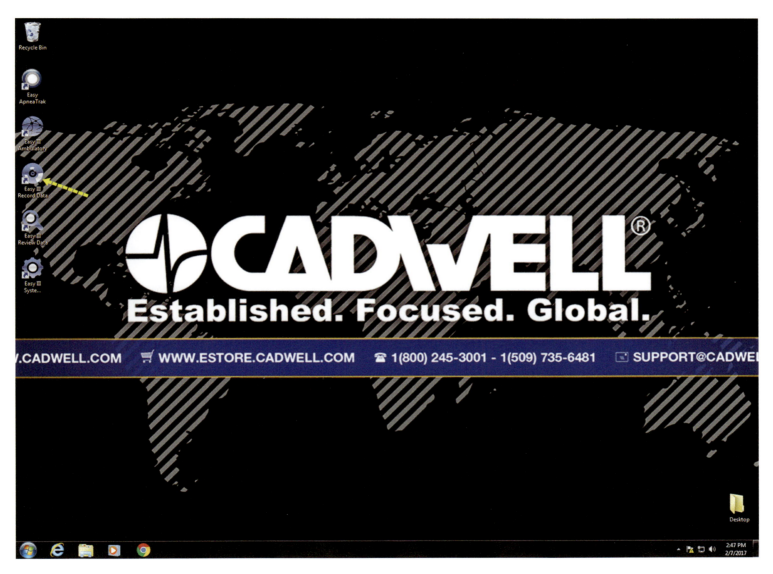

Figure 7.24 Handling Cadwell Software: Click on Easy III Record Data.

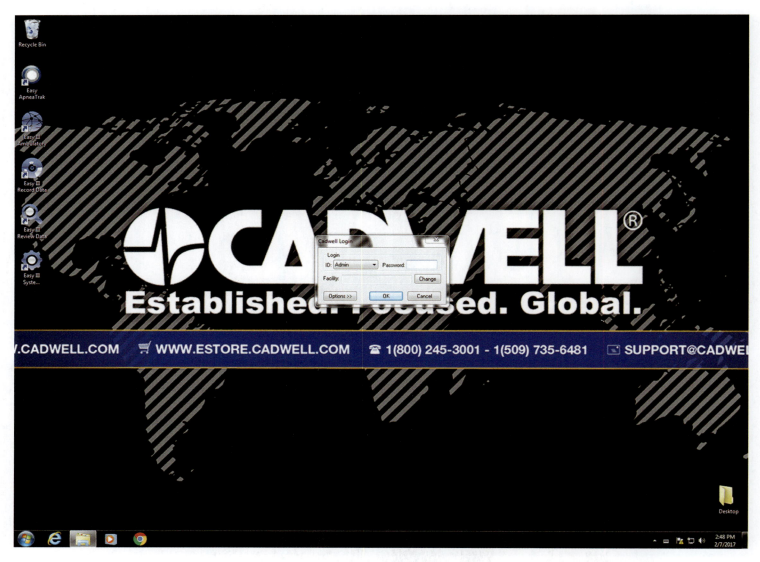

Figure 7.25 Fill the information.

Starting and Closing the Study

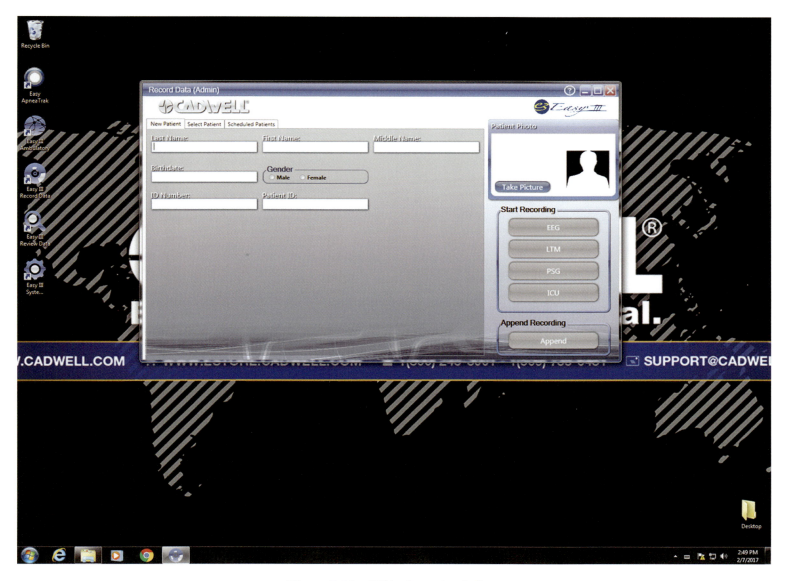

Figure 7.26 Fill in the patient's data.

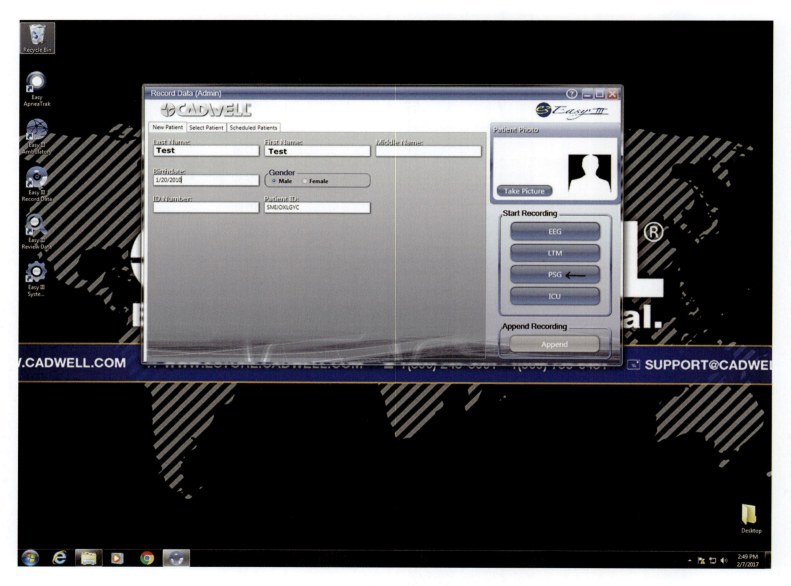

Figure 7.27 Click on PSG.

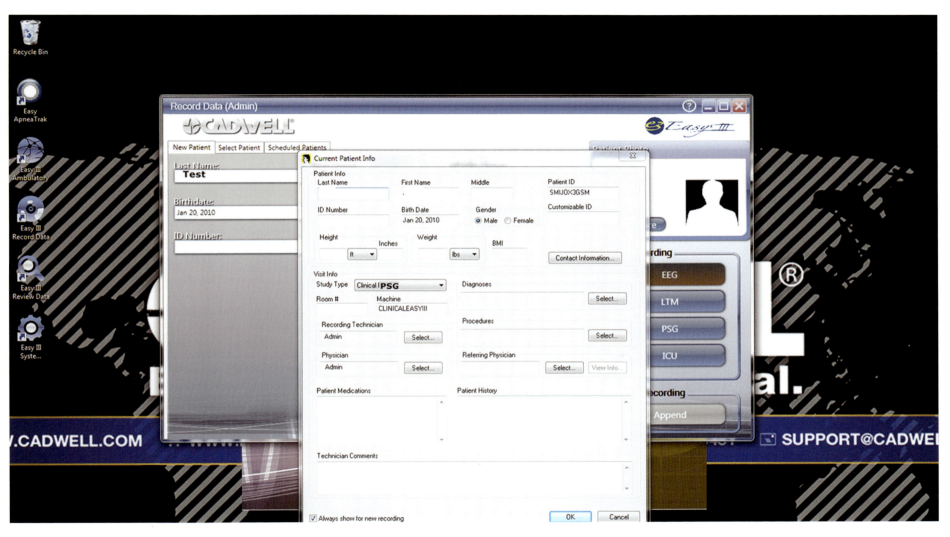

Figure 7.28 Fill in the details.

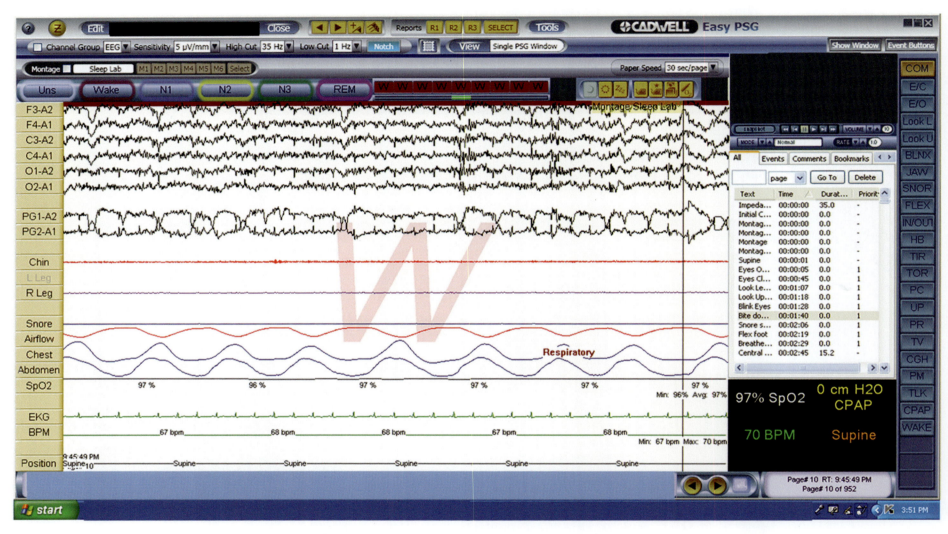

Figure 7.29 This page will appear. Start Calibration (Chapter 8).

Starting and Closing the Study

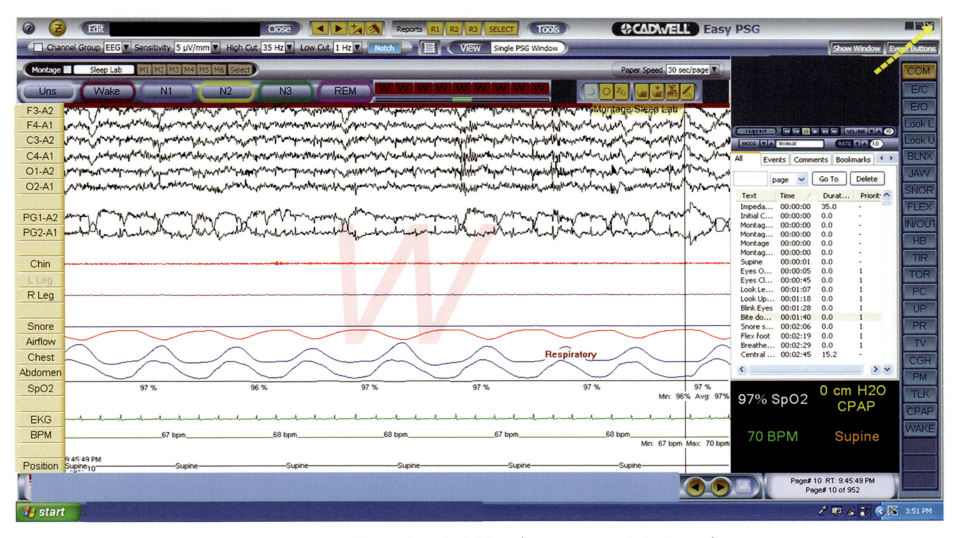

Figure 7.30 To stop the study, click here. (see arrow in top right-hand corner).

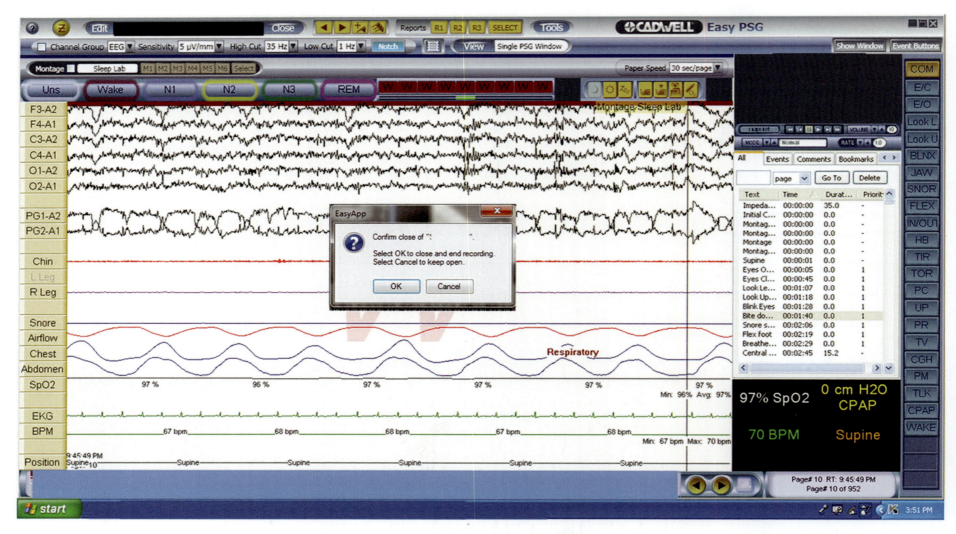

Figure 7.31 This box will appear. Click Yes.

Starting and Closing the Study 177

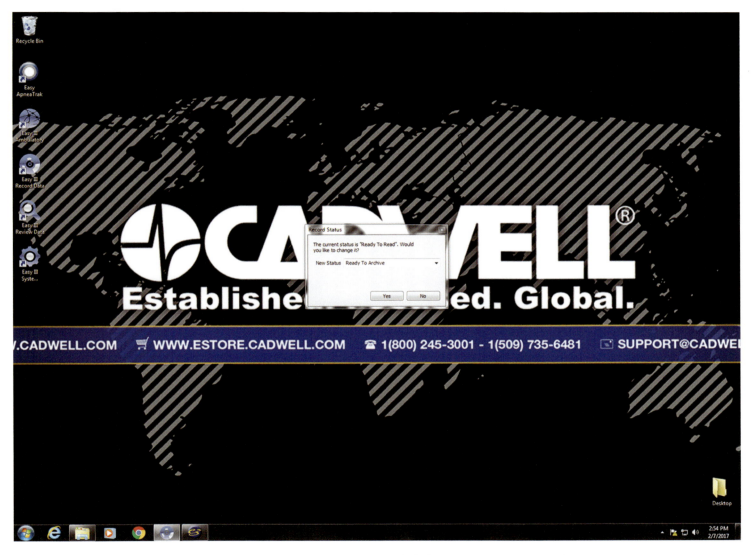

Figure 7.32 This box will appear. Click the appropriate command.

REVIEW QUESTIONS

1. While starting the sleep study, following should be carefully checked and fixed:
   A. Impedance
   B. Lighting of room
   C. Oxygen saturation
   D. Ventilation of the room

2. Data entered during start of study helps in:
   A. Diagnosis of the patient
   B. Management of the patient
   C. Recall of the data with similar characteristics at a later period
   D. Assessing the severity of illness

3. While closing the sleep study:
   A. Make sure that patient is awake
   B. Make sure that data is saved
   C. Make sure that patient is sleeping
   D. Make sure that software is closed

ANSWER KEY

1. A   2. C   3. B

8

CALIBRATION AND BIOCALIBRATION

LEARNING OBJECTIVES

After reading this chapter, the reader should be able to:

1. Discuss the needs for biocalibration.
2. Can recognize the waveforms that appear during biocalibration.
3. Can perform the biocalibration during a sleep study.

CONTENTS

Steps of Biocalbration 179

Further Reading 194

Review Questions 194

Answer Key 195

Calibration is done to check that all the channels are getting adequate information and are free from artifacts. This should be done before every study. The position of calibration command varies among software provided by different manufacturers (Figure 8.1A–C).

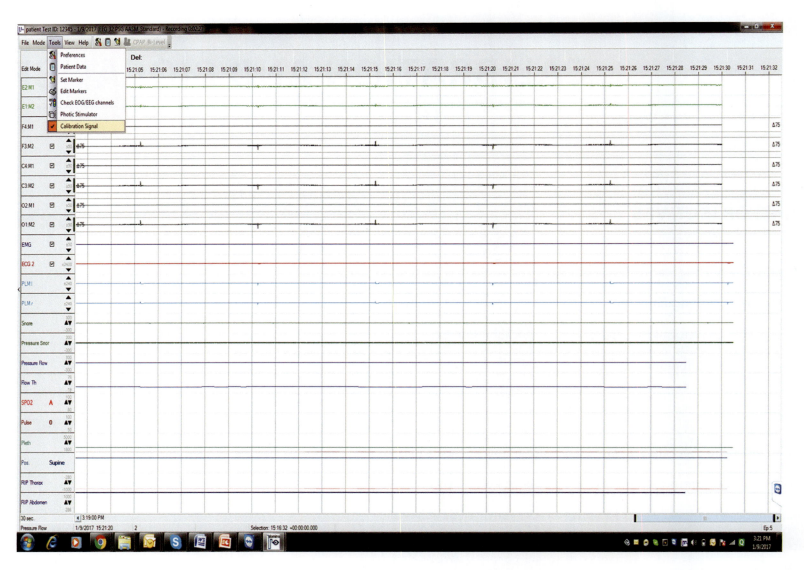

Figure 8.1 A Location of Calibration Command varies among manufacturers. Here, we show how to get Calibration Command in machines from three major manufacturers. A: Calibration Command in Somnoscreen.

Calibration and Bio-Calibration

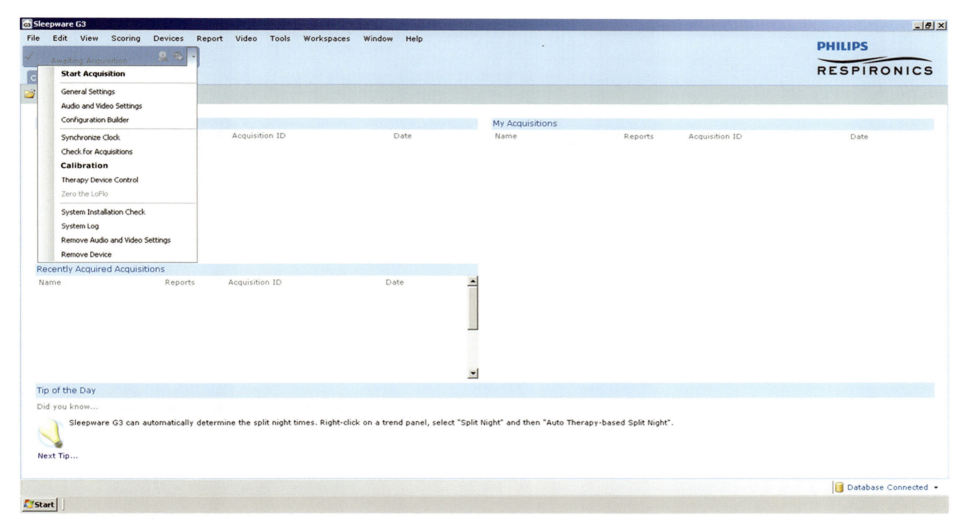

Figure 8.1 B Calibration Command in Alice 5.

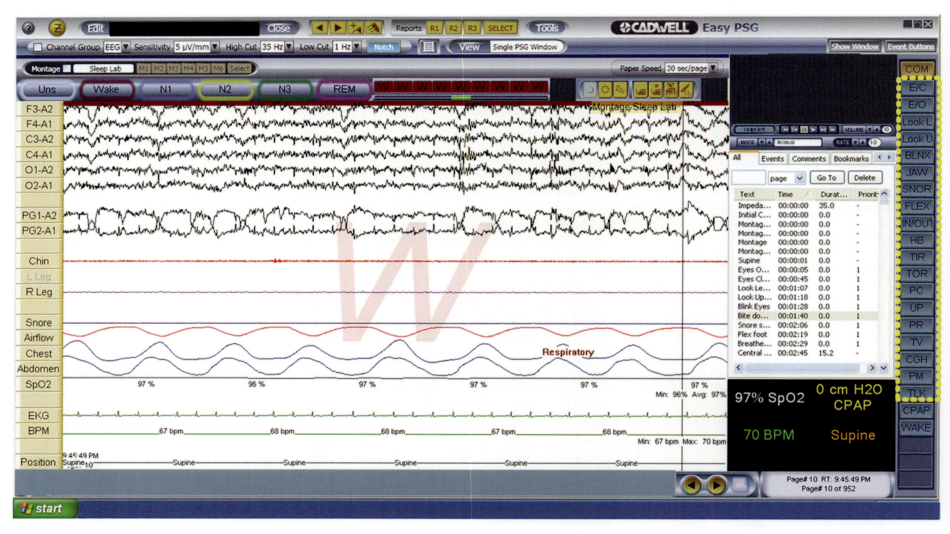

Figure 8.1C Calibration Command in Easy III (in yellow box).

For electrical calibration, we usually pass a current of around 50 μV in the EEG channels and look for the waveform. All the channels must have similar waveform during calibration. This should be done for at least 10 seconds (Figure 8.2). For optimal quality signals, the impedance of the electrode should be below 5 Kohms.

After calibration, it is time to check that all channels are working properly, that is, bio-calibration. This should be done by providing the commands to the patient and checking that the respective channels are working properly. To do this, please give the following commands and look for the changes in the display on the monitor.

1. Record with eyes open and eyes closed for at least 30 seconds duration each.
2. Ask the patient to blink his/her eyes 5 times (Figure 8.3).
3. Ask the patient to look up and down 3 times and then right and left 3 times without moving head (Figure 8.4).
4. Ask the patient to close eyes for 10 seconds and look for change in EEG (appearance of Alpha rhythm in occipital leads) (Figure 8.5).
5. Ask the patient to open his/her eyes.
6. Now ask the patient to stop breathing for 10 seconds (Figure 8.6).
7. Ask the patient to resume breathing and produce a snoring sound (Figure 8.7).
8. Ask the patient to move his/her feet up and down (Figure 8.8).
9. Some machines provide blood pressure in real-time. In those machines, blood pressure has to be calibrated after manual measurement of blood pressure during the start of the recording (Figure 8.9).
10. Ask the patient to clench his/her teeth for 5 seconds (Figure 8.10).
11. Ask the patient to speak something and make sure that you can hear his/her voice through a speaker (Figure 8.11).
12. Body position signals are calibrated by asking the patient to turn left and then turn right in bed. He/She should remain in each position for at least 5 seconds.

Also, check that signals are good in ECG leads, oximeter, capnograph, and video. Once you are satisfied, you may start the study.

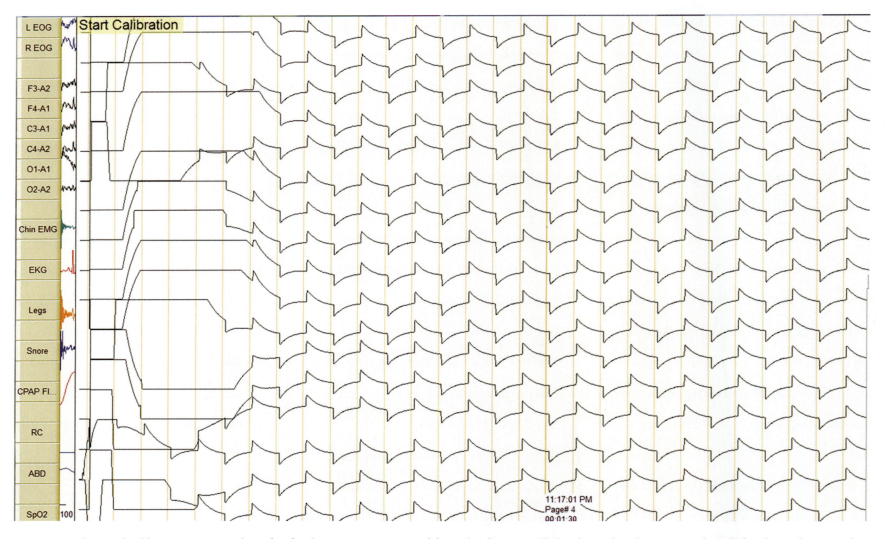

Figure 8.2 Electrical calibration: During this a fixed voltage current is passed from the device to all the channels. This ensures that all the channels are working normally.

Calibration and Biocalibration

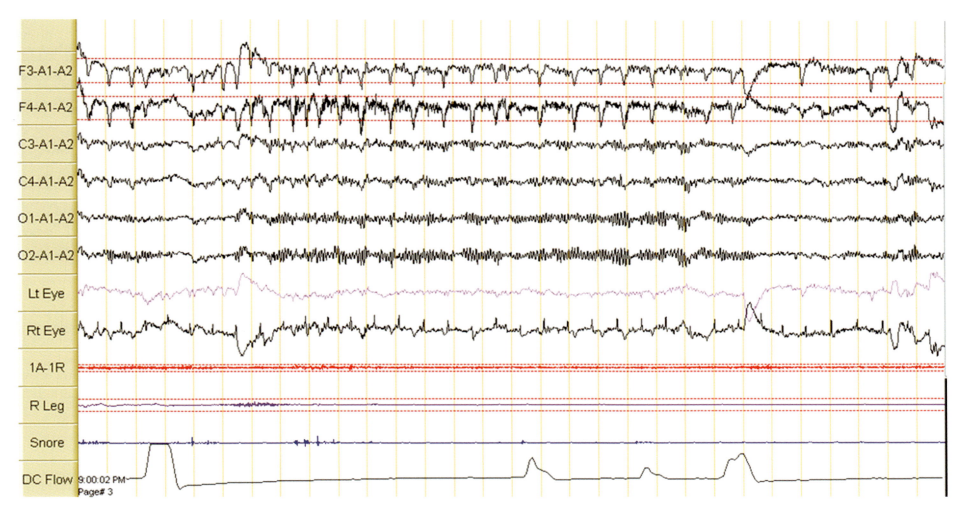

Figure 8.3 Calibration of blinking: Blinking appears as positive wave in frontal derivations as positively charged cornea moves towards frontal derivations due to bells-eye phenomenon during blinking.

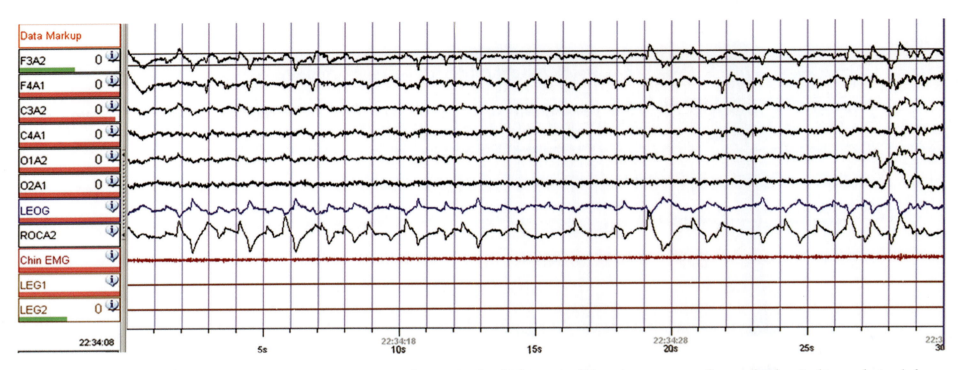

Figure 8.4 Calibration of eye movement: Eye movements appear as darting signals, which are out of phase, i.e., move away from each other. In this epoch signals from left EOG are shorter in amplitude compared to signals from right EOG. This is because of higher impedance in left EOG. Also appreciate spillage of eye movement signals in frontal derivations due to bells-eye phenomenon.

Calibration and Biocalibration

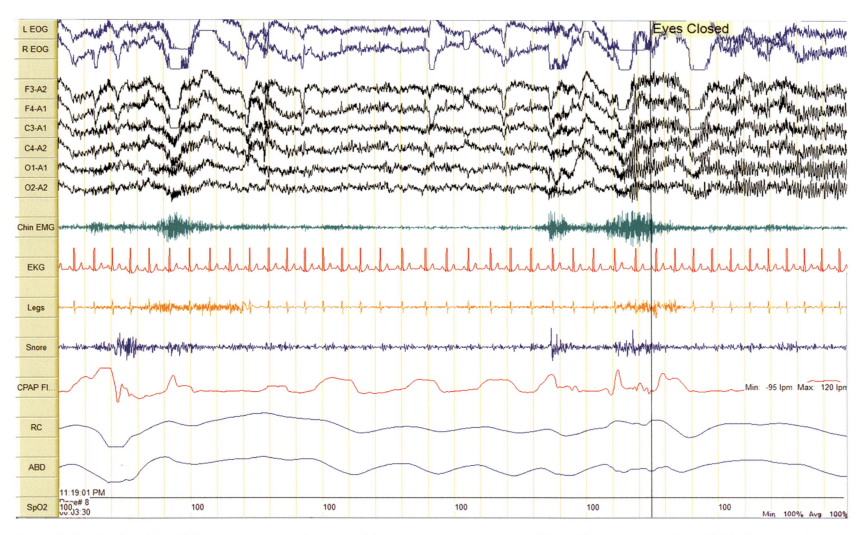

Figure 8.5 Calibration for Alpha: Alpha waves appear in the occipital derivations during quiet wakefulness. However, about 10% individuals are not alpha producers. See the low voltage mixed frequency activity during open eyes. Rapid eye movements suggest that patient is scanning the environment. Alpha appears as soon as he is asked to close his eyes.

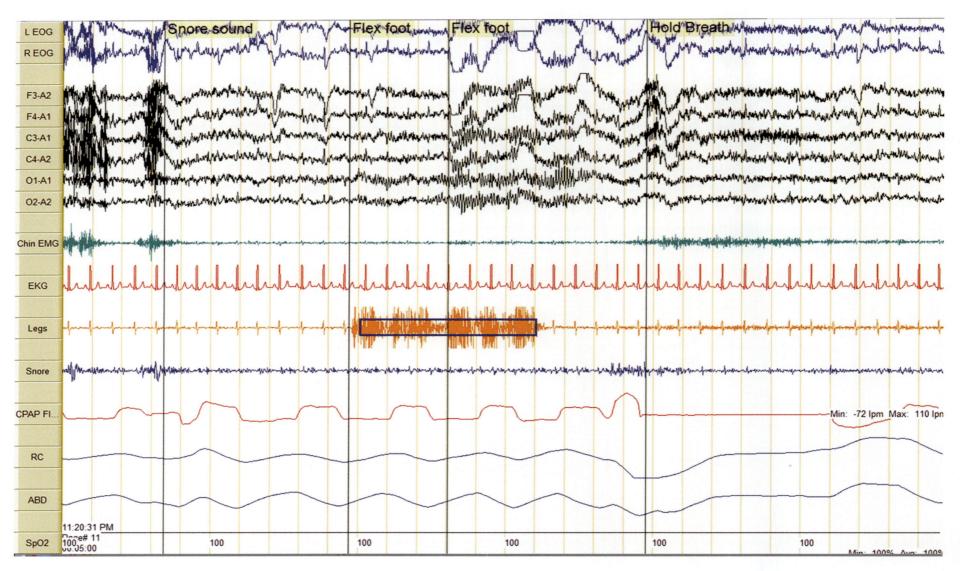

Figure 8.6 Calibration of respiratory signals: Flattening of signals in the flow and respiratory effort sensors appear during wishful breath holding. Look after "Hold Breath."

Calibration and Biocalibration 189

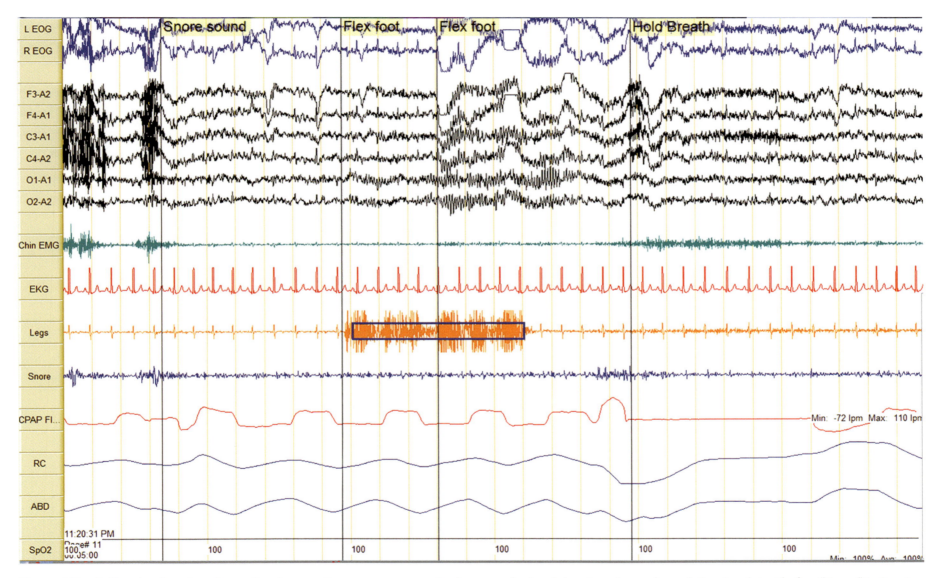

Figure 8.7 Calibration for microphone: Snoring usually appear as crescendo-decrescendo signals in microphone channel. See signals just before "Snore" in Snoring Channel. Wishful snoring sound also creates signals in Chin EMG Channel.

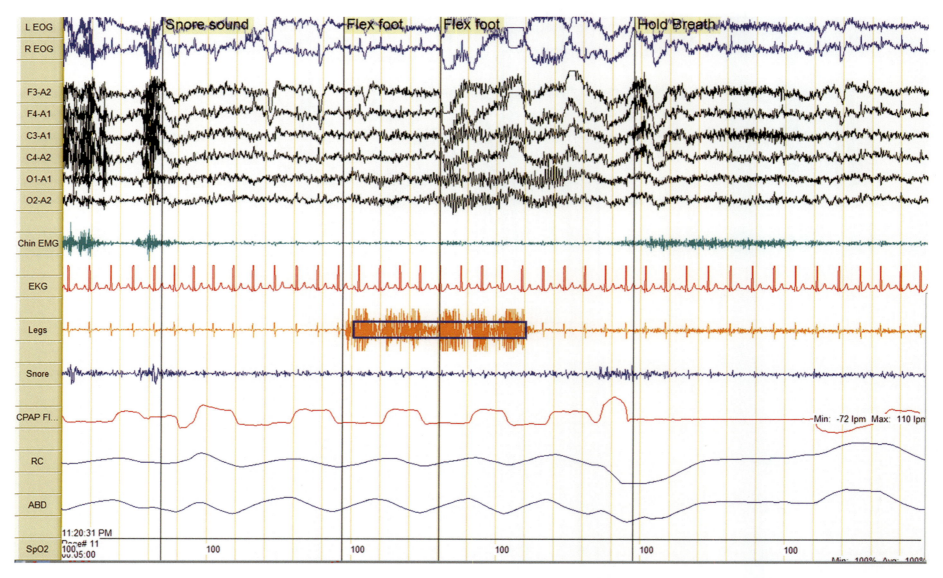

Figure 8.8 Calibration of leg EMG signals: Patient is asked to flex each of his feet up and down without moving thighs or legs. Signals appear in Leg channel.

Calibration and Biocalibration

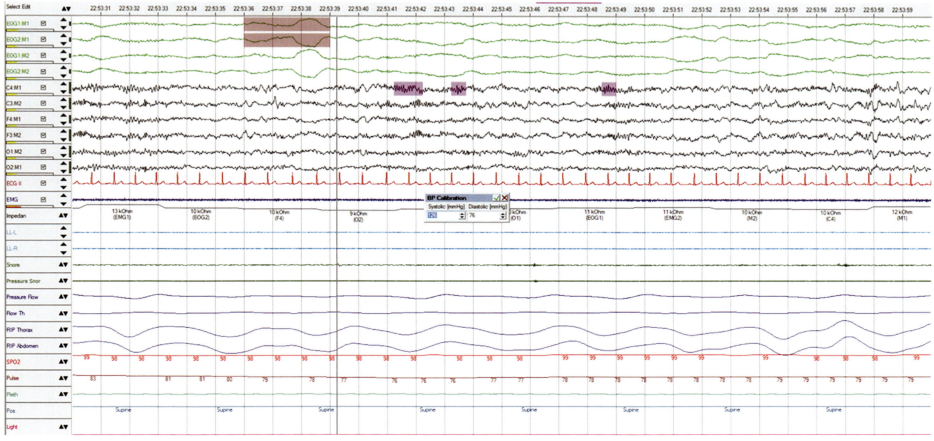

Figure 8.9 Calibration of blood pressure: Some PSG machines provide real time BP based upon the surrogate signals. In those machines, BP calibration is done by measuring the blood pressure at the start of study and entering the values in software. Software then calibrates the signals itself and shows blood pressure over the duration of study.

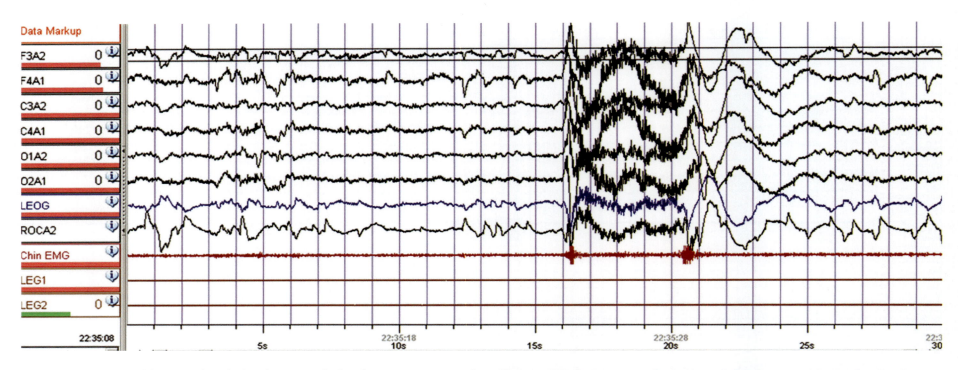

Figure 8.10 Calibration of teeth clenching: Teeth clenching appear as muscle artifacts in EEG derivations and as sustained muscle contraction in chin EMG.

Calibration and Biocalibration

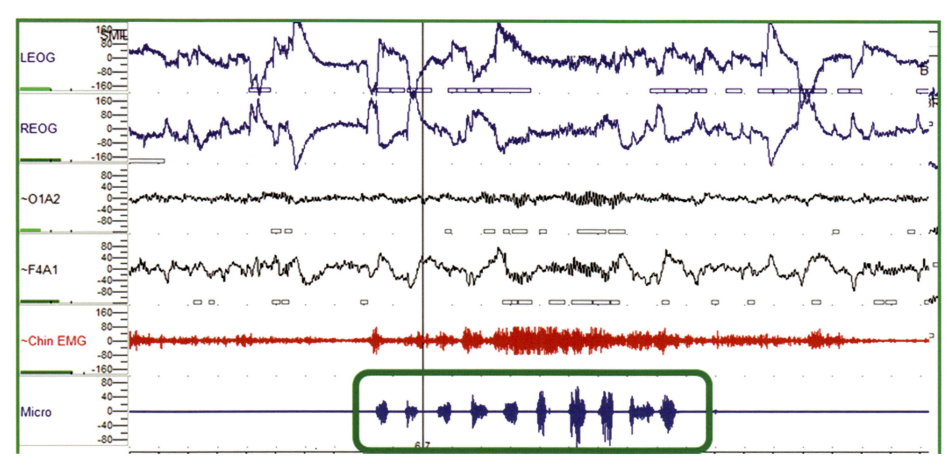

Figure 8.11 Calibration of voice signals in microphone: This is done by asking the patient to speak something. Check that you can hear his voice in your room. This ensures that the sound system is working properly. Signals are also seen on the microphone lead.

Table 8.1 Utility of Biocalibration

S.No.	Command	Look for?	Output
1.	Blink eyes	Blink artifacts in frontal leads	Frontal channels are working properly
2.	Look up down, right left	Deflection in opposite phase in EOG	EOG working properly
3.	Close eyes	Alpha in occipital leads	Occipital leads working properly
4.	Stop breathing	Loss of signals in thermistor, pressure transducer, and RIP belts	Respiratory measures working
5.	Snore sound	Signal in microphone	Microphone working
6.	Feet movement	Signals in leg EMG channels	leg EMG working properly
7.	Turn in bed	Change in signals in body position	Sensor working properly
8.	Clench teeth	EMG artifacts in EEG and chin EMG	Sensors working properly
9.	Speaking	Sound in speaker	Sound system working properly

FURTHER READING

1. Berry, R. B., Brooks, R., Gamaldo, C. E., Harding, S. M., Lloyd, R. M., Marcus, C. L., & Vaughan, B. V., (2017). For the American Academy of Sleep Medicine. The AASM manual for scoring of sleep and associated events: rules, terminology and technical specifications. Version 2.4. www.aasmnet.org. Darian, Illinois: *American Academy of Sleep Medicine.*

REVIEW QUESTIONS

1. Calibration is done to:

    A. Ensure that filter and amplifiers are working properly

    B. Ensure that all sensors are properly placed and are working

    C. Ensure that computer is function properly

Calibration and Biocalibration

D. Ensure that patient is having normal vital parameters

2. During biocalibration of EEG, following waves appear in the EEG:

   A. Delta
   B. Theta
   C. Beta
   D. Alpha

3. Waveforms during the biocalibration of eye movements appear due to:

   A. Electrical activity of extra-ocular muscles
   B. Activity of the motor areas of brain
   C. Conjugate movements of eyes
   D. Neural activity in the facial nerve

4. During biocalibration of the respiratory channels following will be apparent:

   A. Obstructive Sleep Apnea
   B. Hypopnea
   C. Mixed Apnea
   D. Central Sleep Apnea

5. Activity in which of the following channels does not alter during biocalibration:

   A. EEG
   B. EMG
   C. EKG
   D. EOG

ANSWER KEY

1. B   2. A   3. C   4. D   5. C

9

MINIMAL RECORDING PARAMETERS AND EXTENDED MONTAGE

LEARNING OBJECTIVES

After reading this chapter, the reader should be able to:

1. Discuss the minimum recordable parameters during an attended sleep study.
2. Reason the importance for including these parameters.
3. Enumerate the optional parameters and their utility.
4. Discuss the extended parameters in special situations.

CONTENTS

9.1　Optional Parameters 214

9.2　Extended Parameters for Special Circumstances 219

Further Reading 219

Review Questions 219

Answer Key 220

The aim of the polysomnography is to obtain the data to arrive at a diagnosis, which, in turn, can help in the formulation of an optimal treatment plan. It translates into symptom resolution, improvement in the quality of life, and increased productivity of the patient.

Sleep disorders rarely occur in isolation and many times co-morbid disorders influence the courses of a sleep disorder, or at times, two-sleep disorders are seen together. As a case point, certain medical disorders, for example, chronic renal failure and Parkinson's disease increase the risk for obstructive sleep apnea as well as the restless legs syndrome. Thus, a patient who has been referred for OSA may have co-morbid periodic limb movement during sleep. Similarly, periodic limb movement during sleep as well as insomnia, are not uncommon with obstructive sleep apnea; certain parasomnias, for example, night terrors and sleep-walking are common in patients with obstructive sleep apnea. In such cases, if patients are subjected to home sleep testing with limited parameters, other sleep disorders may be missed, culminating in a compromised treatment planning.

To overcome it, the AASM has proposed that a minimal number of parameters must be recorded. These recommendations were made considering that:

1. Main sleep disorder and co-morbid disorder can be picked after the scoring.
2. Optimal technology is used to pick the signals related to a given parameter.
3. Signals should help the scorer/clinician to reach a conclusion.
4. Prevention of data that may be lost due to issues arising during sleep study, for example, falling leads, malfunction of channels.

The AASM has proposed that during a sleep study, the following parameters must be recorded:

1. **Electroencephalogram (EEG):** EEG must be recorded using at least six channels. Three of these channels are placed on right side of the head and the other three on the left side. These channels are placed at frontal, central and occipital areas on both sides and are referred to the opposite mastoid. These places are important as they help in picking specific waveforms to score a sleep stage. Alpha waves that are seen during quiet wakefulness, are best seen in the occipital area, hence, this area was chosen (Figures 9.1 and 9.2). Central leads show vertex waves that characterize Stage 1 sleep, sleep spindles that characterize stage 2 sleep, and sawtooth waves seen during REM sleep (Figures 9.3–9.5). Similarly, K-complexes of stage 2 sleep and delta waves that characterize stage 3 sleep (deep

Minimal Recording Parameters and Extended Montage 199

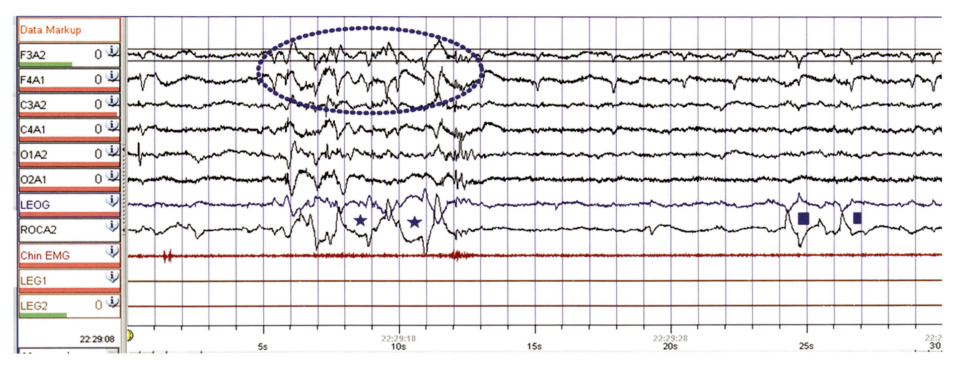

Figure 9.1 Determination of active wakefulness: Active wakefulness can be determined by EEG, EOG, and chin activity together. EEG shows mixed alpha and beta activity blinking artifacts (in the ellipse), rapid eye movements (waveforms around stars) and relatively high chin tone.

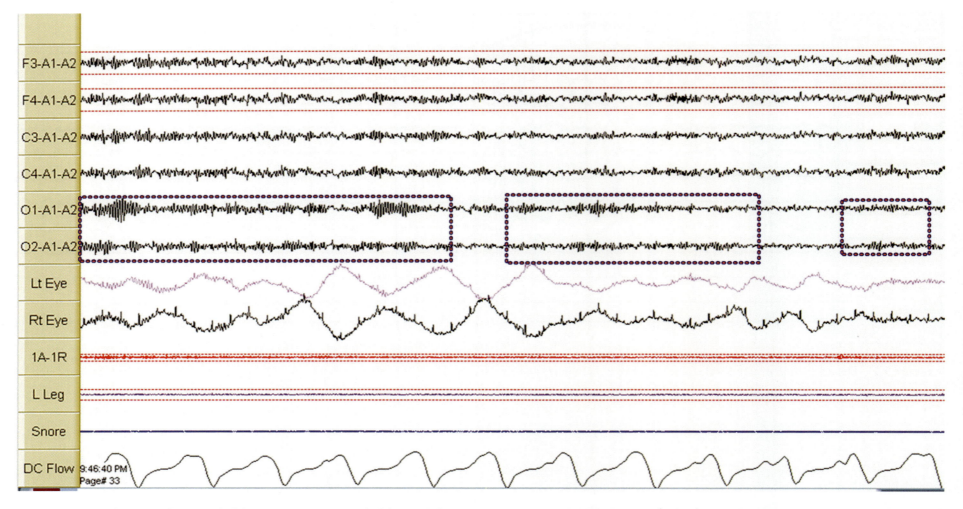

Figure 9.2 Alpha waves during wakefulness: During quiet wakefulness, alpha waves appear in occipital derivations (in box). In addition, eye movements are slower and chin tone reduces. Thus composite data from all three gives an idea regarding sleep-wake state.

Minimal Recording Parameters and Extended Montage

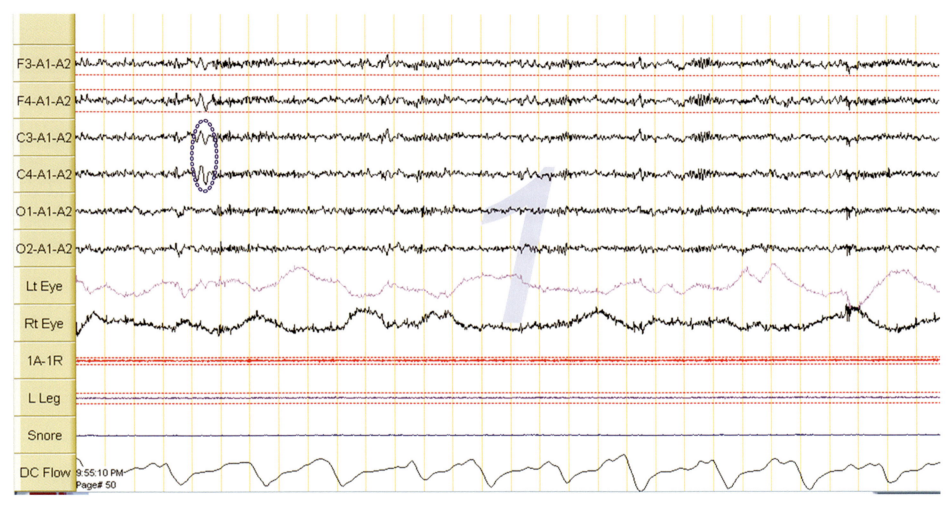

Figure 9.3 Vertex waves in central derivations: Vertex waves (circle) are best seen in central derivations during stage 1 sleep.

202　Clinical Atlas of Polysomnography

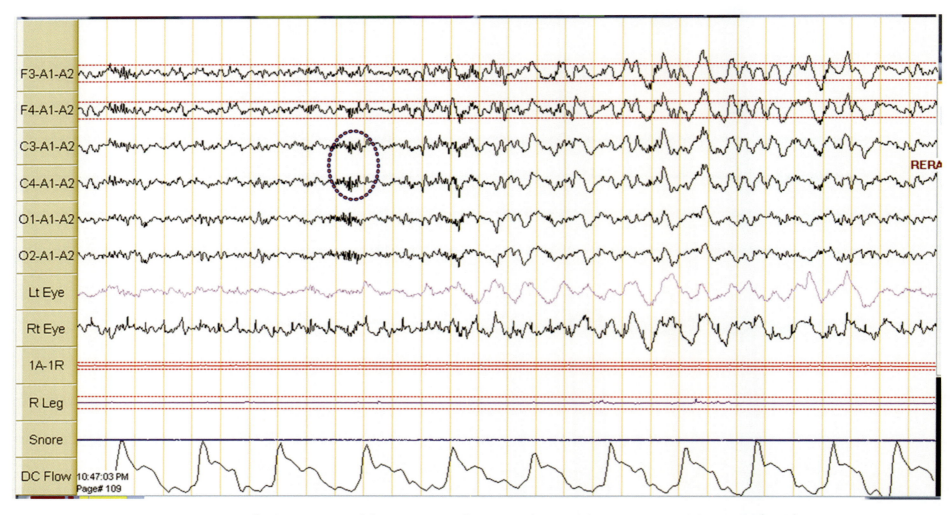

Figure 9.4 Spindle activity in central derivations: Spindle activity of stage 2 is best seen in central derivations (circle).

Minimal Recording Parameters and Extended Montage 203

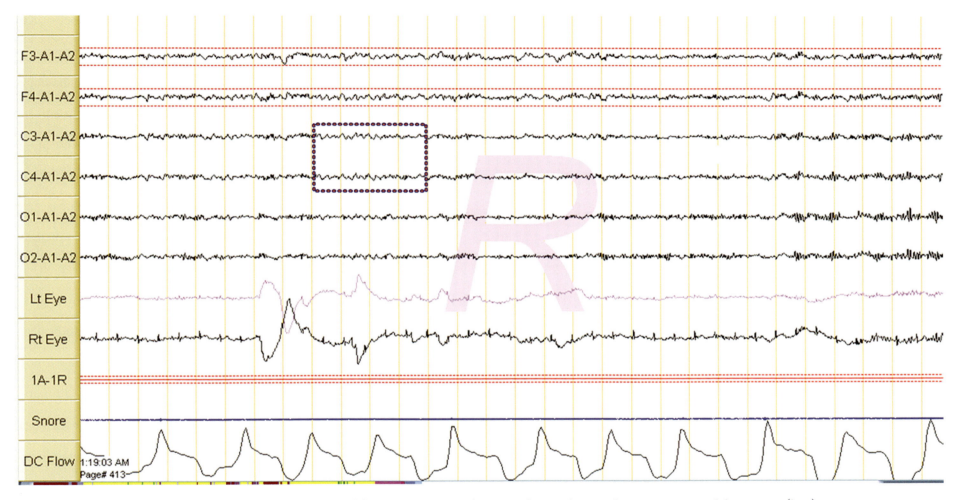

Figure 9.5 Sawtooth waves in central derivations: Sawtooth waves of REM sleep are best seen in central derivations (box).

sleep) are best seen in frontal leads (Figures 9.6 and 9.7). Electrodes are placed on both sides of the head, corresponding to both hemispheres of the brain. During the study, the patient may turn in the bed, resulting in fall of one or more leads, resulting in loss of signals. If the electrodes are placed on both sides, data from the other side can still be scored to reach a conclusion.

Some evidence suggests that a single EEG channel, that is, placement of central electrodes helps in optimal scoring of sleep stages, still the placement of three channels (along with three back-up channels) is considered gold standard.

2. **Electrooculogram (EOG):** An EOG that records the movement of the eyeball is also recorded from both eyes. Having the data from both eyes not only prevents the data loss in case of fall-off of lead but also clearly delineates the eye-movement. For the reasons discussed in Chapter 3 (section on EOG), true eye movement is always in the opposite phase, that is, if one channel deflects towards the negative side, the other deflects towards the positive side (Figure 9.8). Since these channels are placed close to the frontal area, sometimes, delta waves may be spilled in these channels, giving the impression of eye movement (Figure 9.9). Eye movements, along with EEG, help in scoring wakefulness, quiet wakefulness, and REM sleep.

3. **Electromyogram (EMG):**
   a. **Muscle tone:** An EMG is required to score the sleep stage as well as the abnormal movements during sleep. During wakefulness, muscles have a basal tone that reduces as well fall asleep. REM sleep is characterized by profound atonia, however, in patients with REM sleep behavior disorder, this atonia is not seen. Submentalis muscle has been chosen because it is a skeletal muscle, lies just beneath the skin, thus, provides good signals, and normally remains inactive during sleep, in contrast to other skeletal muscles that may get activated episodically during sleep. To prevent data loss, signals are recorded from sub-mentalis/mentalis muscles of both sides. Along with EEG, these signals help in recognizing REM sleep stage and REM sleep without atonia.

   b. **Limb movements:** To record the periodic limb movement during sleep

Minimal Recording Parameters and Extended Montage

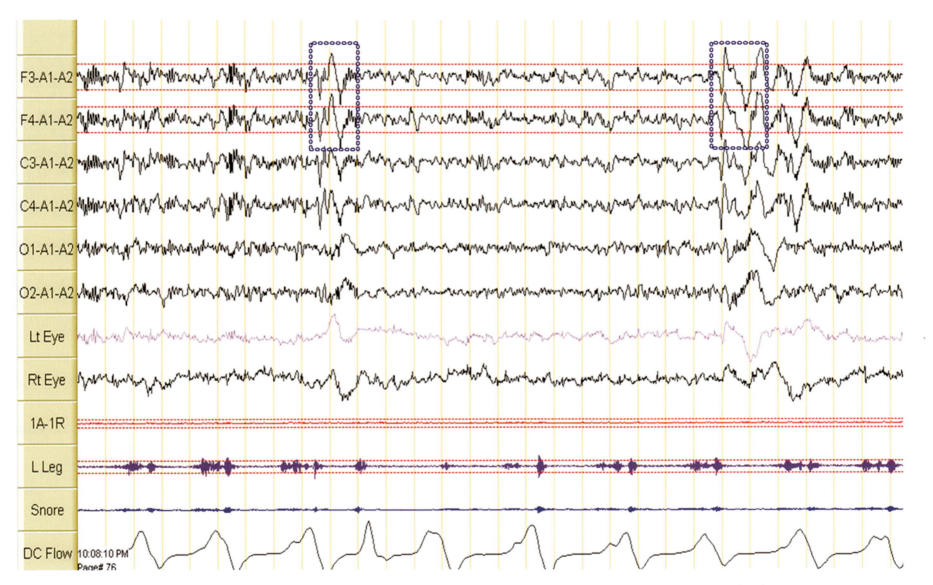

Figure 9.6 K complexes in frontal derivations: K-complexes of stage 2 are best seen in frontal derivations (box).

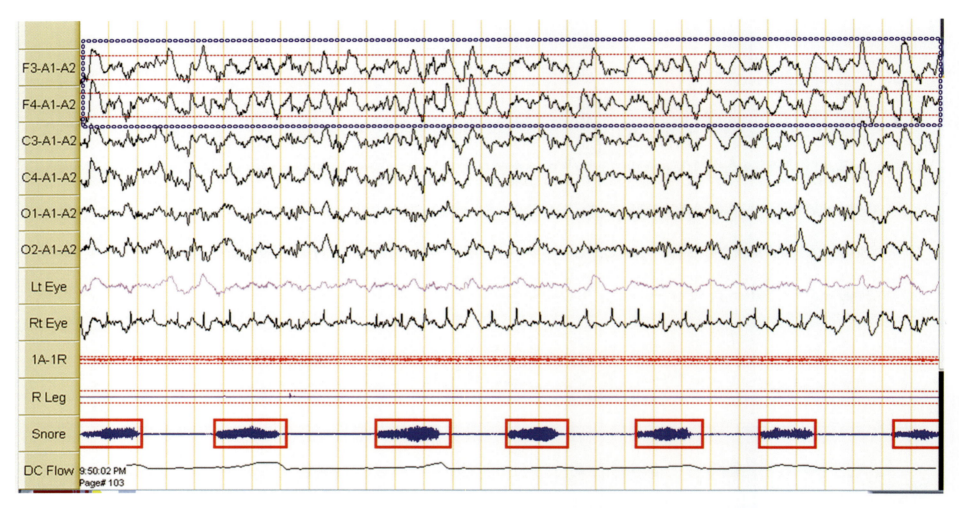

Figure 9.7 Delta waves in frontal derivations: Delta waves (in broken box) of stage 3 sleep are best seen in frontal derivations.

Minimal Recording Parameters and Extended Montage

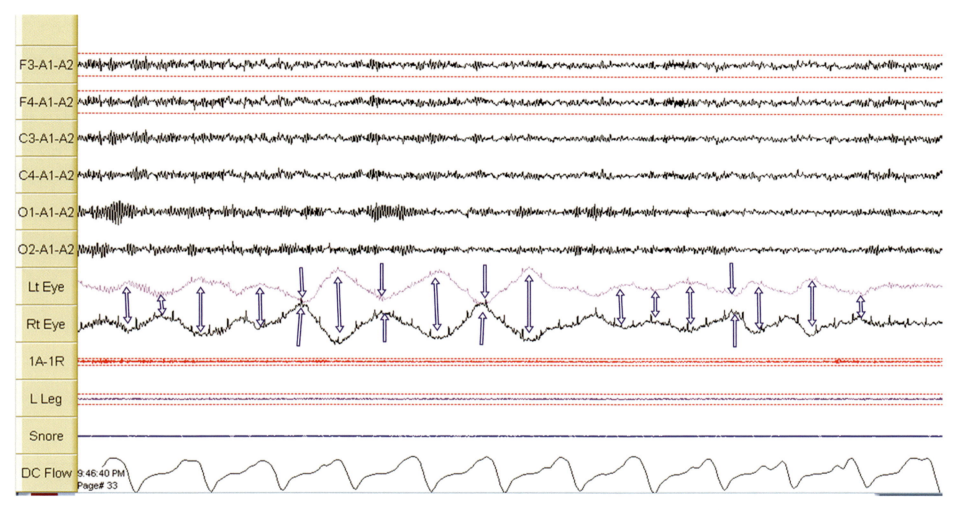

Figure 9.8 True eye movement: True eye movements are in opposite phase (see arrows).

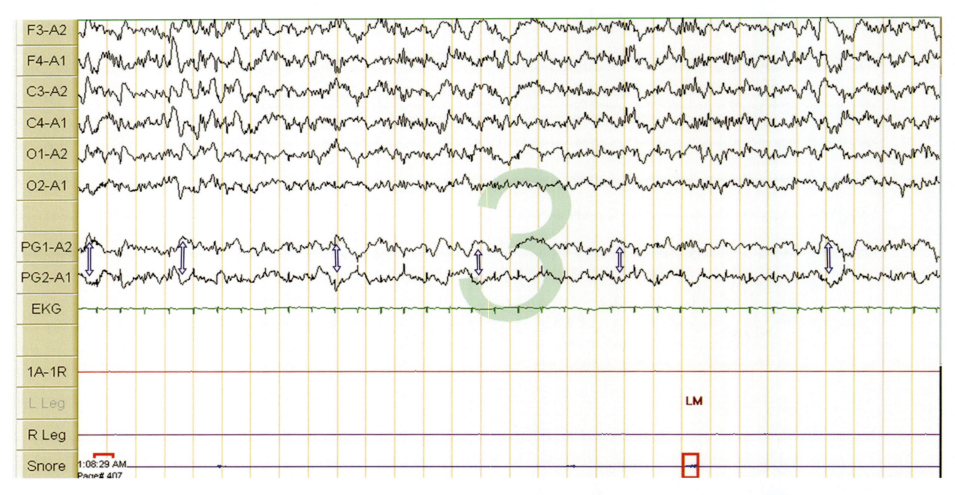

Figure 9.9 Delta waves may be mistaken for eye movements: Frontal delta activity or K complexes may appear in eye channels, but these deflections are in-phase, as opposite to true eye movements. In this epoch some of them appear out of phase (arrows), however, their waveform is similar to right and left frontal derivations respectively.

(PLMS), signals are recorded from anterior tibialis muscles of both legs. To minimize data loss, signals are recorded from both legs separately (Figure 9.10). In addition, signals from both sides should not be combined as it may reduce the detectable numbers of limb movements. These signals, along with the EEG, help in scoring PLMS-associated arousals. In addition, these signals also help in scoring "alternate leg muscle activity" (ALMA), Hypnogogic foot tremors (HFT), and excessive fragmentary myoclonus (EFM).

4. **Respiratory Flow:** Respiratory flow is an important signal to score apnea and hypopnea. It is measured using a thermistor and a pressure transducer (Figure 9.11). As discussed in chapter 3, the thermistor is important for the diagnosis of apnea while a pressure transducer helps in recognizing hypopnea, Cheyne-Stokes breathing, and periodic breathing.

   However, during titration with PAP machine, signals must be taken from the PAP device, as placing a cannula or thermistor below the mask may not provide good signal due to a continuous flow of air and also increase the leakage from the mask owing to incomplete sealing.

5. **Respiratory Effort:** Analysis of respiratory effort is important for differentiating between obstructive and central respiratory events (both apneas and hypopneas). In obstructive sleep apnea and hypopnea, the respiratory efforts are present in absence of respiratory flow, while in central apnea, the effort is also absent. Though it can be measured using strain gauges, RIP belts are considered superior and are recommended for the measurement of respiratory efforts (Chapter 3). In addition, sum of the chest is abdominal efforts is calculated and depicted as a waveform. During obstructive apnea and hypopnea, the chest and abdomen breath "out of phase" and thus, their sum shows a flat line (figure out of phase movement during apnea and sum showing flattening). Thus, the waveform of the "sum" of efforts makes the scoring of respiratory events easier (Figure 9.12).

6. **Oxygen Saturation:** Oxygen saturation is measured using a pulse oximeter that provides a value that is an average of 3 seconds values (Figure 9.13).

7. **Electrocardiogram:** Some patients may have arrhythmia during sleep, especially

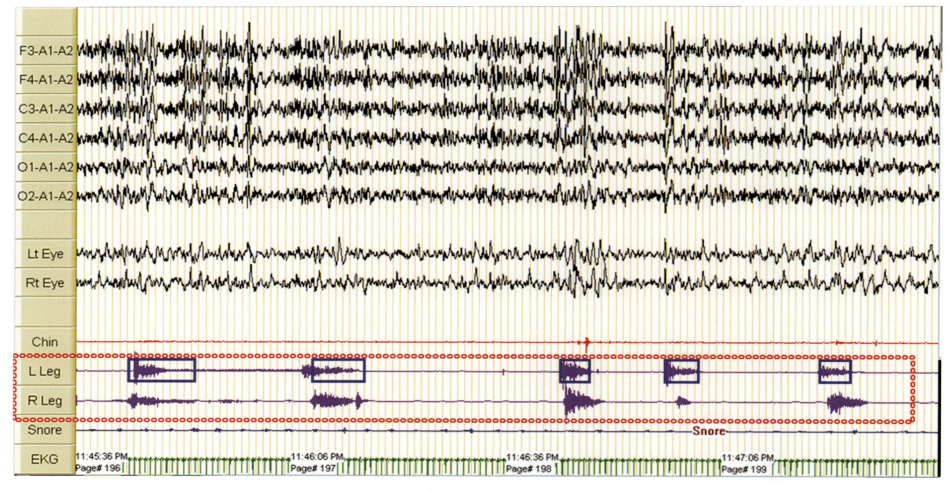

Figure 9.10 EMG signals from legs: Signals from both legs are recorded separately.

Minimal Recording Parameters and Extended Montage 211

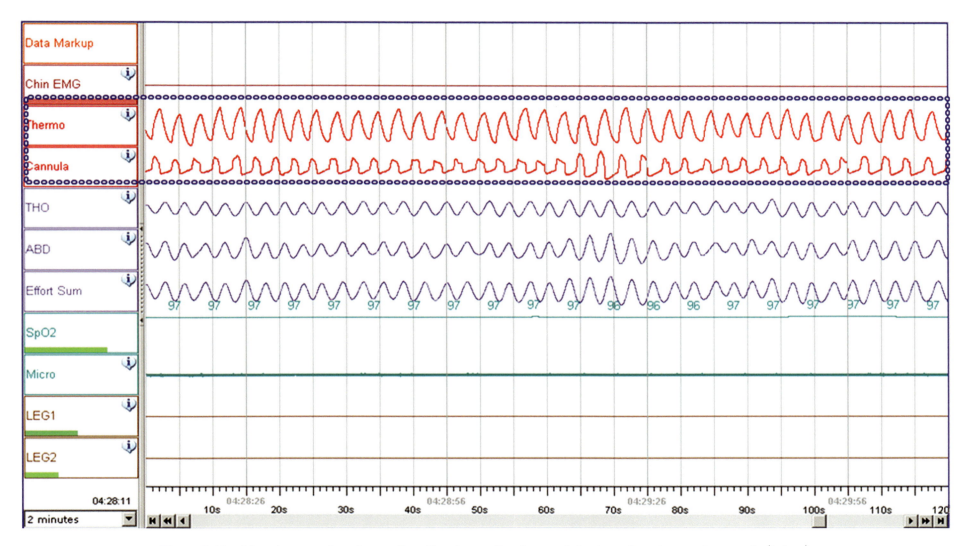

Figure 9.11 Measurement of respiratory flow: Respiratory flow is recorded using a thermistor and a cannula (in box).

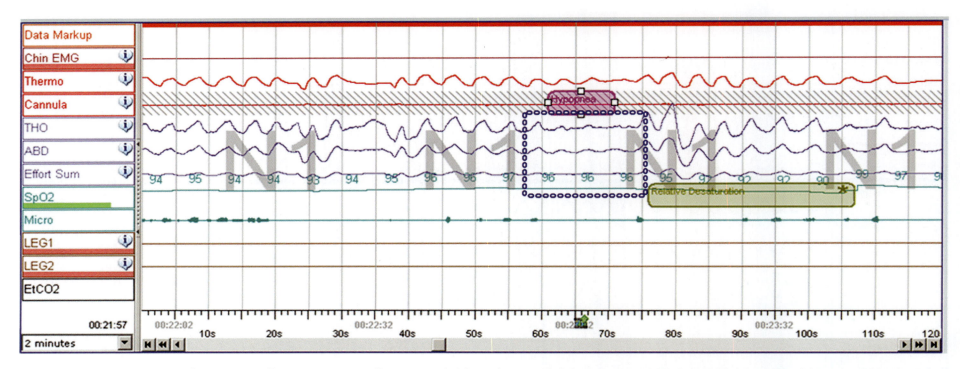

Figure 9.12 Measurement of respiratory efforts: Respiratory effort is recorded from thorax and abdomen using RIP belt. In addition a sum of them should be displayed. During apnea, paradoxical movement is seen between thorax and abdomen and thus the RIP shows flattening (in box).

Minimal Recording Parameters and Extended Montage

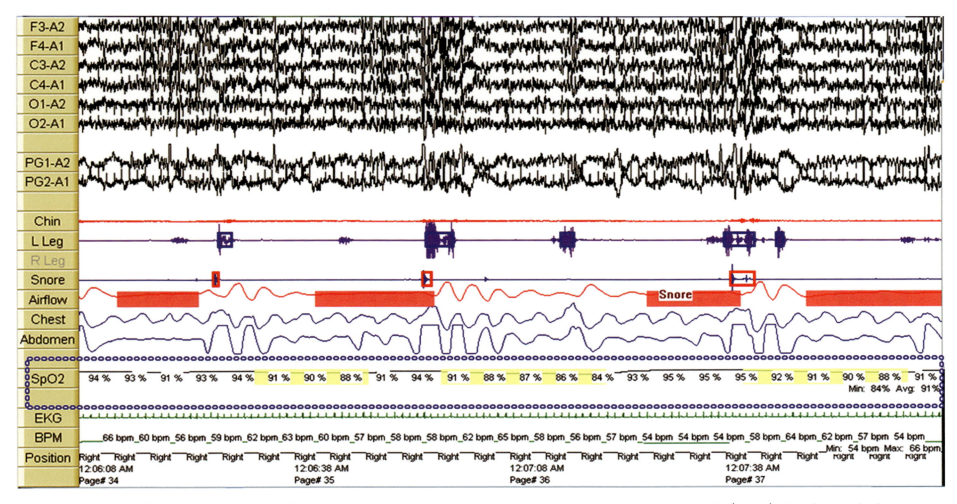

Figure 9.13 Recording of oxygen saturation: Oxygen saturation is recorded as 3 seconds averaged value during the sleep study (in box). Also shown is body position.

during obstructive sleep apnea, or may have sleep-related asystole that worsens the sleep quality, or may develop ischemia during sleep due to hypoxemia. Hence, cardiac electrical activity has to be recorded. It is recommended that it uses at least two leads, preferably limb lead I and II (Figure 9.14).

8. **Body Position:** A number of patients have sleep apnea that is specific to a body position, for example, supine-dependent. In such cases, at least theoretically, positional therapy may help to ameliorate the symptoms. Hence, recording of body position is recommended throughout the study (Figure 9.13).

9.1 OPTIONAL PARAMETERS

1. **Synchronized Audio-Video Recording:** This will help to differentiate between the parasomnia and seizure activity. A laboratory may choose to have the provision for synchronized video recording using an infrared imaging camera (that captures images in the dark as well) and audio recording by placing a microphone near the patient's head end. With the help of a microphone, one can also discern the verbal content spoken during the study.

2. **Capnography:** This is important for the diagnosis of hypoventilation during sleep. One may choose to use the transcutaneous sensor, however, the value in the transcutaneous sensor lags behind by approximately two minutes to that of $PaCO_2$ and it requires regular calibration. Another sensor is to use the End-tidal CO_2, through a nasal cannula, however, it may provide false low values in cases of nasal obstruction and when supplemental oxygen is administered (Figure 9.15).

3. **Snoring:** Although snoring can be recorded through pressure transducer as mentioned in Chapter 3, it is recommended that it is recorded using a snore microphone that is placed on the neck close to the larynx (Figure 9.16). This microphone also helps in detecting sleep talking and bruxism.

4. **Espohageal Manometery:** This measures the esophageal pressure that changes during the respiration and is the most reliable method to differentiate obstructive events from central events. However, being an invasive method, it is uncomfortable and its use is limited to research only.

5. **Pharyngeal pH:** This is measured in cases of nocturnal gastro-esophageal reflux

Minimal Recording Parameters and Extended Montage 215

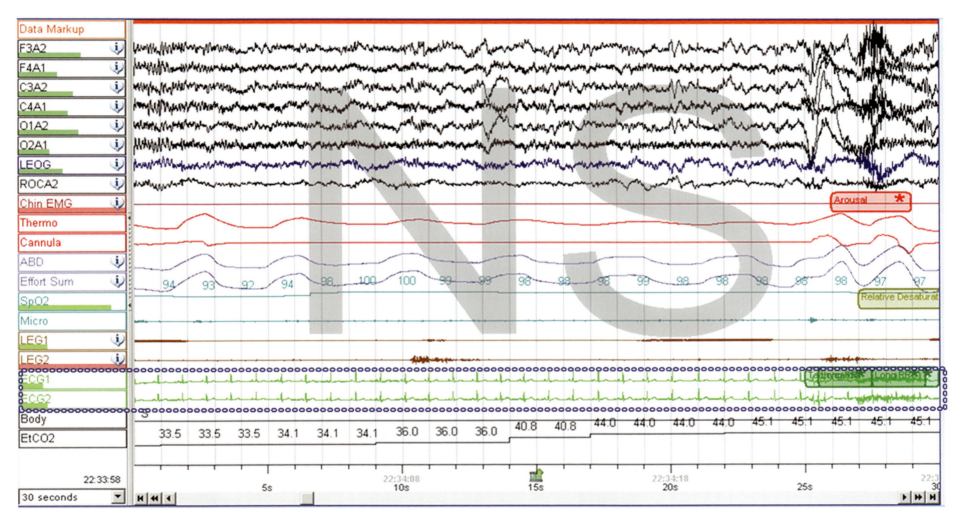

Figure 9.14 Recording of electrical activity of heart: Two channels of electrocardiogram are recorded during the sleep study (in box).

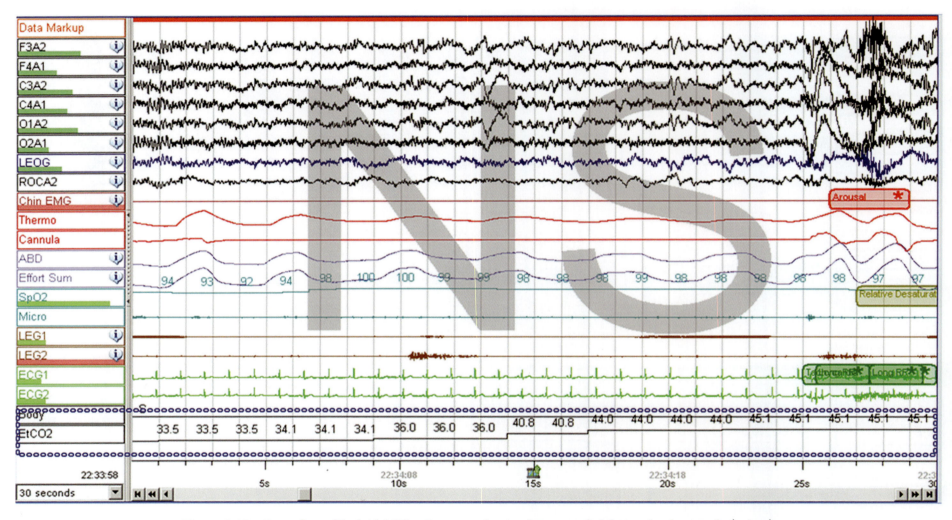

Figure 9.15 Recording of End tidal CO_2: Capnography signals as recorded during the sleep study (in box).

Minimal Recording Parameters and Extended Montage 217

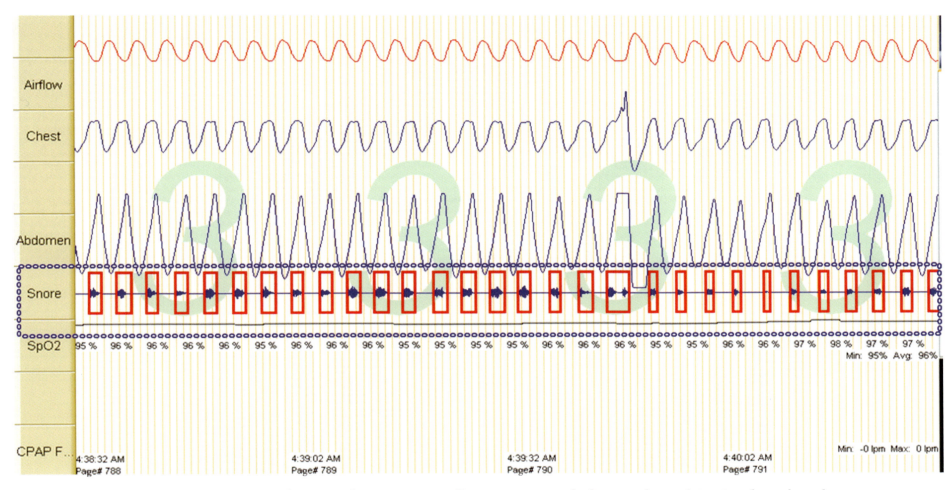

Figure 9.16 Snoring signals in microphone: Snoring usually appear as crescendo-decrescendo signals in microphone channel.

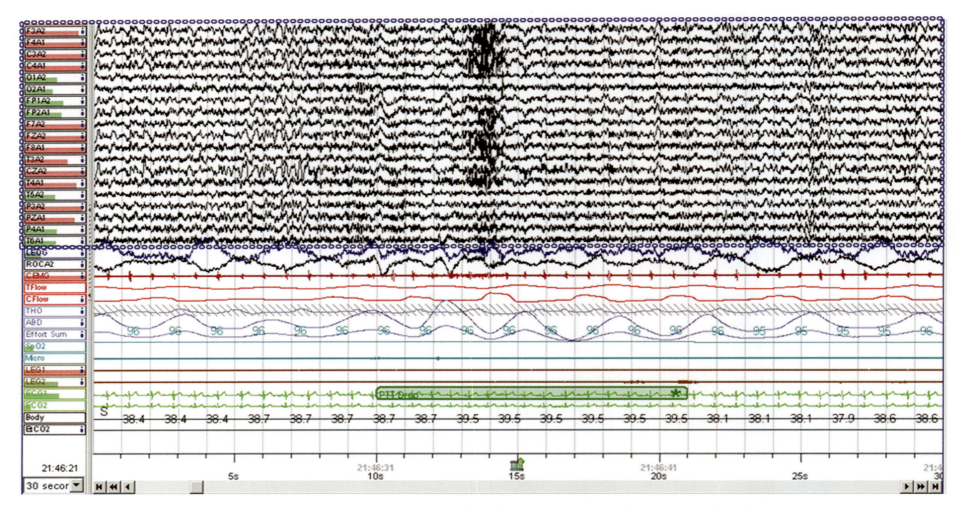

Figure 9.17 Extended montage for seizures: Extended Seizure montage includes all the EEG channels.

disease and helps in differentiating sleep-related laryngospasm from chocking during obstructive sleep apnea. However, its use is also limited to research purpose only.

9.2 EXTENDED PARAMETERS FOR SPECIAL CIRCUMSTANCES

1. **Sleep-Related Seizure:** Sometimes, sleep-related seizures are difficult to differentiate from parasomnia. In such cases, instead of six recommended leads, their number may be extended to 24 or 32 so that seizure activity from any of the cortical area can be recorded (Figure 9.17).
2. **Bruxism/REM Sleep Behavior Disorder:** Bruxism is characterized by phasic or sustained contraction of the temporalis and masseter muscles. To record their activity, additional leads may be placed on these muscles with the EMG derivation setting. Similarly, in cases of REM sleep behavior disorder, additional leads may be placed on the upper limbs, so as to catch their activity with EMG derivation settings.

FURTHER READING

1. Silber, M. H., Ancoli-Israel, S., Bonnet, M. H., et al. (2007). The visual scoring of sleep in adults. *J Clin Sleep Med.* 3(2), 121–131.
2. Berry, R. B., Brooks, R., Gamaldo, C. E., Harding, S. M., Lloyd, R. M., Marcus, C. L., et al., (2017). The AASM manual for scoring of sleep and associated events: rules, terminology and technical specifications. Version 2.4. www.aasmnet.org. Darian, Illinois: American Academy of Sleep Medicine.

REVIEW QUESTIONS

1. Determination of sleep wake stage is dependent upon all of the following EXCEPT:
   A. EEG
   B. EMG
   C. EKG
   D. EOG
2. Scoring of the REM sleep is dependent upon change in amplitude in:
   A. Chin EMG
   B. EOG
   C. Leg EMG
   D. EKG
3. Placement of EMG on one leg will:
   A. Interfere with diagnosis of ALMA
   B. Increase the number of PLMs
   C. Provide the accurate measure of PLMs

D. Can be done in known case of restless legs syndrome

4. During Home Sleep Testing, AASM recommends use of following parameters, EXCEPT:

A. Thermal sensor or pressure transducer

B. At least one of thoracoabdominal RIP belts

C. Pulse oximeter

D. Acoustic sensor

5. Most reliable scoring of respiratory effort can be done by:

A. RIP belts

B. Esophageal manometry

C. Piezoelectric belts

D. Thermistor

ANSWER KEY

1. C   2. A   3. A   4. D   5. B

10
MONTAGES

LEARNING OBJECTIVES

After reading this chapter, the reader should be able to:

1. Discuss various montages used during polysomnography.

CONTENTS

Configuration of Various Montages 221

Review Questions 234

Answer Key 234

Lead or electrode refers to a sensor that captures electrical signals. Derivation or channel refers to the combination of two electrodes that depict a potential difference between two areas. On the other hand, montage refers to the constellation of electrophysiological channels that provide us adequate information that helps us decipher the underlying pathology. Although manufacturers provide some montages, still you may design your own montage in the software, depending upon your need (Figure 10.1).

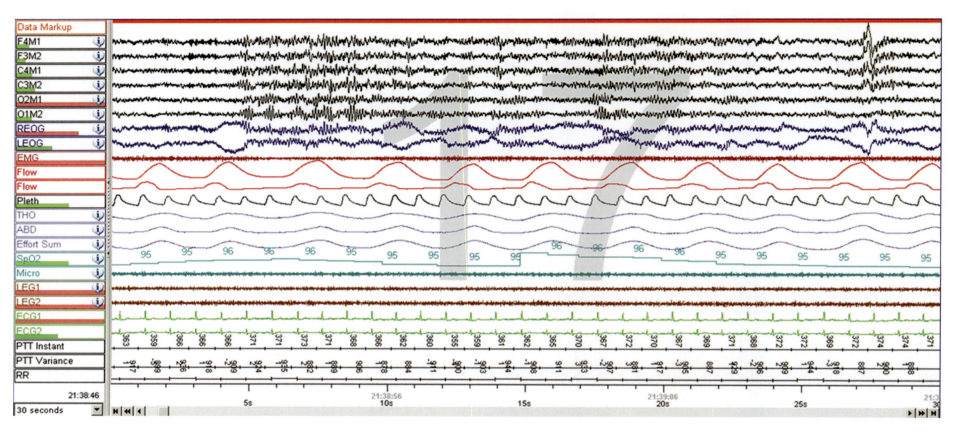

Figure 10.1 Supplied montage.

Montages

In general, you may need the following montages:

1. Recording of diagnostic sleep study (Figure 10.2)
2. Recording of PAP titration study (Figure 10.3)
3. Seizure/parasomnia montage
   a. Referential (Figure 10.4A-C)
   b. Banana (Figure 10.5)
4. For scoring of polysomnography data:
   a. Sleep Stages (Figure 10.6)
   b. Respiratory parameters (Figure 10.7)
   c. Movement (Figure 10.8)
   d. EKG (Figure 10.9)

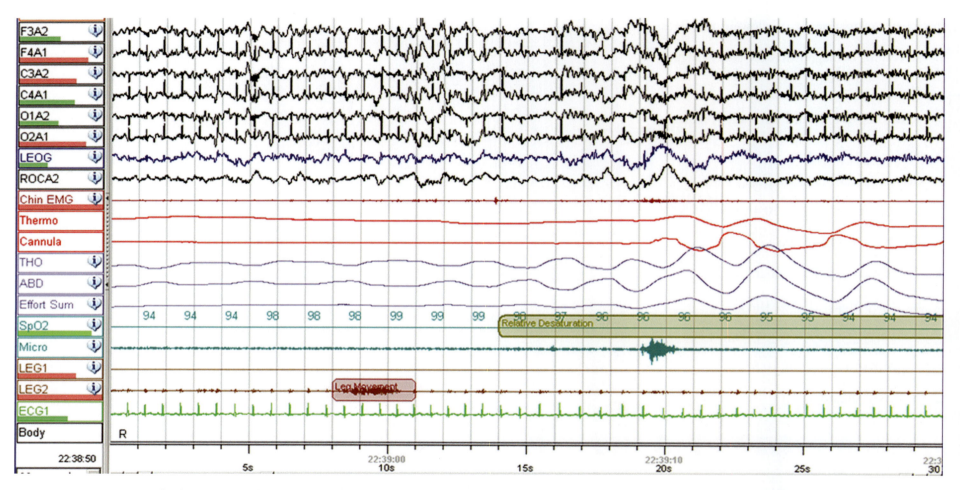

Figure 10.2 Montage for diagnostic study: This montage contains all the data for diagnostic study except video. Signals from Video may be added to this montage to make it complete.

Montages

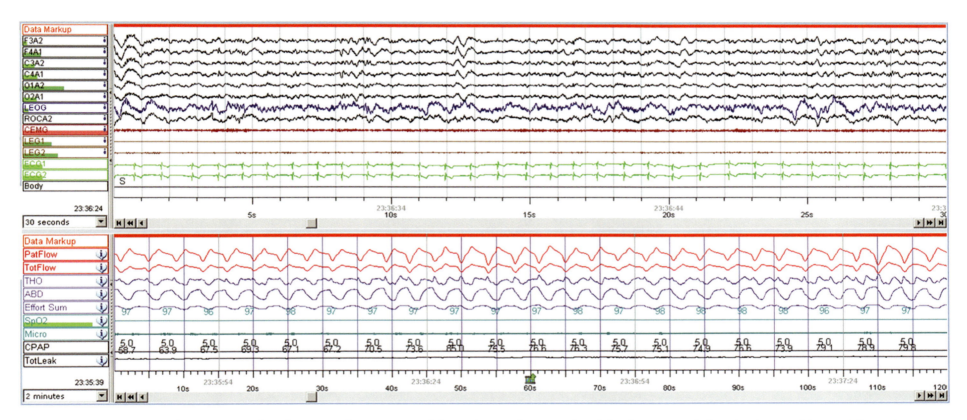

Figure 10.3 Montage for titration study: Please note that upper panel shows sleep-wake state related data hence, epoch duration is set to 30 sec, while respiratory parameters are depicted in lower panel. Epoch duration setting for this panel is 120 seconds data for better recognition.

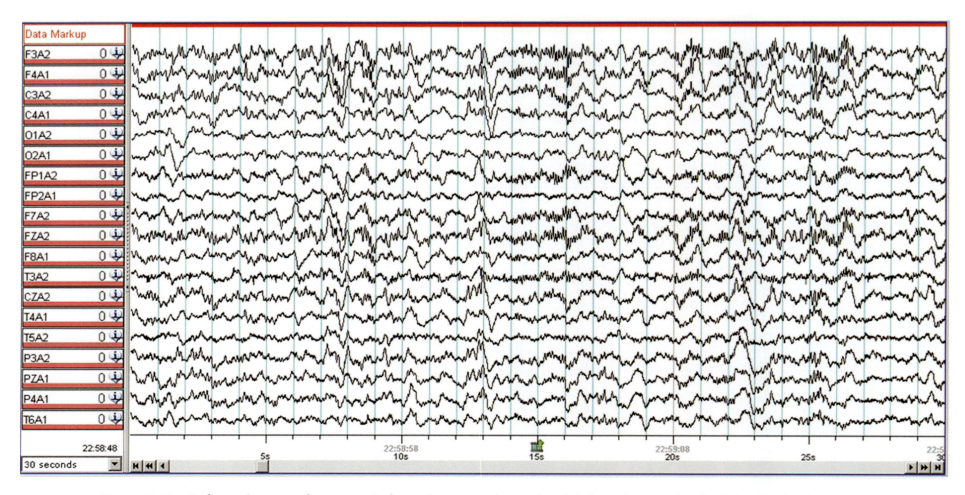

Figure 10.4A Referential montage for seizures. Referential montage where right sided electrodes are referred to left mastoid and vice versa.

Montages

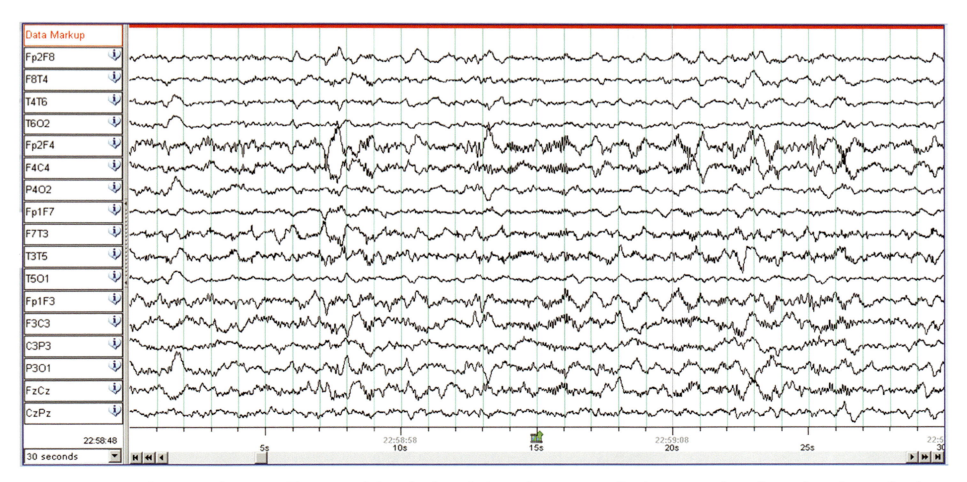

Figure 10.4B Bipolar montage for seizures. This montage helps in localizing the focus of seizure activity. Bipolar montage where adjacent electrodes are referred to each other.

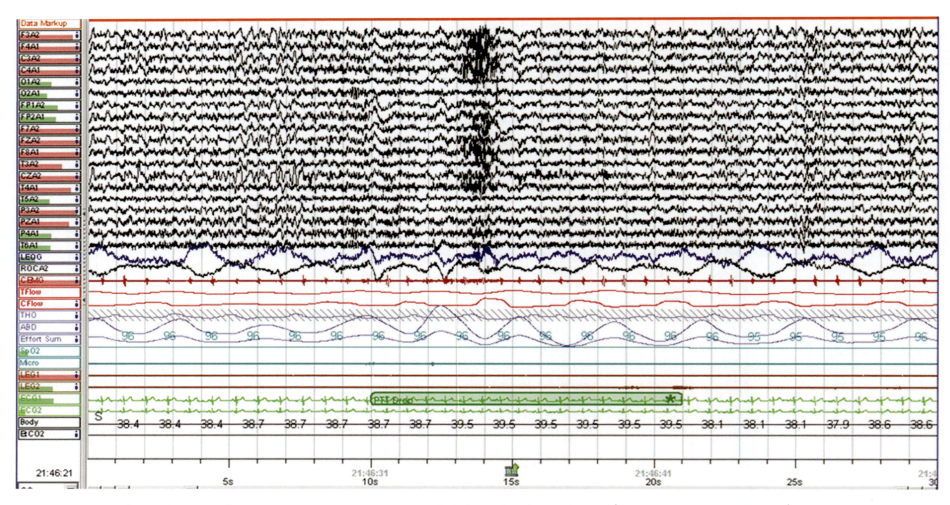

Figure 10.4C Referential montage for parasomnia. Parasomnia montage should include other parameters (respiratory, cardiac, movement) also so as to differentiate between seizure and parasomnia.

Montages

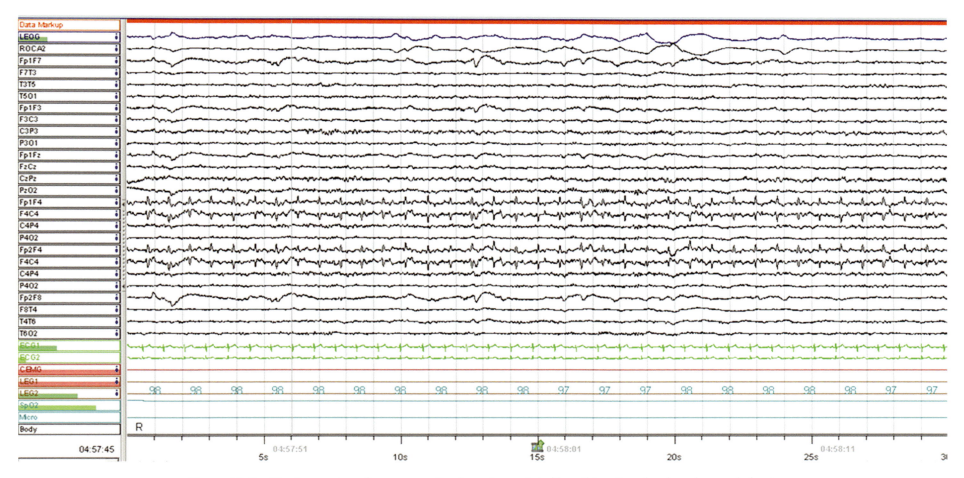

Figure 10.5 Bipolar montage for parasomnia where sleep related breathing disorders have been ruled out.

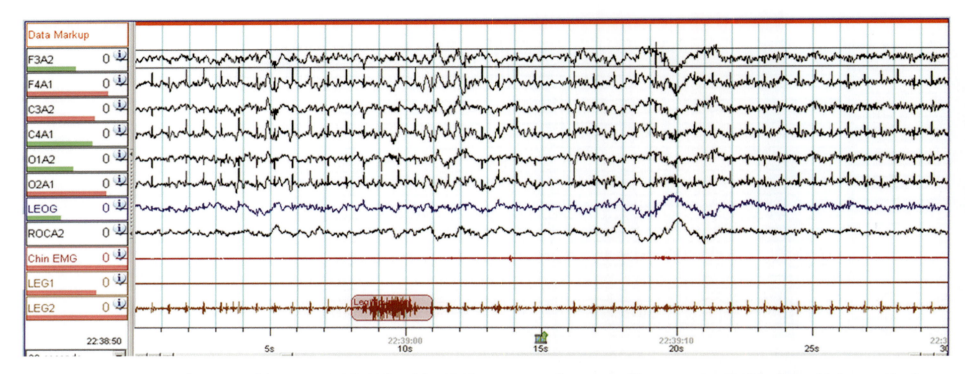

Figure 10.6 Montage for scoring of sleep stages: EEG, EOG and chin EMG are important for scoring of sleep stages. Leg EMG has been added to score Limb Movement related arousals.

Montages

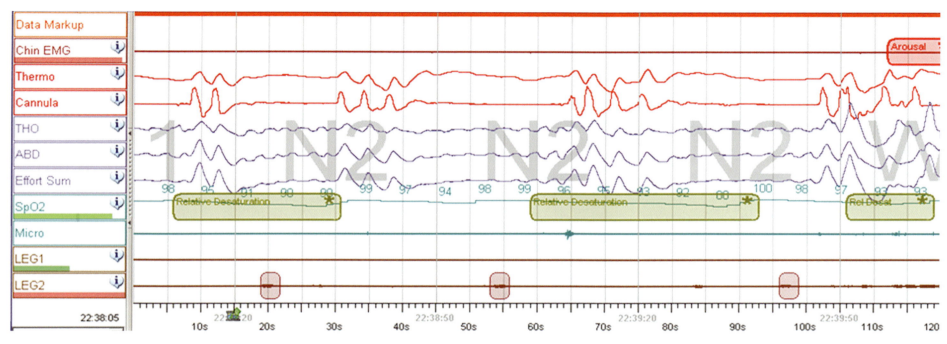

Figure 10.7 Montage for scoring of respiratory data.

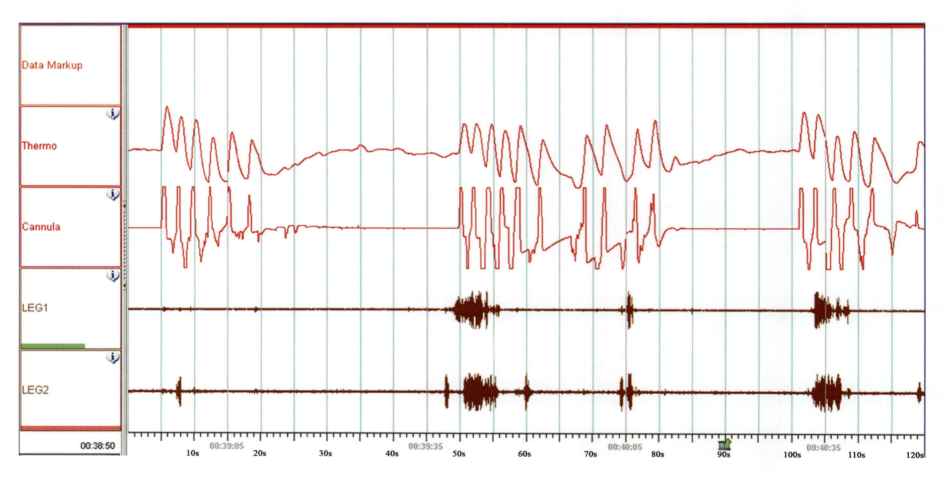

Figure 10.8 Montage for scoring leg movement.

Montages

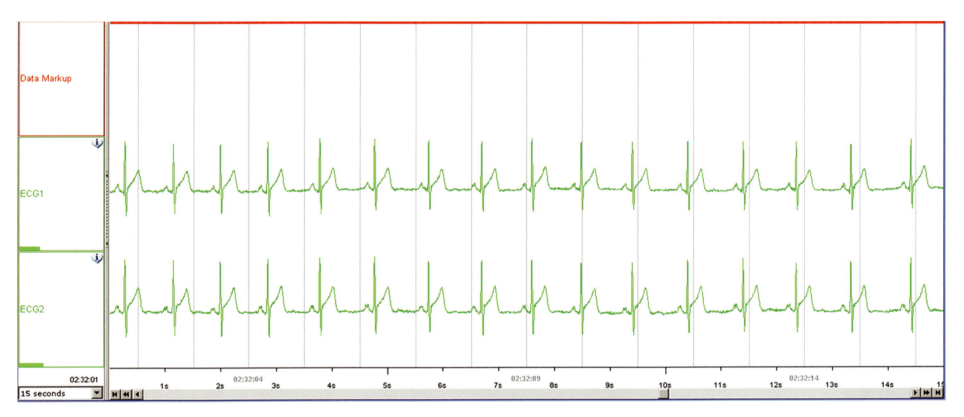

Figure 10.9 Montage for scoring electrocardiogram.

REVIEW QUESTIONS

1. Designing of montage is important for:
   A. Looking at the relevant data at a given point of time
   B. Scoring of the data
   C. Paying attention to the information
   D. Nothing, but done as it is customary

2. Montage for parasomnia or seizure should include:
   A. Six EEG derivations as described in AASM manual
   B. Twelve EEG derivations including that described in AASM manual
   C. All EEG channels
   D. All EEG channels with referential and banana montages

3. Localization of seizure activity can be done best in:
   A. Referential montage
   B. Banana montage
   C. Sleep staging montage
   D. Limb movement along with referential montage

4. For better scoring of limb movements, leg EMG channels must be included in:
   A. Sleep Stage montage
   B. Respiratory montage
   C. EKG montage
   D. Seizure montage

5. For correct scoring of RERA following should be done:
   A. Include chin EMG in respiratory montage
   B. Include leg EMG in sleep stage montage
   C. Include pressure transducer signals in sleep staging montage
   D. Include EKG in respiratory montage

ANSWER KEY

1. A   2. D   3. A   4. B   5. C

11. THE CONCEPT OF EPOCHS

LEARNING OBJECTIVES

After reading this chapter, the reader should be able to:

1. Discuss the concept of epochs.
2. Enumerate the duration of epochs for scoring different parameters during sleep study.

CONTENTS

Concepts of Epochs 235
Further Reading 236
Review Questions 251
Answer Key 251

Usually, we record the 6–8 hours sleep data during the diagnostic study. However, we cannot analyze whole of the data just at one glance. There are many reasons for that—first, as we have mentioned, sleep consists of different stages that give discrete appearance on an EEG (see below); second, such a large raw data cannot be concise to a single page; third, transition between wakefulness and sleep and among different sleep stages is gradual. Hence, we divide the data into small parts called epochs.

Most of the polysomnography machines provide the independence to set the epoch duration varying from as short as 5 seconds to as long as 300 seconds. According to standard guidelines, for scoring various parameters, epochs are set at different durations (Table 11.1).

With the illustrations (Figures 11.1–11.12), we will depict why such setting is necessary.

Table 11.1 Standard Duration of An Epoch

S. No.	Parameter to be scored	Duration of epoch
1	Sleep stage	30 sec
2.	EEG for seizures	15 sec
3.	Respiratory	2 min
4.	Cheyne-Stokes breathing	5 min
5.	Leg movements	5 min
6.	EKG	15 sec

FURTHER READING

1. Berry, R. B., Brooks, R., Gamaldo, C. E., Harding, S. M., Lloyd, R. M., Marcus, C. L., et al., (2017). The AASM manual for scoring of sleep and associated events: rules, terminology and technical specifications. Version 2.4. www.aasmnet.org. Darian, Illinois: American Academy of Sleep Medicine.

The Concept of Epochs

237

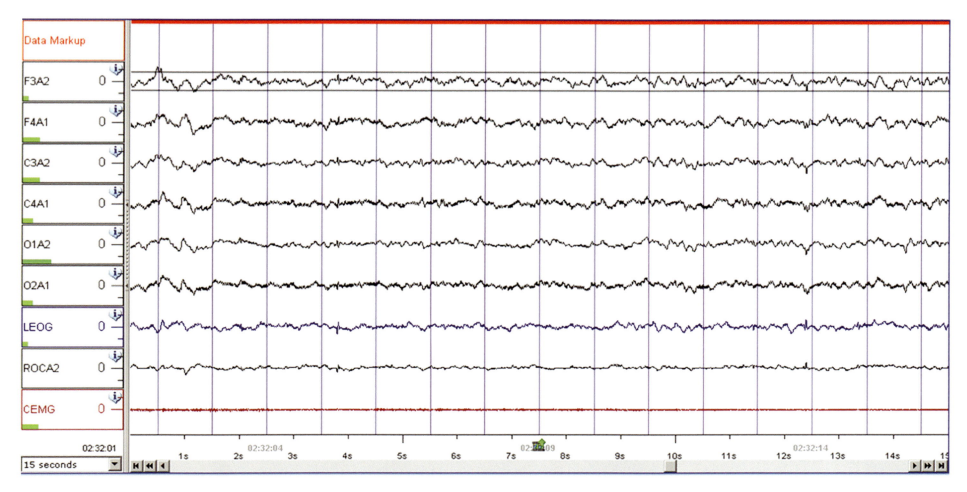

Figure 11.1 Sleep stage in 15 seconds. This epoch allows to identify each waveform separately and clearly.

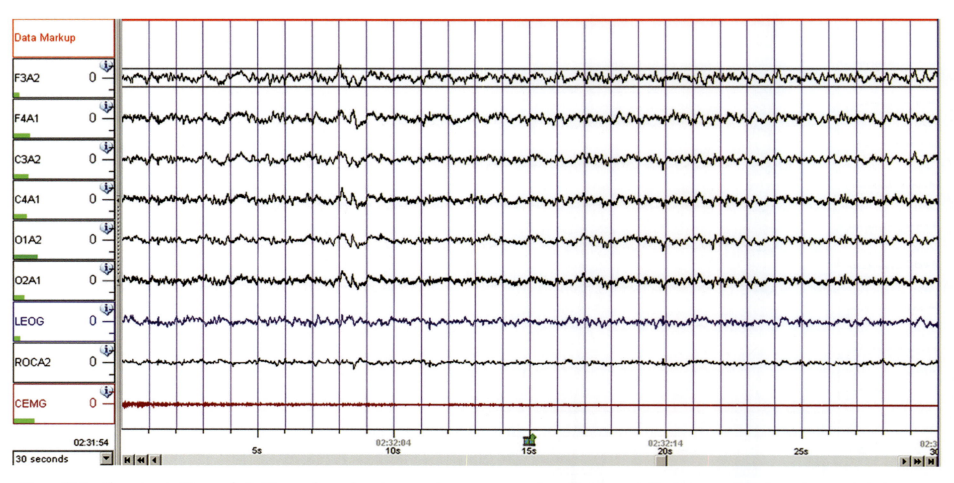

Figure 11.2 Sleep stage in 30 seconds: In 30 seconds epoch, various waveforms appears condensed but recognizable. 30 sec epoch has been chosen as it depicts most of waveforms important for sleep-wake stage scoring clearly (compare with 60 sec epoch) and at the same time, reduces time spent in scoring the data (compare 15 sec epoch).

The Concept of Epochs

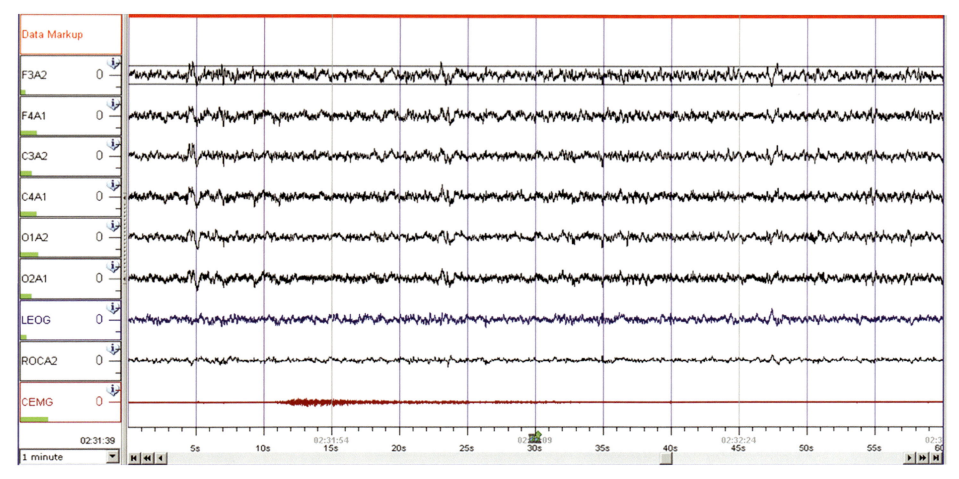

Figure 11.3 Sleep stage in 60 seconds: In 60 seconds epoch, waveforms are so condensed that they cant be recognized.

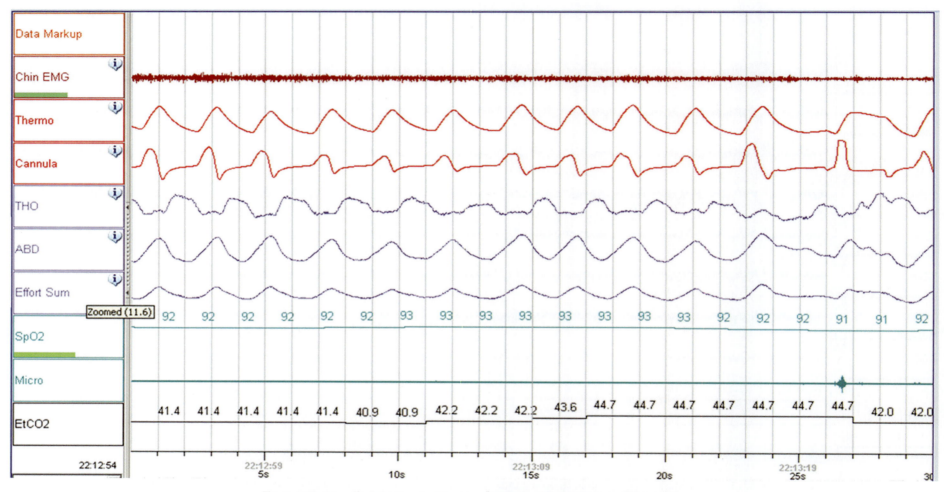

Figure 11.4A Respiration in 30 seconds. Normal respiration in 30 seconds.

The Concept of Epochs

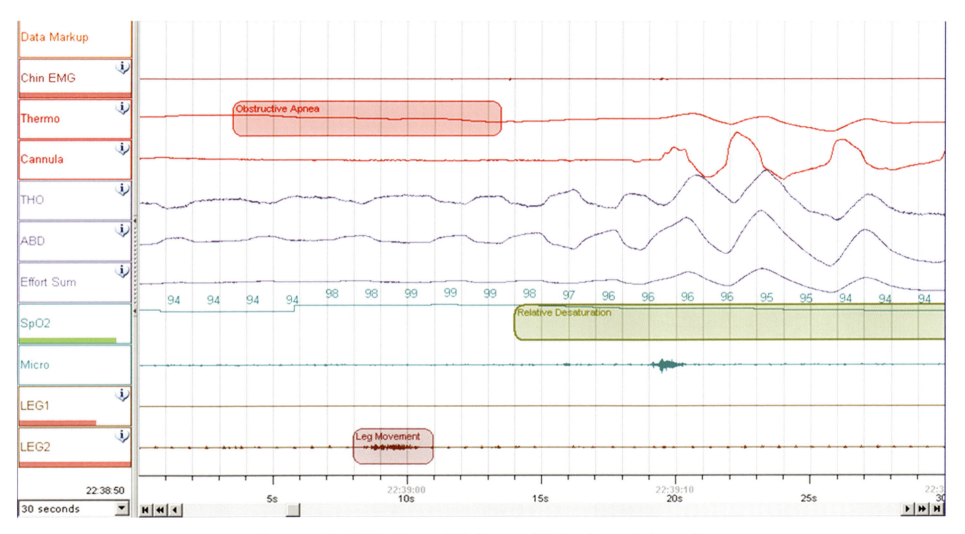

Figure 11.4B OSA in 30 seconds. Only a part of OSA can be seen in the epoch.

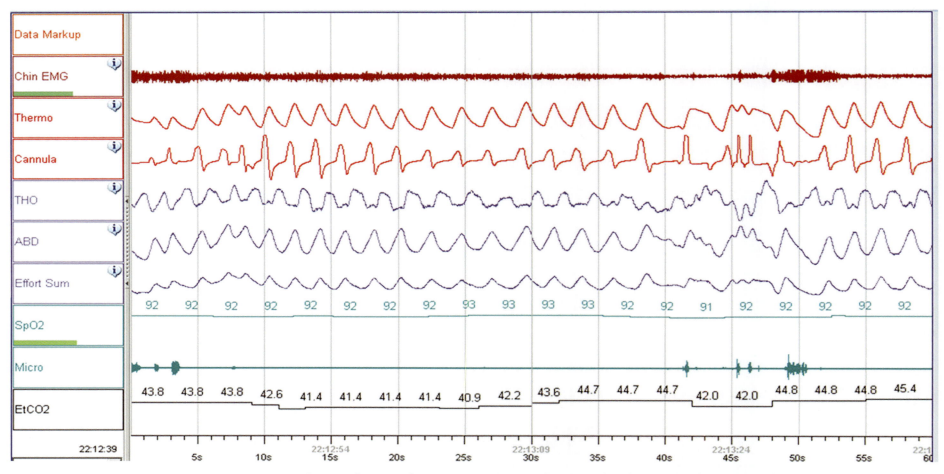

Figure 11.5 Respiration in 60 seconds: Normal respiration in 60 seconds. It provides a better view than 30 seconds epoch.

The Concept of Epochs 243

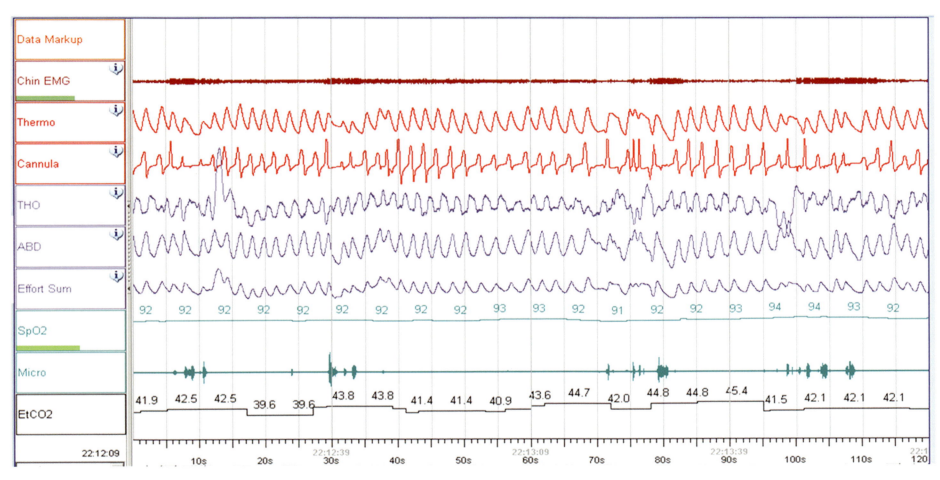

Figure 11.6A Respiration in 120 seconds. Normal respiration in 120 seconds.

244 Clinical Atlas of Polysomnography

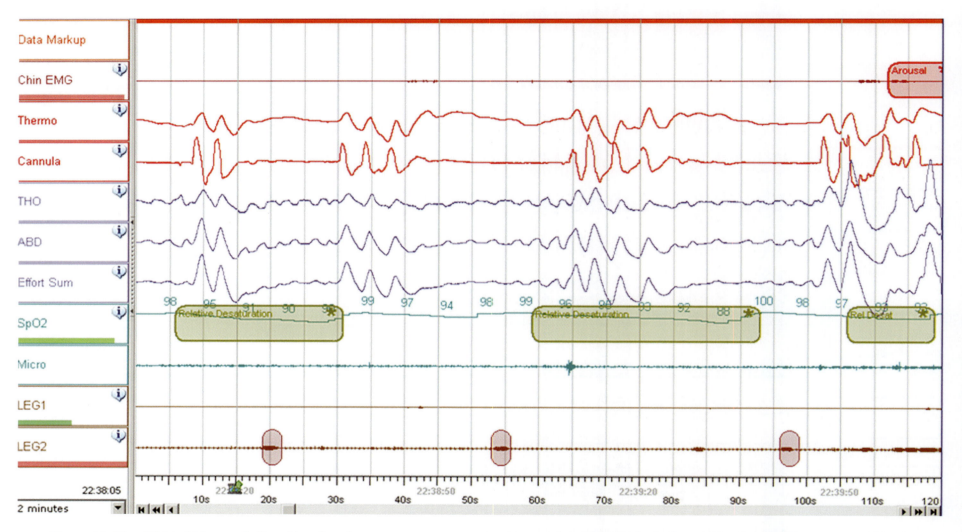

Figure 11.6B OSA in 120 seconds. Breathing disorders are easy to recognize in 120 seconds. In this epoch, rhythm and amplitude of the respiration can be recognized easily.

The Concept of Epochs

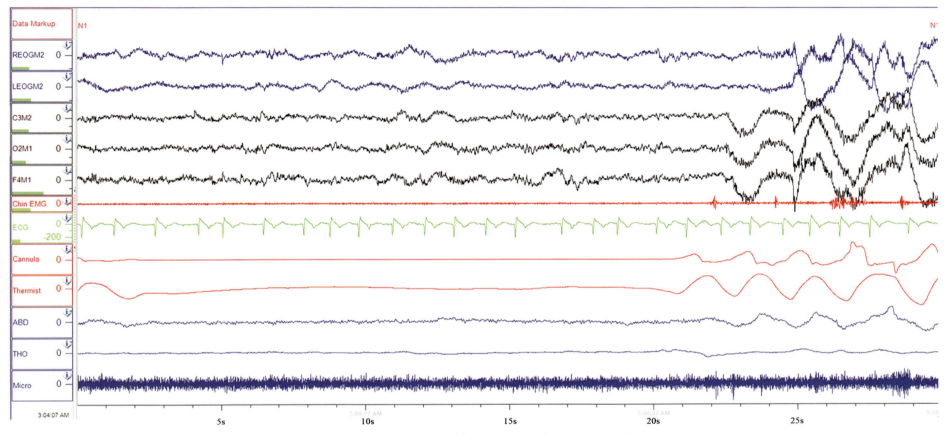

Figure 11.7 CSB in 30 seconds: In 30 seconds, it appears as hypopnea.

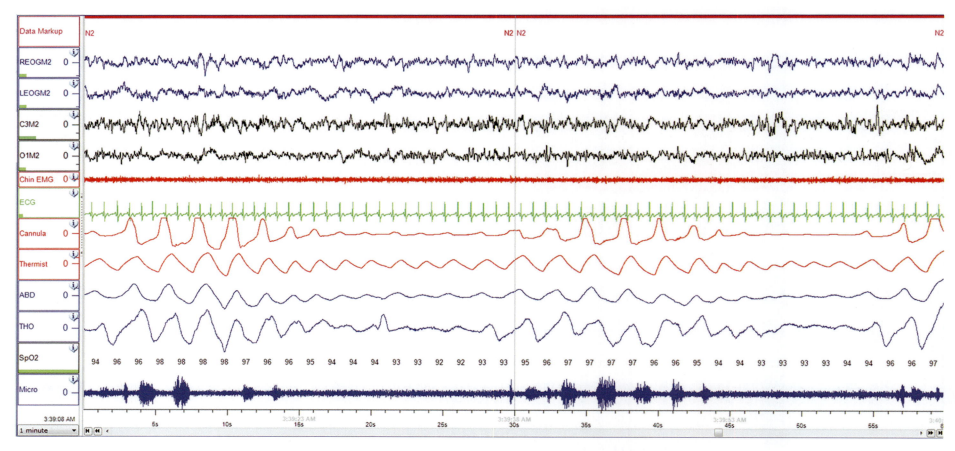

Figure 11.8 CSB in 60 seconds: In 60 seconds, appearance is better, still CSB cannot be assessed.

The Concept of Epochs

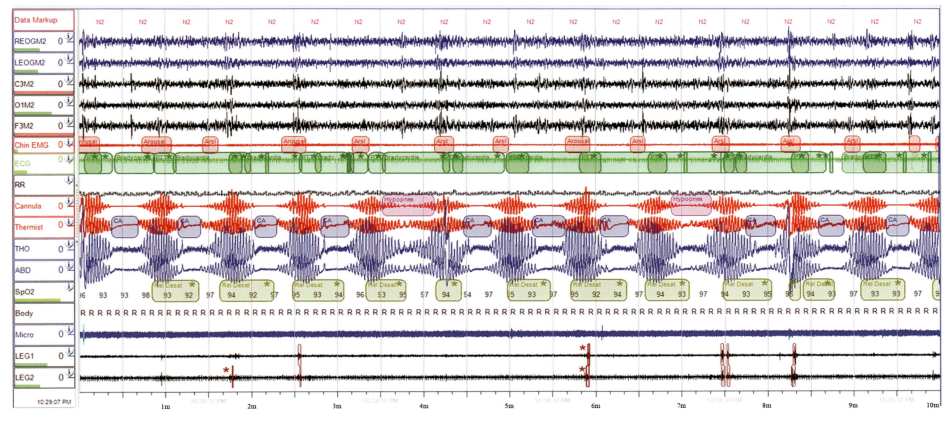

Figure 11.9 CSB in 300 seconds: 300 seconds epoch shows that central apnea/ hypopnea is actually the parts of Cheyne-Stokes breathing. This should be differentiated from respiratory pattern seen in OSA in 300 seconds epoch (Figure 11.12).

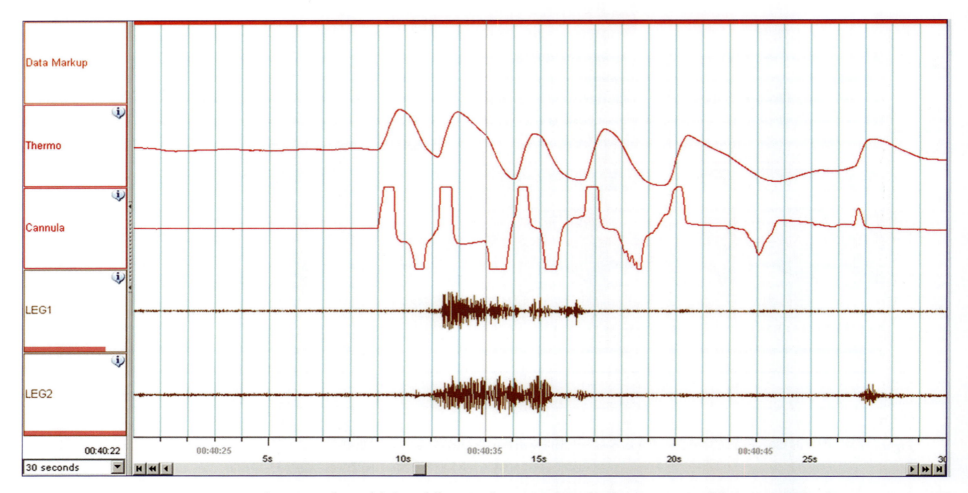

Figure 11.10 Leg movement in 30 seconds: 30 seconds epoch help to differentiate between independent leg movement and those associated with respiratory events. If a movement starts within 1 seconds of respiratory event, it is not scored as movement.

The Concept of Epochs

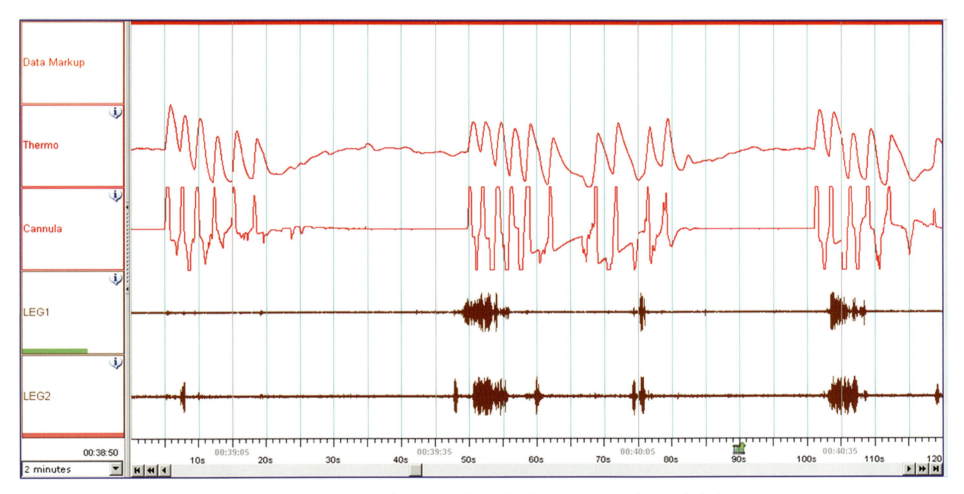

Figure 11.11 Leg movement in 120 seconds: 120 seconds epoch makes it easier to see the periodic limb movements series.

250 Clinical Atlas of Polysomnography

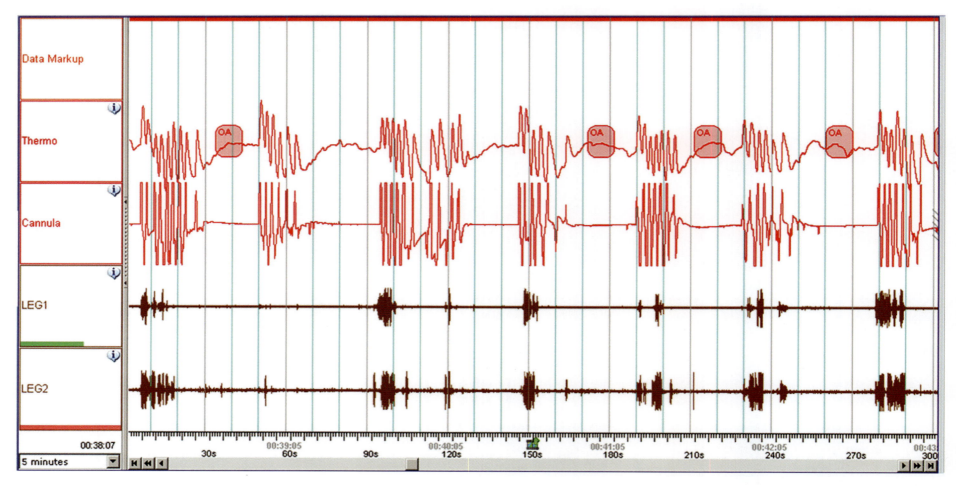

Figure 11.12 Leg movement in 300 seconds: 300 seconds epoch makes PLM series prominent.

REVIEW QUESTIONS

1. Epoch of a sleep study refers to:
   A. A small portion of the data recorded during whole night
   B. The depiction of a fixed number of channels
   C. Capturing the video data
   D. The time when the recording starts

2. Best epoch setting for Cheyne-Stokes Breathing is:
   A. 30 seconds
   B. 120 seconds
   C. 300 seconds
   D. 360 seconds

3. Epoch setting of the following is greater than that for the sleep wake staging:
   A. EKG
   B. Leg movement
   C. Seizures
   D. Suspected parasomnia

4. Following is best seen in 120 seconds epoch:
   A. Sleep stage
   B. Cardiac activity
   C. Eye movements
   D. Respiration

5. Digital recording of sleep data is advantageous over paper recording as:
   A. Provides clearer waveforms during analysis
   B. Epochs can be set differently for different parameters during analysis
   C. It is cheaper than the paper-based recording
   D. Report can be prepared quickly

ANSWER KEY

1. A B. C 3. B 4. D 5. B

12
ARTIFACTS

LEARNING OBJECTIVES

After reading this chapter, the reader should be able to:

1. Define artifacts.
2. Recognize the artifacts during a sleep study.
3. Fix the artifacts during the sleep study.

CONTENTS

Various Artifacts	253
Further Reading	268
Review Questions	268
Answer Key	268

In polysomnography, the artifact is defined as any abnormal signal that is neither physiological nor pathological but still appears in the recording because of various reasons impedance problems, movements, electrical spilling, poor filter setting, etc.

Signals for sleep staging are picked through electrodes placed on the scalp, face, and sub-mentalis muscles. EEG signals originate in the brain (cortex and thalamocortial circuit) and before being picked up by the electrode, they have to travel through the skull, scalp and other anatomical structures in the vicinity. These signals are electrical in nature. Because of these issues, these are liable to alter in a variety of situations. Many times, a number of artifacts may appear in any of the PSG channels.

The artifact is defined as any abnormal activity appearing in any diagnostic aid that is not pathognomic of any disease but can be ascribed to the alteration in the signals because of undesired input from other sources.

Commonly following artifacts may appear in the PSG channels:

1. Electrode pop artifact: because of poor contact between electrode and skin (Figure 12.1);
2. Respiration artifact also is known as sweating artifact (Figure 12.2);
3. EKG artifact (Figure 12.3);
4. Movement artifact (Figure 12.4);
5. Artifact due to physiological activities of orofacial structures (Figure 12.5);
6. Electrical artifact (Figure 12.6);
7. Muscle artifact (Figure 12.7);
8. Cardioballistic artifact (Figure 12.8).

The scorer must be able to differentiate the artifact from the true signals so that a definitive conclusion may be reached.

As we have discussed above, the source of the signal may be the patient, the electrodes, head-box, or the amplifier. Artifacts arising secondary to the head-box or amplifier problems are widespread. Localized artifacts are either attributed to a patient's physiological activity spilling in the channel (e.g., pulse artifact, chewing artifact) or to a problem localized to one or few leads (sweating artifact or 60 Hz artifact secondary to increased impedance, a pop artifact due to poor contact of the lead with skin). In addition, physiological activities that involve the whole body often produce widespread artifacts, for example, ECG artifact, sweating artifact or movement artifact.

It is of paramount importance to recognize these artifacts while recording the study and also while scoring the recorded data. If they are recognized during data acquisition, they can be fixed to obtain a good quality recording (as in Level I polysomnography). Similarly, a scorer has to be vigilant of the artifacts should they remain unattended (as in Level II study or level III study) to prepare a good quality report.

In this chapter, we will discuss how to recognize them and fix them during data acquisition and during scoring of data (Table 12.1).

Artifacts

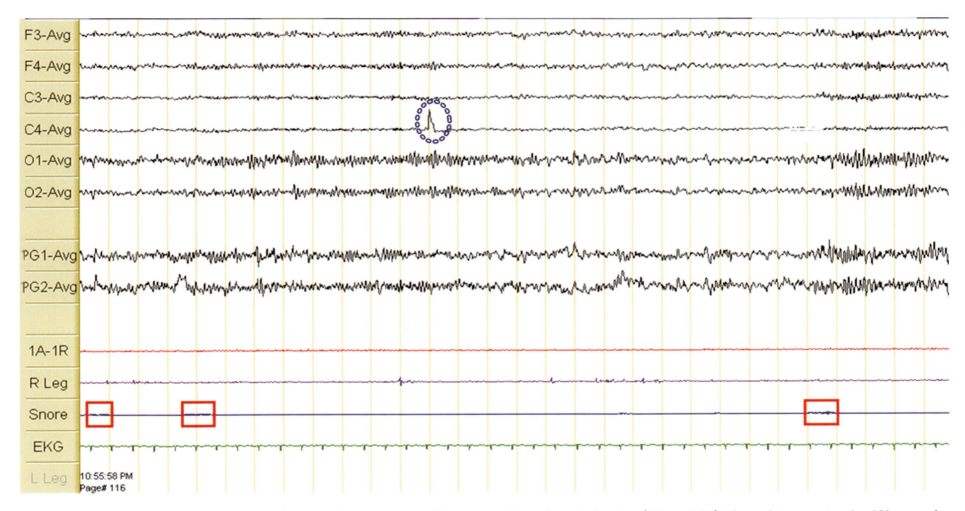

Figure 12.1 Electrode pop artifact: Electrode pop artifacts appear as sudden uprise of wave from the baseline (Channel C4). Electrode in question should be pasted properly in this case.

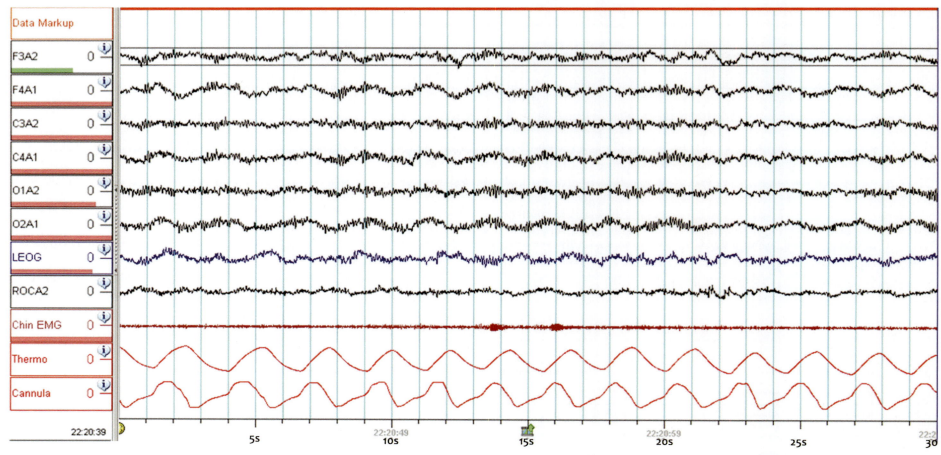

Figure 12.2A Sweating artifacts or poor contact artifacts: Sweating artifacts appear due to poor contact between the skin and electrode owing to presence of sweat in between. A: Note the undulating waves in F4 and O2 with respiration.

Artifacts

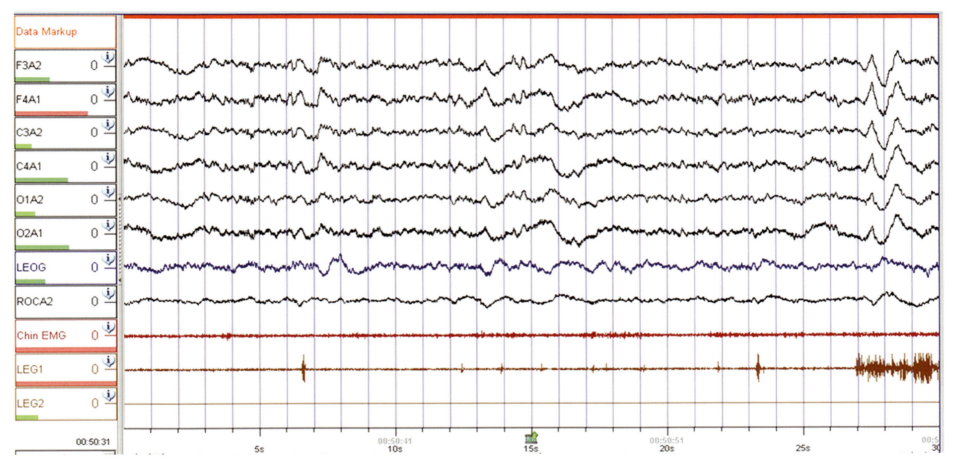

Figure 12.2B Sweating artifacts without respiratory waveform. In such cases, room should be cooled using an air conditioner.

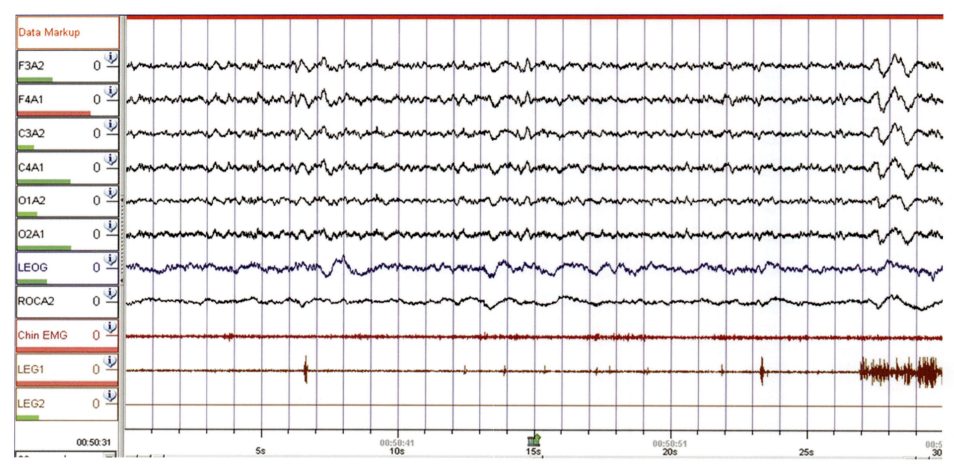

Figure 12.2C If not fixed during recording of data, sweating artifacts can also be removed by increasing the LFF to 1 Hz from 0.3. However, it may compromise appearance of slow waves in EEG.

Artifacts

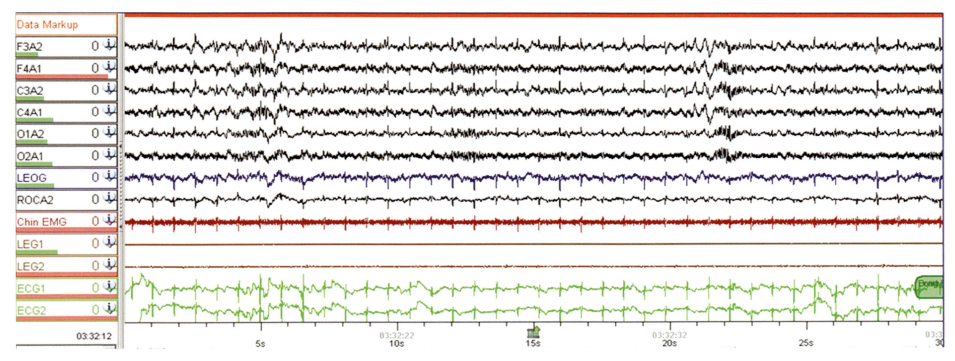

Figure 12.3A EKG artifacts: EKG artifacts appear when an electrode is placed over a blood-vessel. Electrical activity from the heart is carried through blood-vessels. They may be recognized by looking at the EEG and EEG simultaneously.

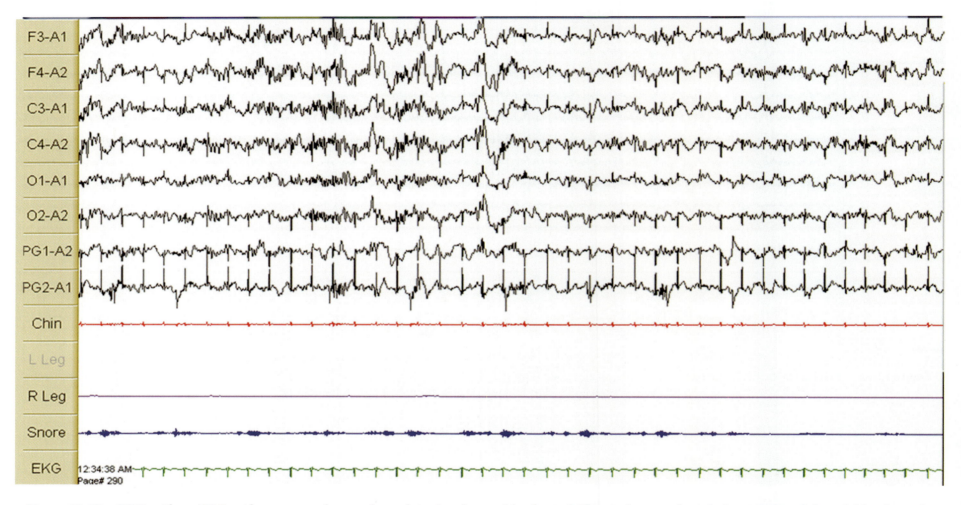

Figure 12.3B EKG artifacts: EKG artifacts appear when an electrode is placed over a blood-vessel. Electrical activity from the heart is carried through blood-vessels. They may be recognized by looking at the EEG and EEG simultaneously.

Artifacts

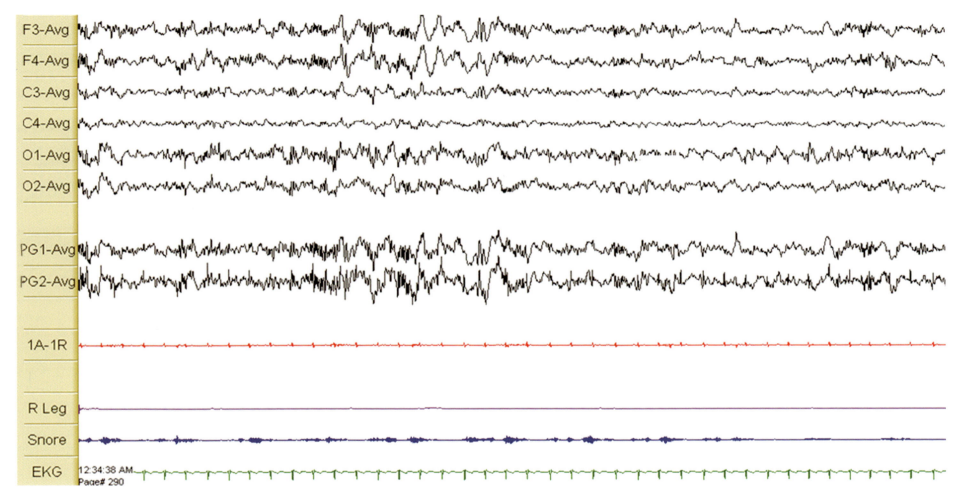

Figure 12.3C EKG artifacts may be removed by changing the place of electrode or using the EKG filter in the software or referring the scalp electrodes to M1 and M2 simultaneously. This illustration shows that short-circuiting M1 and M2 to form an average reference has removed the EKG artifacts.

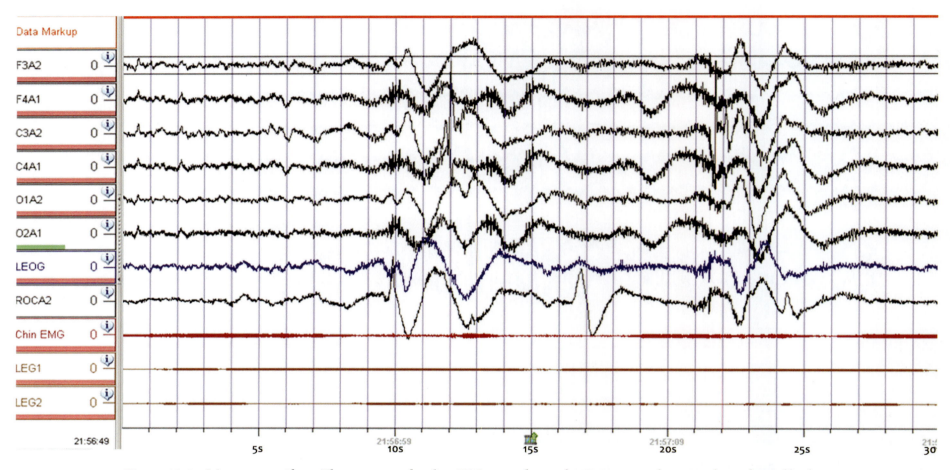

Figure 12.4 Movement artifacts: They appear as dangling EEG waves due to change in contact between electrodes and body.

Artifacts

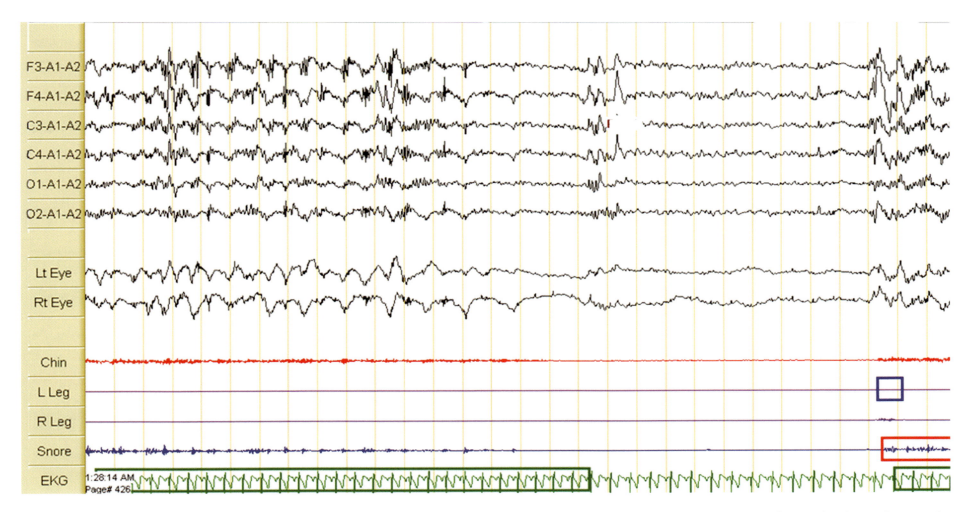

Figure 12.5 Artifacts due to orofacial structures: Like any other organ, tongue is also a dipole and its movement inside the oral cavity changes the electrical potential beneath some electrodes. In this epoch patient is talking during sleep (signals in microphone). Also shown that movement of tongue is showing deflections in EEG and EOG derivations due to change in electrical potentials close to the electrodes.

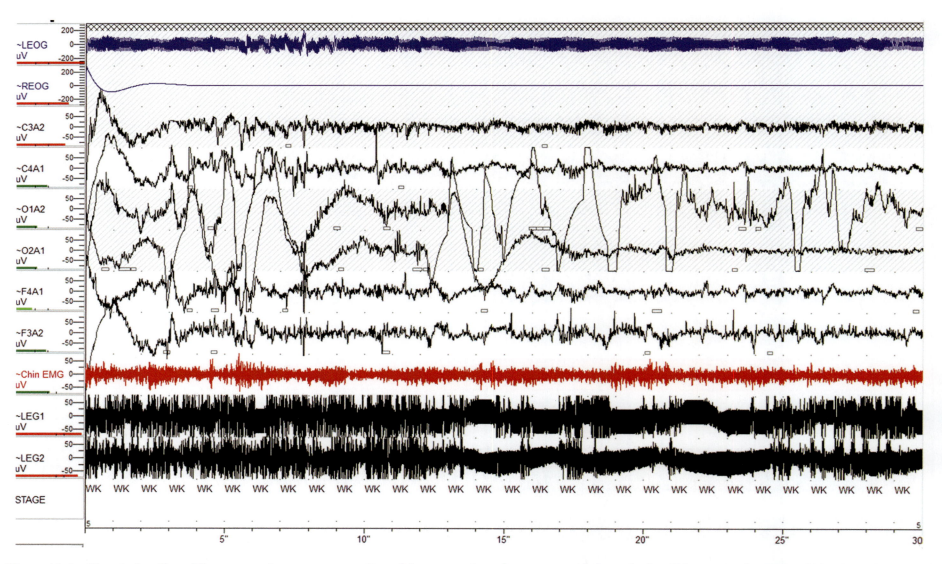

Figure 12.6 Electrical artifacts: They appear due to poor grounding of the patient. In such cases, ground electrode should be removed, area should be cleaned again and then electrode should be placed.

Artifacts

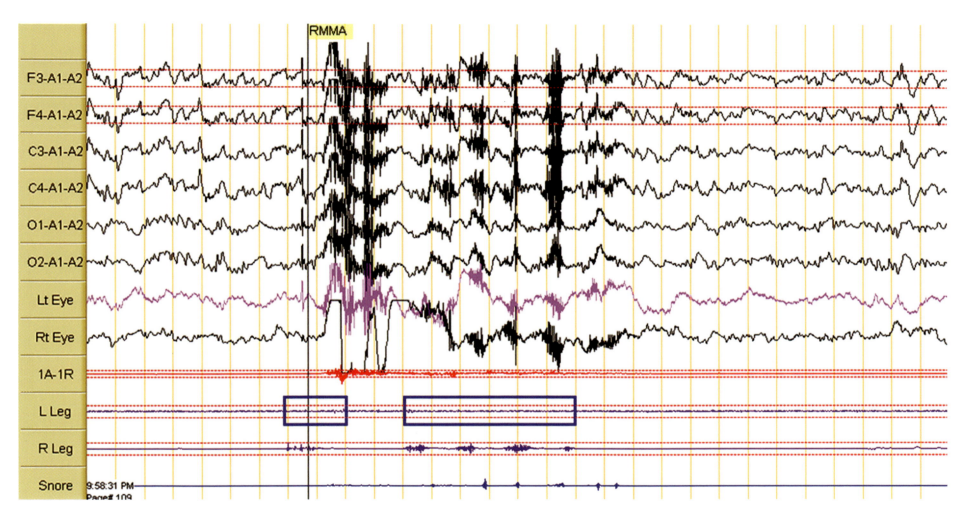

Figure 12.7 Muscle artifacts: In this epoch you can see muscle artifacts due to grinding of the teeth. Noise produced during teeth-grinding can be seen in the microphone channel as well. This is rhythmic activity is and is known as Rhythmic Masticatory Movement Activity (RMMA).

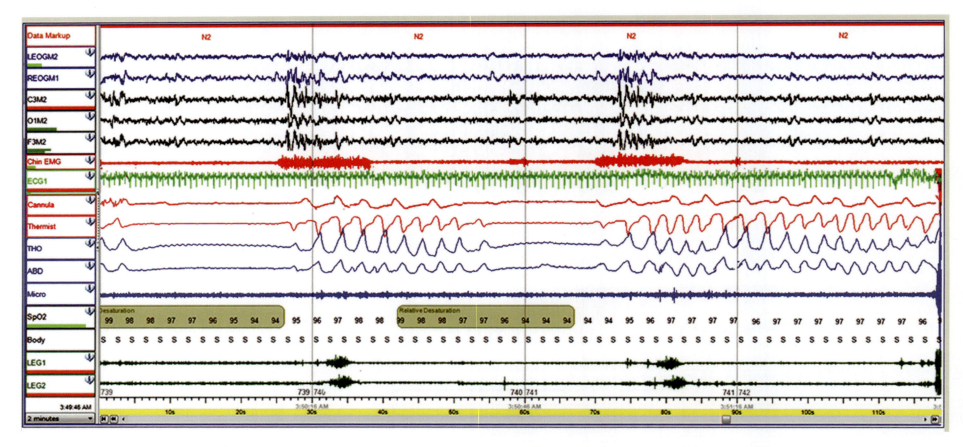

Figure 12.8 Cardioballistic artifacts: They may create problem in scoring central sleep apnea. Due to heart beat, chest belt records some movements in this epoch cardioballistic artifacts may be seen in THO channel..

Artifacts

Table 12.1 Fixing the Artifact

Artifact	Fixing during data acquisition	Remarks	Fixing during scoring	Remarks
Electrode pop artifact	Clean the area under electrode after removing it. Fix with adequate amount of conductive gel and surgical tape	Improves the signal quality for the rest of the study	Can not be fixed	N/A
Sweating artifact	• Look for the reason of sweating • Reduce the ambient temperature by resetting the temperature of air-conditioner, if possible	Improves the quality of recording and signal quality for rest of the study	Can be fixed by increasing LFF from 0.3 Hz to 1 Hz	May result in loss or alteration of waveforms below the LFF settings.
EKG artifact	• Instead of referring the lead to contralateral mastoid, refer it to both mastoids (e.g., F4-M1M2)	Improves the quality of signals for rest of the study	In some machines, may be fixed afterward by applying "QRS filter"	May be mistaken for spikes or sharp wave if attention not paid to. Results in error during reporting.
Movement artifact	Look if the patient is uneasy in the bed. The reason for discomfort should be sorted out.	Improves the quality of signals for rest of the study	Fixing not possible during scoring	Results in loss of data
Artifacts due to physiological activities of orofacial structures	Can be ignored if infrequent. If not already sought, next day enquire for the sleep-related bruxism, sleep-related laryngospasm	Improves the quality of signals for rest of the study	Fixing not possible during scoring	Loss of data or diagnosis of sleep related bruxism can be made.
Electrical artifact	Suggests poor filtering or poor grounding. Look for the impedance in ground electrode, clean the area again and fix the ground electrode. If still not fixed, turn-on Notch filter	Improves quality of signals for rest of the study	Fixing not possible	Loss of data
Muscle artifact	Mainly by contraction of scalp muscles. Try to relax the patient	Improves quality of signals for rset of the study	Fixing not possible	Loss of data

FURTHER READING

1. Patil, S. P. (2010). What every clinician should know about polysomnography? *Respiratory Care.* 55, 1179–1193.
2. Beine, B. (2005). Troubleshooting and elimination of artifact in polysomnography. *Respiratory Care.* 11, 617–634.

REVIEW QUESTIONS

1. Sweating artifacts may be removed by all of the following EXCEPT:
   A. Improving the cooling of the room
   B. Changing the low frequency filter setting
   C. Changing the amplitude of the waveform
   D. Replacing the electrode after cleaning the area
2. EKG artifact appear due to all of the following EXCEPT:
   A. Placement of electrical lead close to an artery
   B. Unequal impedance in the electrodes
   C. Travelling of cardiac activity in the body
   D. Poor setting of filters
3. Following artifact can be fixed during the scoring of data:
   A. Electrode pop
   B. EKG
   C. Movement artifact
   D. Poor grounding
4. Cardioballistic artifact may interfere with the scoring of:
   A. Hypopnea
   B. Obstructive Sleep Apnea
   C. Central Sleep Apnea
   D. Hyperventilation
5. Following should be avoided close to the head box to prevent appearance of artifact:
   A. Intravenous fluid
   B. Mobile phone
   C. Air-conditioner
   D. Fan

ANSWER KEY

1. C 2. D 3. A 4. C 5. B

13

SCORING OF DATA IN ADULTS

LEARNING OBJECTIVES

After reading this chapter, the reader should be able to:

1. Recognize and score sleep-wake stages in the recording of an adult patient.
2. Recognize and score respiratory events in the recording of an adult patient.
3. Recognize and score leg movements in the recording of an adult patient.
4. Recognize and score cardiac rhythms.
5. Recognize and score REM sleep behavior disorder, sleep-related bruxism and rhythmic movement disorder.

CONTENTS

13.1	Scoring of Sleep Stage	270
13.2	Scoring of Respiratory Data	340
13.3	Leg Movements	350
13.4	Bruxism	363
13.5	REM Sleep Behavior Disorder	363
13.6	Rhythmic Movement Disorder	363
13.7	Scoring of EKG Data	382
Further Reading		382
Review Questions		382
Answer Key		385

Sleep and wakefulness lie on a continuum. They can be absolute states, once fully progressed, but the switch from one state to another occurs through slow transition. These transitions are gradual, especially from wakefulness to sleep as compared to one from sleep to wakefulness. This can be easily seen during scoring of polysomnographic data. Similar to the EEG changes, other physiological parameters, for example, respiration, muscle tone, heart rate, and blood pressure also change from wakefulness to sleep and then across different sleep stages.

To maintain uniformity across different scorers and across different sleep laboratories, some rules have been proposed for the scoring of data. Globally, rules proposed by the American Academy of Sleep Medicine are followed for the scoring of polysomnography data. In this chapter, we will discuss how to score the data obtained after a full night of recording.

13.1 SCORING OF SLEEP STAGE

Before we go for the scoring of sleep stage, it would be better to understand the progression of sleep cycles through the night. This information will be helpful in understanding the extent of normality. This will also help you to understand why you are getting a particular type of sleep stage during a certain part of the night.

As an adult person progresses from wakefulness to sleep, he/she passes through various stages of NREM sleep. The NREM sleep lasts for approximately 100 minutes and then the REM sleep appears. After spending a certain period in REM sleep, the NREM sleep reappears. This cycle of NREM-REM transition continues whole of the night, as explained below, with reduction in amounts of N3 and progressive lengthening of REM towards the morning. This can be easily understood by following a hypnogram (Figure 1.8).

NREM can be further divided into three different stages—N1, N2, and N3, depending upon the characteristic of EEG waveforms obtained during sleep. Any person who is falling asleep first attains the N1 stage (which can behaviorally be equated with drowsiness), then passes through the N2 stage to enjoy the deep sleep—the N3 sleep. Usually, the change to the less deep stage (i.e., from N3 to N2 and N2 to N1) in normal persons is also gradual throughout the night, unless there is sudden arousal. Also, remember that earlier rules (R and K rules) proposed two different stages of deep sleep (S3 and S4). However, rules have been modified and now we score deep sleep as N3. This change was brought as scientific literature failed to show any advantage of dividing deep sleep (N3) into two stages (S3 and S4).

Scoring of Data in Adults

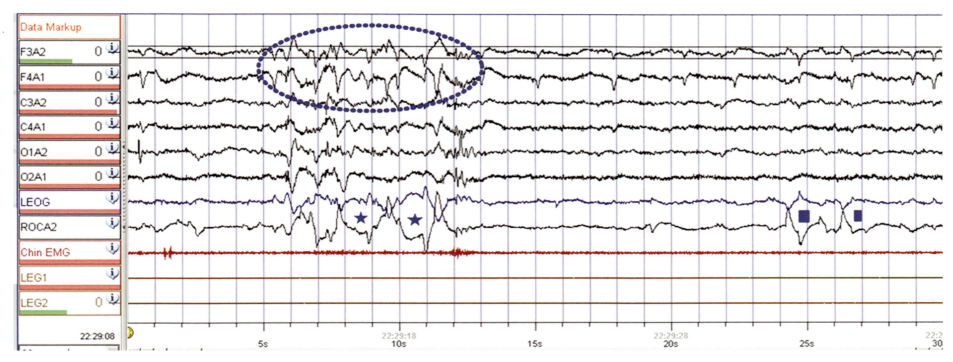

Figure 13.1 Active wakefulness is characterized by electrical activity in EEG channels that is low voltage, mixed frequency (Gamma-beta-theta), Reading eye movements (See Star) in opposite phase having initial slow phase followed by a rapid phase (LEOG and ROCA2) associated with eye blinks having frequency of 0.5–2 Hz (Frontal derivations) (see inside circle), and rapid eye movements, where initial deflection is less than 0.5 seconds (see Box) (30 seconds epoch).

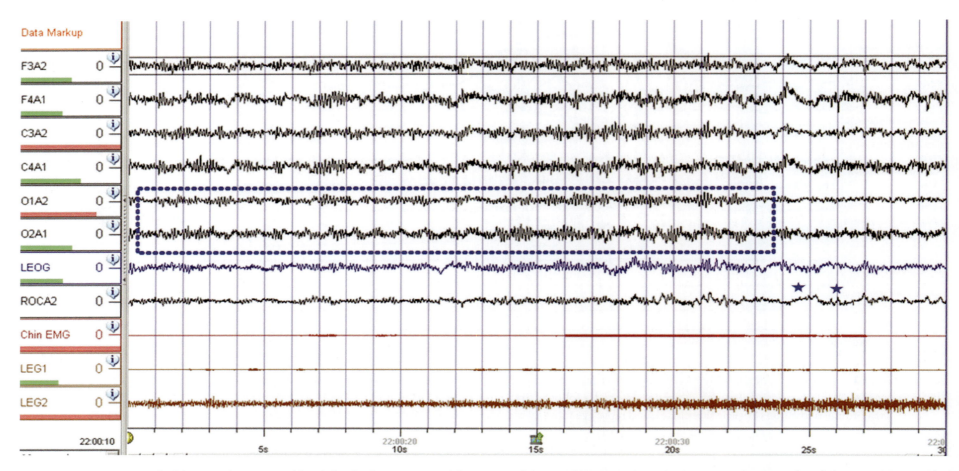

Figure 13.2 Quiet wakefulness is characterized by alpha rhythm in occipital derivations for more than 50% of epoch, slow eye movements (Star) (30 seconds epoch).

Scoring of Data in Adults

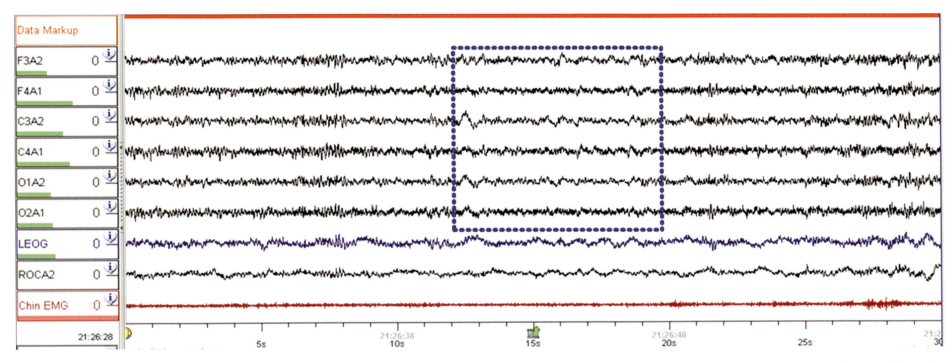

Figure 13.3 Microsleep: Please note that transition from wakefulness is gradual. Before a scorable N1 epoch, few epochs of microsleep are seen. In these epochs, EEG shows islands of theta rhythm (in the box) in between alpha. However, this epoch will be scored as Wake as alpha occupies more than 50% of epoch.

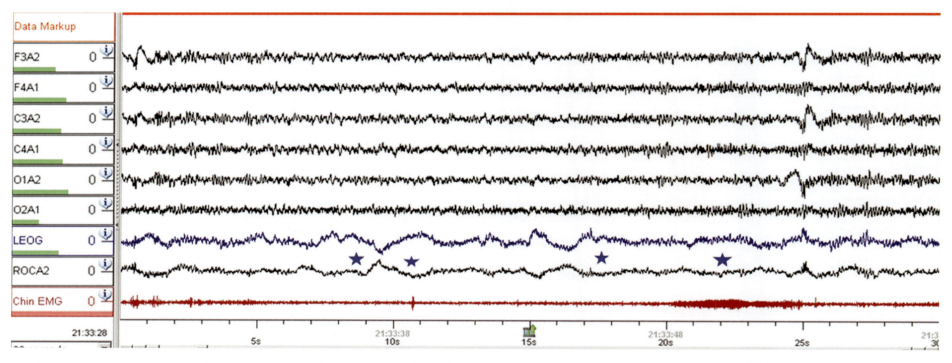

Figure 13.4 A Microsleep: Slow eye movements (star) are sinusoidal and duration of initial deflection is more than 0.5 seconds. They may persist during quiet wakefulness as well as during N1. This epoch shows SEMs while a person is drifting into sleep. Shall be scored as Wake as alpha occupies more than 15 seconds of epoch.

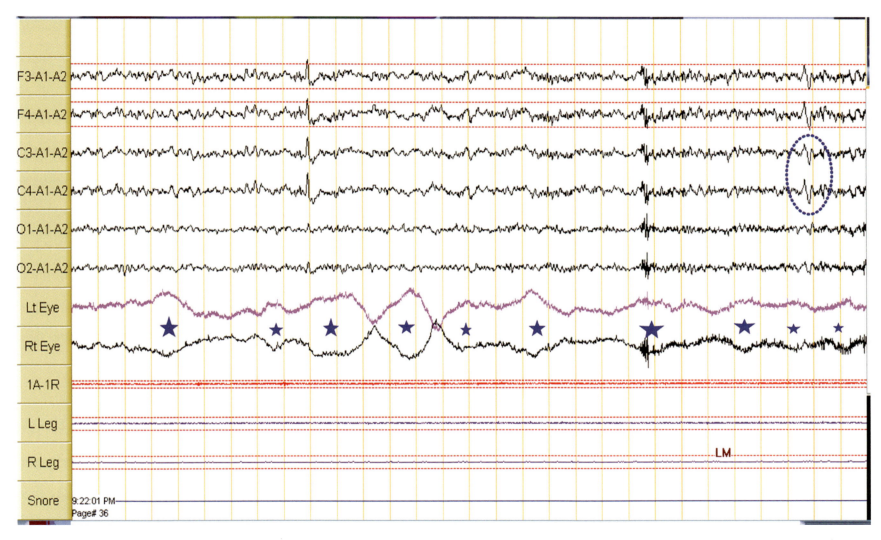

Figure 13.4B N1 is scored when alpha activity in EEG is replaced by low voltage mixed frequency activity in theta range in more than half of the epoch (> 15 seconds in an epoch of 30 seconds) and slow eye movements (satr). Chin tone goes down with N1. Vertex sharp waves (duration < 0.5 seconds and in circle) appear maximally in central derivations. First scorable N1 epoch also marks the sleep onset (30 seconds epoch)

Earlier rules (R and K rules) proposed four different stages of normal NREM sleep – S1, S2, S3 and S4. In new rules, S1 has been replaced by N1, S2 by N2 while S3 and S4 together formed stage N3.. This change was brought as scientific literature failed to show any advantage of dividing deep sleep (N3) into two stages (S3 and S4).

With this information in the background, we can now proceed with the scoring of sleep stages.

For the scoring of sleep staging, we depend upon the EEG, EOG, and chin EMG information. When we are scoring the sleep stage, we need to determine arousals also (see section on the scoring of arousals). These arousals can be physiological, viz., at the transition of sleep stages, or pathological, for example, associated with the respiratory event or leg movement. These arousals determine not only the epoch in which they are seen but also the scoring of the subsequent epoch, as we shall see. As we have already discussed, sleep stages are scored in an epoch of 30 seconds. Remember, if waveforms for two different stages are seen in an epoch, the epoch is assigned the stage whose waveforms occupy more than 50% of the epoch. At times, waveforms suggesting three or more stages may be seen in an epoch. In such case, first, decide whether it represents "wake" or "sleep" depending upon the relative amount of respective waveforms in the epoch. Once it is decided that it should be scored as "sleep," assign the sleep stage (N1, N2, N3 or REM) whose waveforms represent most of the part of the epoch. During the study, patients may be disconnected from the PSG as they may need to go to the washroom. However, recording continues during this period and this should be scored as awake.

Wakefulness is scored when the epoch shows blinking; rapid eye movements with high tone in chin EMG or saccadic eye movements are seen (Figure 13.1). Alpha activity (8–13 Hz) is seen during wakefulness if the subjects close their eyes in the occipital leads, also known as posterior dominant rhythm (Figures 13.2 and 13.3). However, nearly 10% of individuals are not alpha producers and their EEG activity remains same during eyes open and eyes closed. Rules for scoring of N1 are different between alpha producers and alpha-nonproducers (Figures 13.4 A,B and 13.5 C-G).

It must be remembered that in alpha producers, vertex waves and slow eye movements, though generally seen during N1, are not required for scoring of N1. In alpha producers, the only attenuation of alpha to low-voltage, mixed-frequency activity is sufficient for scoring of N1 (Figures 13.5A and 13.8B). In alpha non-producers, N1 is

Scoring of Data in Adults

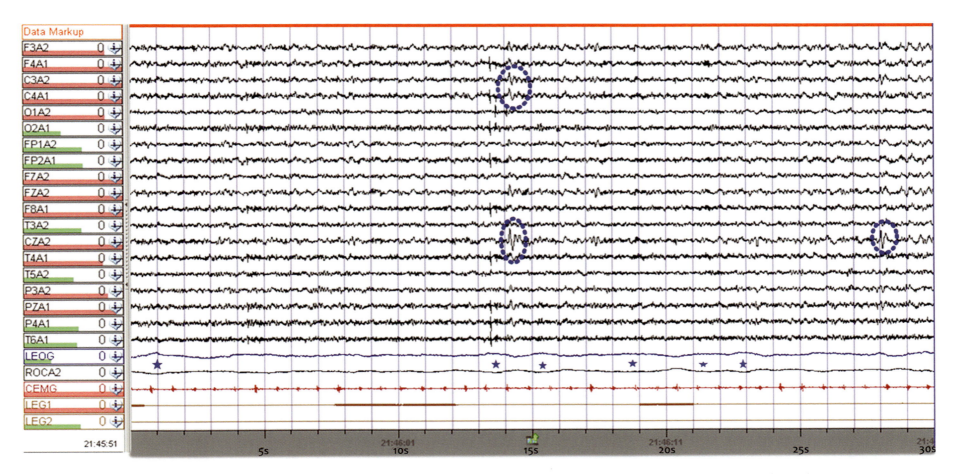

Figure 13.5 A Vertex waves in N1: Vertex waves are sharp waves that appear distinct to the background EEG in central derivations (Circle) and have duration less than half a second. Slow eye movements are also seen (star) and background EEG activity is low voltage mixed frequency in theta range during whole of epoch. So this is scored as N1.

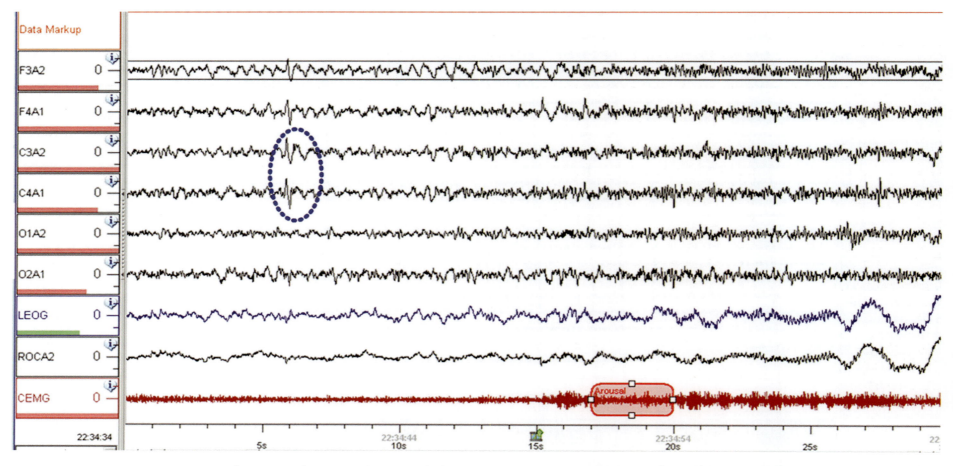

Figure 13.5 B N1: Vertex waves are sharp waves that appear distinct to the background EEG in central derivations (Ellipse) and have duration less than half a second. Background EEG activity is low voltage mixed frequency in theta range for more than 15 seconds. Though alpha is seen in the last part of epoch; but it occupies less than 50% of epoch, hence this epoch will be scored as N1.

Scoring of Data in Adults

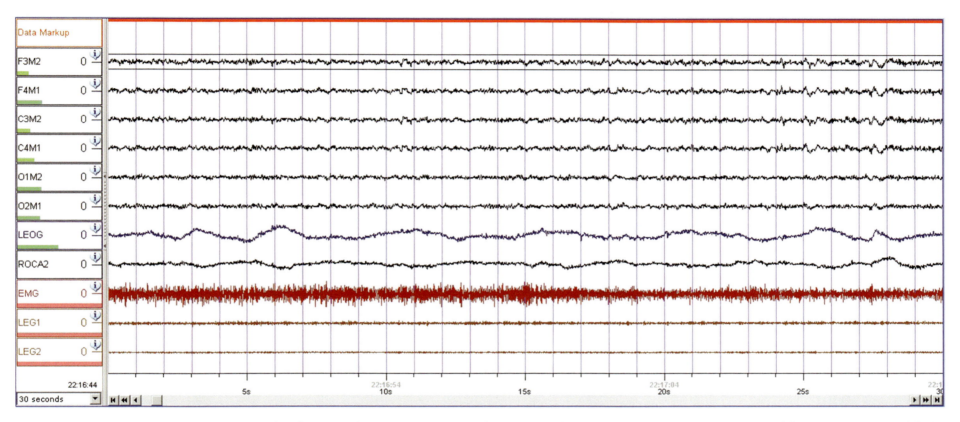

Figure 13.5C–G Scoring of N1 in subjects that do not produce alpha as posterior dominant rhythm: This epoch shows low voltage mixed frequency activity with high chin tone and slow eye movements. This is quiet wakefulness.

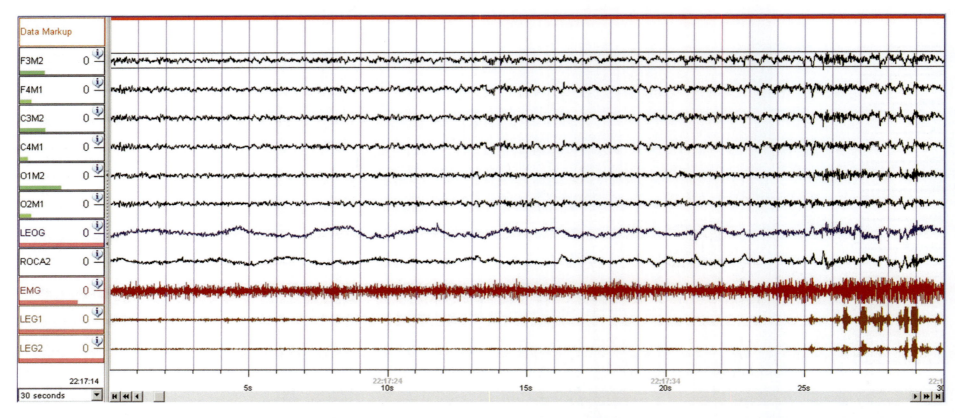

Figure 13.5D This epoch shows low voltage mixed frequency activity with high chin tone and slow eye movement. Last part of epoch shows eye blink artifacts, movement artifacts and further increase in chin tone. EEG activity is just like previous epoch. This will be scored as wake.

Scoring of Data in Adults

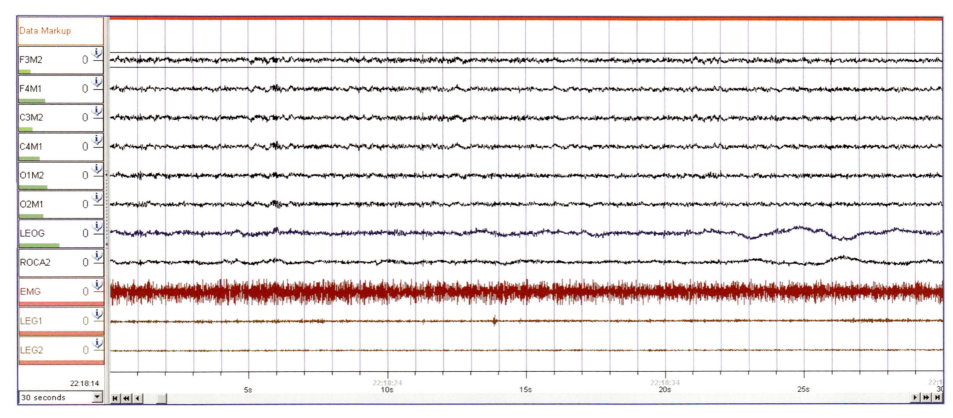

Figure 13.5E This epoch does not show any change in EEG, however, eye movements are lesser pronounced. This will also be scored as wake.

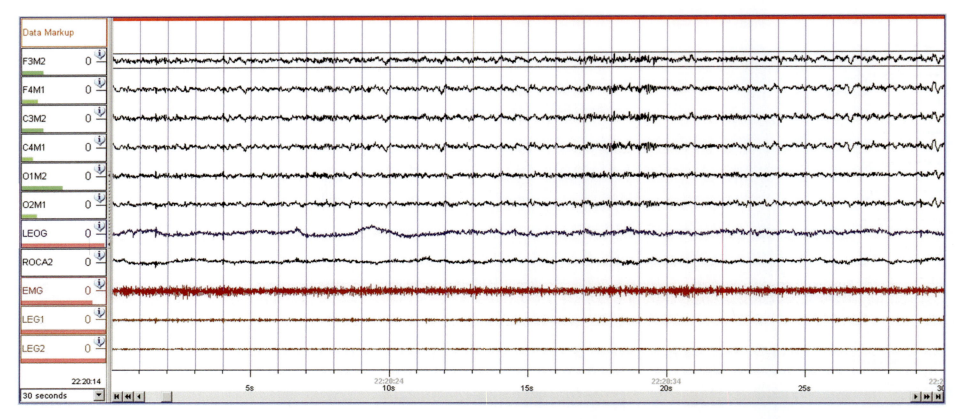

Figure 13.5F This epoch shows slowing of background EEG activity compared to previous epoch with lowering of chin tone and reduction of eye movement. This should be scored as N1.

Scoring of Data in Adults

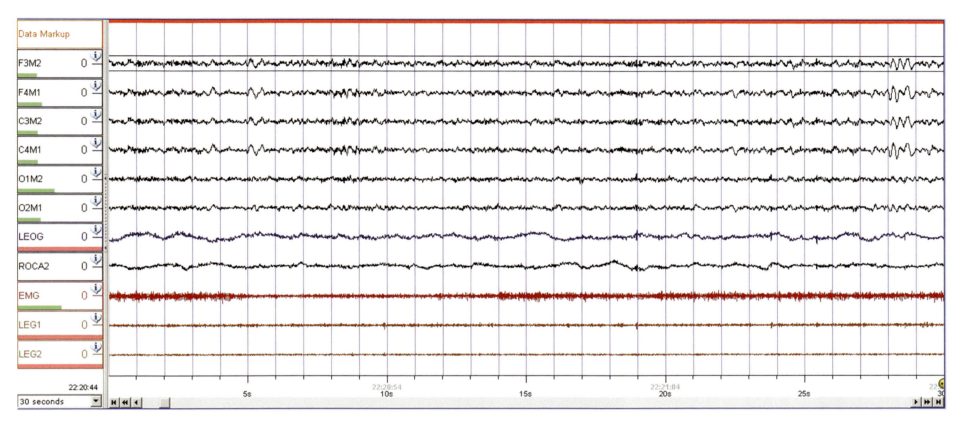

Figure 13.5G This epoch is like previous epoch showing N1 sleep.

scored when the EEG activity shows theta waves (4–7.99 Hz) with slowing of background EEG activity by 1 Hz or greater than the awake EEG activity, or vertex sharp waves appear, or slow eye movements appear. Any one of these three features is sufficient to score N1 in alpha-non-producers (Figure 13.5 C-G).

The patient may have arousals during any stage of sleep. An arousal in NREM sleep is scored when the EEG activity abruptly shifts for at least 3 seconds to alpha or it becomes higher than the earlier frequency. However, this must be differentiated from sleep spindles as they may be prolonged. At least 10 seconds of stable sleep must be present before arousal (Figures 13.5B and 13.6). If the arousal lasts more than 50% of the epoch, that epoch is scored as wake (Figure 13.7).

Stage 2 (N2) sleep is characterized by K-complexes and sleep spindles. If any of them is seen during the first half of the present epoch or second half of the previous epoch, the present epoch should be scored as N2.

A sleep spindle is a train of high frequency (11–16 Hz) waveform having a duration of at least 0.5 seconds with maximum amplitude in central derivations (Figure 13.8A and 13.8B).

K complexes are large waves that appear distinct from the background activity with initial negative followed by positive deflection having the duration of at least 0.5 seconds and seen with maximum amplitude in frontal leads (Figure 13.9). An epoch having either sleep spindle or K complex in first half is scored as N2. If an epoch shows these waveforms in second half, then subsequent epoch should be scored as N2, unless it confirms waveforms depicting any other sleep-wake stage (Figure 13.10). Any K complex associated with arousal is not considered as K complex (Figure 13.11).

Continue to score epoch with low voltage mixed frequency activity as N2, till any other sleep stage appears, even in absence of K-complexes and sleep spindles (Figure 13.12). Although eye movements are absent during N2, sometimes they may be seen during N2 (Figure 13.13). A microarousal during N2 is shown in Figure 13.14A. Majority of this epoch shows waveforms suggesting N2, hence it will be scored as N2. End scoring N2, if an epoch of K-arousal in the first half appears (Figure 13.11) or an epoch of wakefulness appears (Figure 13.14 B), or an epoch of major body movement (Figures 13.16; 13.17 A,B,C and 13.18 A, B) followed by an epoch with slow eye movement appears (Figure 13.18 A, B) or N2 converts to N3 (Figure 13.19) or REM (Figure 13.20A).

Stage 3 (N3) sleep is scored when slow-wave activity, defined as waveforms having frequency

Scoring of Data in Adults 285

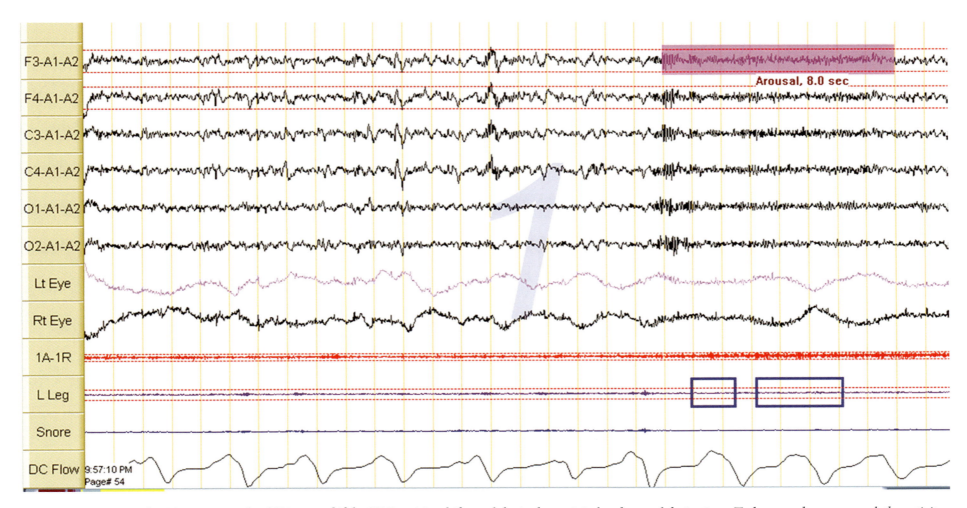

Figure 13.6 Arousal in N1: An arousal in N1 is scored if the EEG activity shifts to alpha in the occipital and central derivations. To be scored as an arousal, the activity must last at least 3 seconds but not more than 15 seconds. If it is more than half the epoch, score the epoch as wake. At least 10 seconds of uninterrupted sleep must be present before an arousal.

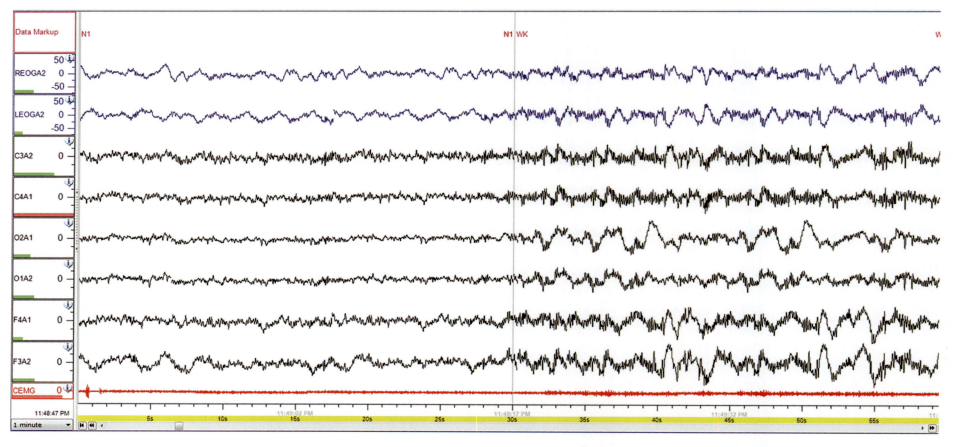

Figure 13.7 N1 to wakefulness: If the arousal lasts more than 15 seconds in an epoch, score it as wake (60 seconds epoch) Subsequent epoch must be scored based upon the waveforms present in epoch, as discussed in Figures 13.4 and 13.10.

Scoring of Data in Adults

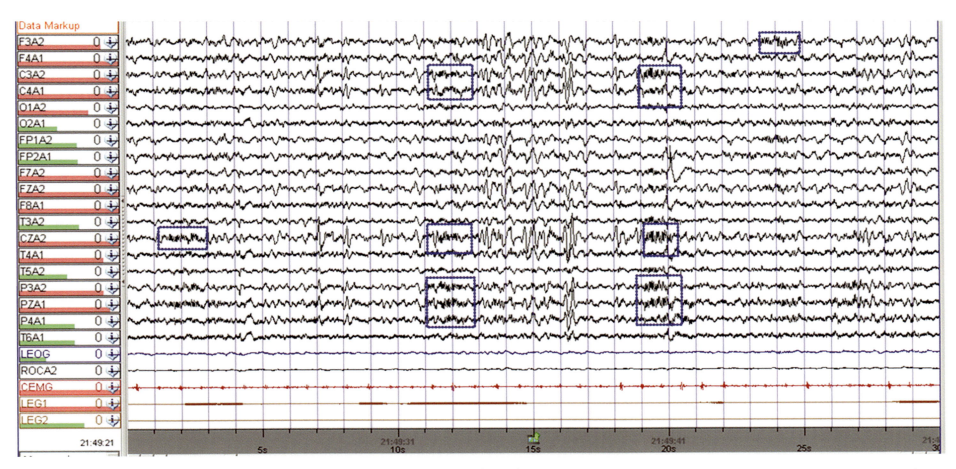

Figure 13.8 A Distribution of Sleep spindles in EEG derivations: Sleep spindles (in box) are distinct waves of 11–16 Hz frequency that are seen most prominently in central derivations, having duration of at least 0.5 seconds and are easily distinguishable from the background. In this epoch (30 seconds) they appear in first half, hence this will be scored as N2.

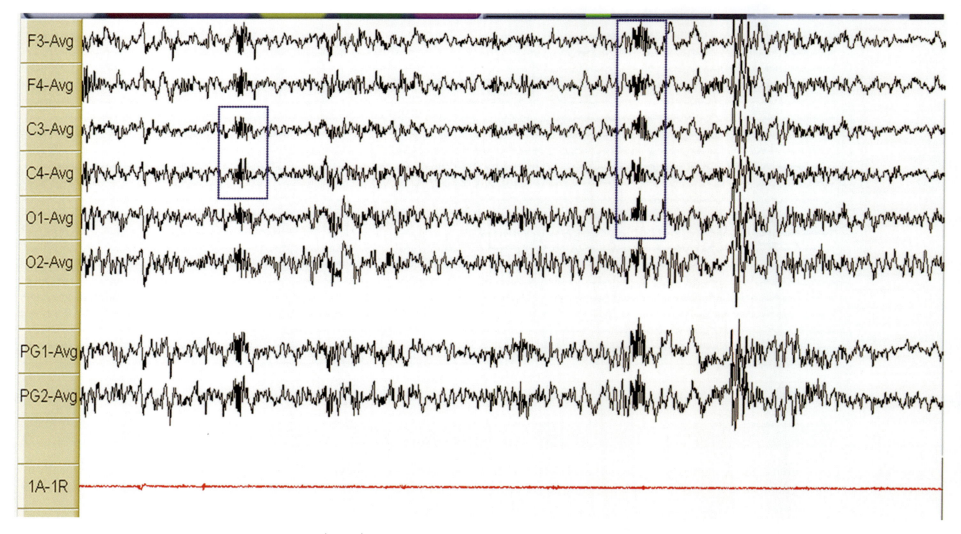

Figure 13.8 B Sleep spindles in N2: Sleep spindles (in box) are distinct waves of 11–16 Hz frequency that are seen most prominently in central derivations, having duration of at least 0.5 seconds and are easily distinguishable from the background. In this epoch (30 seconds) they appear in first half, hence this will be scored as N2.

Scoring of Data in Adults

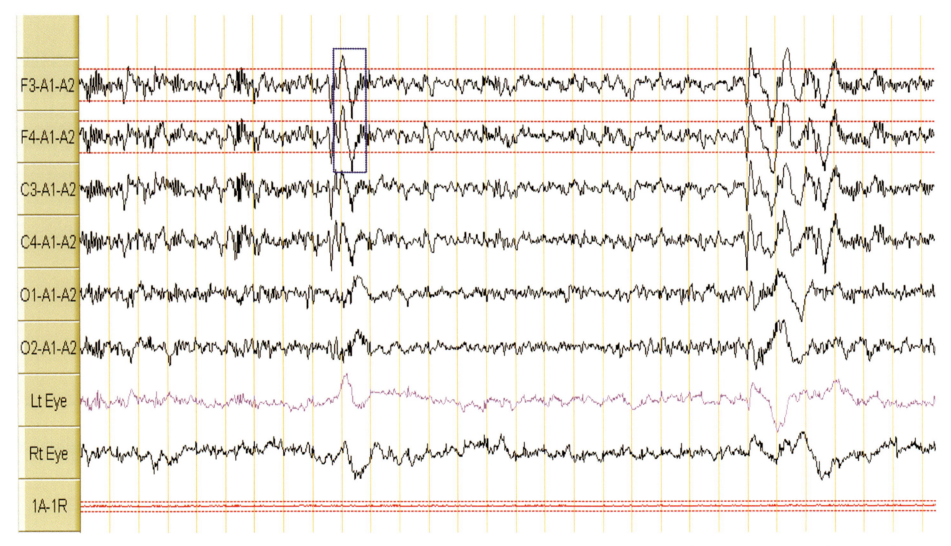

Figure 13.9 K Complexes in N2: K complex are prominently seen in frontal derivations and sometimes in EOG as well due to their proximity to frontal area (Box). Having a duration of more than half a second, they have initial negative deflection, followed by a positive deflection and the coming to baseline. They are quiet distinct from the background.

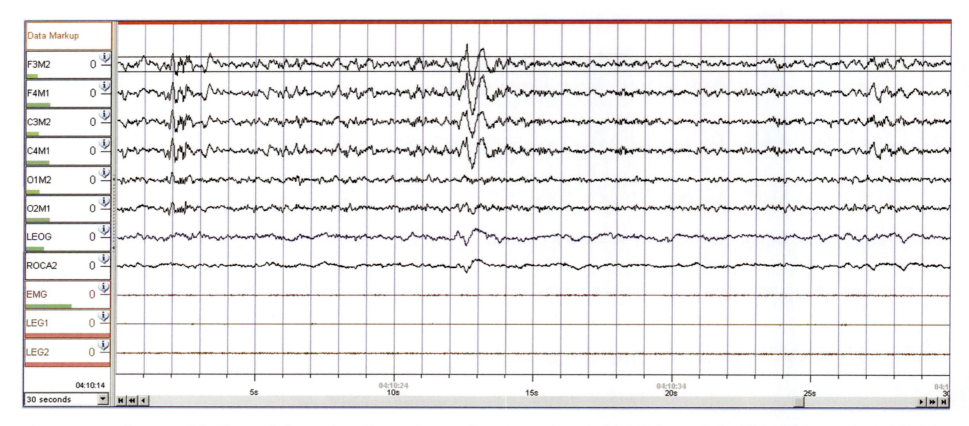

Figure 13.10A Initiation of N2: If an epoch shows either of the two-sleep spindles or K-complex in the first half, that epoch should be scored as N2. If it is seen in later half, score sleep stage in present epoch as that of previous epoch or whichever waveform dominates the present epoch. Here K complexes are seen in first half, no arousal is seen, hence tis will be scored as N2.

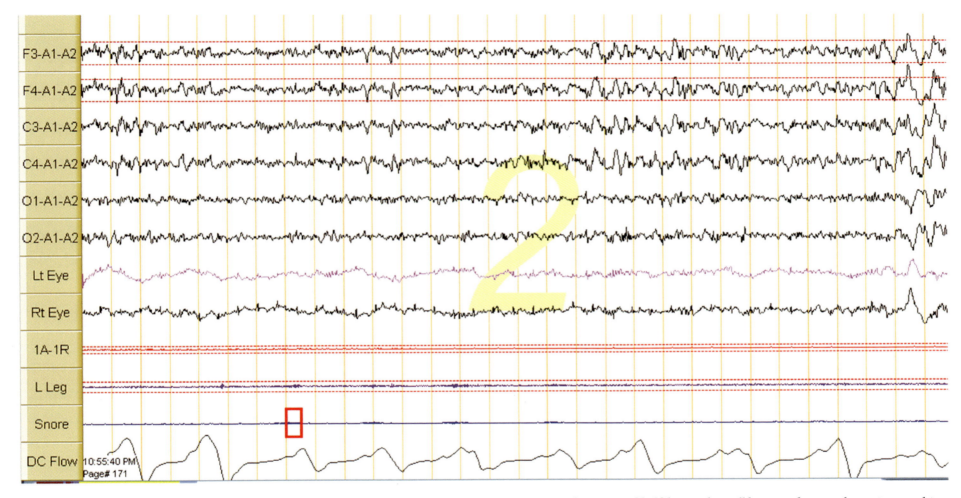

Figure 13.10B Figure shows low voltage mixed frequency activity, however, K complex is seen only in second half, hence this will be scored as epoch previous to this, i.e., N1. Epoch subsequent to this will be scored as N2, unless it is interrupted by arousal or shows waveforms confirming any other stage.

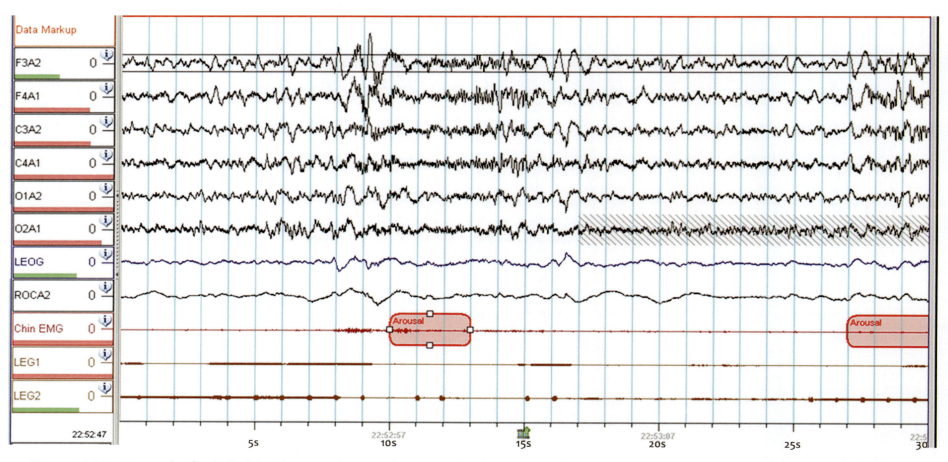

Figure 13.11 K arousal in first half of N2: If an arousal starts within 1 seconds of termination of K complex, it is not considered as K complex. Here K-arousal is apparent in first half of epoch, and sleep spindle or K complex are not seen before or after K-arousal, hence it should be scored as N1. Change the scoring of subsequent epochs to N2 according to Figure 13.10 A, B.

Scoring of Data in Adults

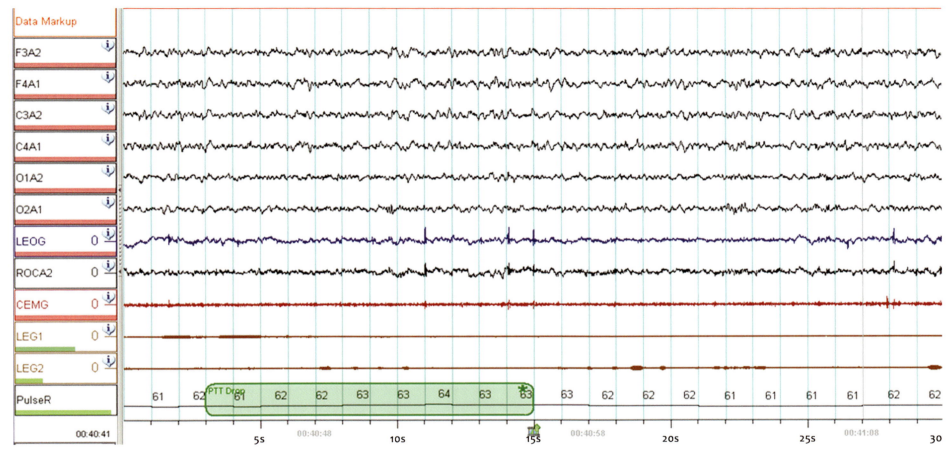

Figure 13.12 N2 scoring without spindle of K complex: Once started, scoring of N2 should be continued if low voltage mixed frequency activity is seen in EEG even in absence of K complex or sleep spindles. However, the previous epochs must have been scored as N2 and did not have intervening arousals.

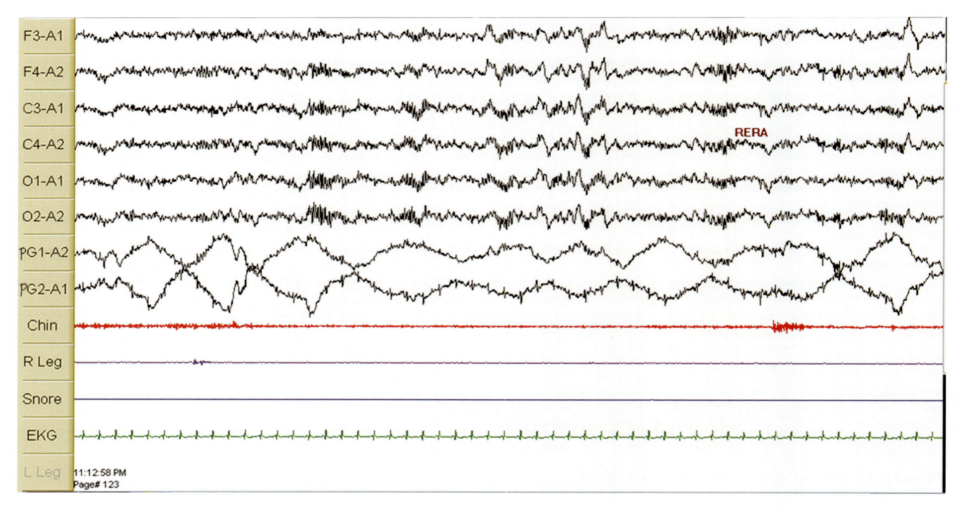

Figure 13.13 SEM during N2: Although eye movements are absent during N2, in some cases, slow eye movements may be seen during N2.

Scoring of Data in Adults

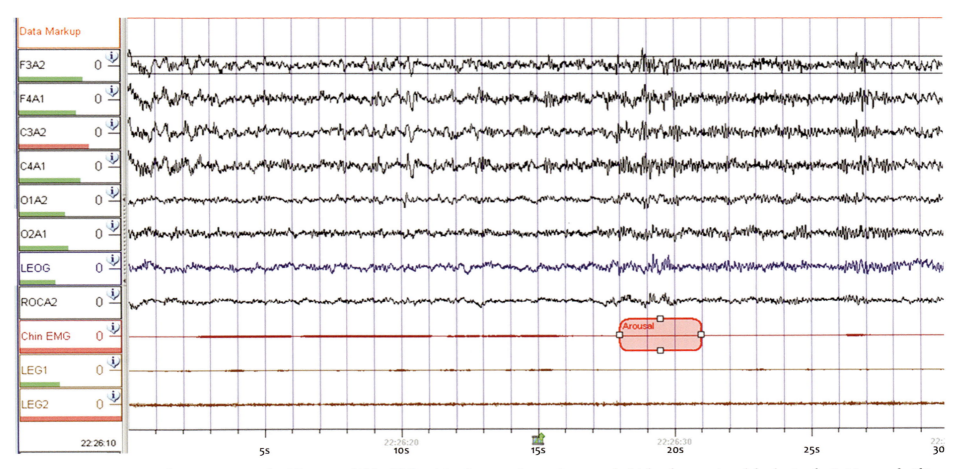

Figure 13.14 A Arousal in N2: An arousal in N2 is scored if the EEG activity shows an abrupt change to the higher frequencies- alpha, lasting for 3–15 seconds. If it occupies more than half of the epoch, score as wake. If less, score as N2. See that theta activity appears after alpha at the end of epoch, following epoch shall be scored as N1 anc continue to score N1 till an epoch with sleep spindle or K complex appears as mentioned in Figure 13.10.

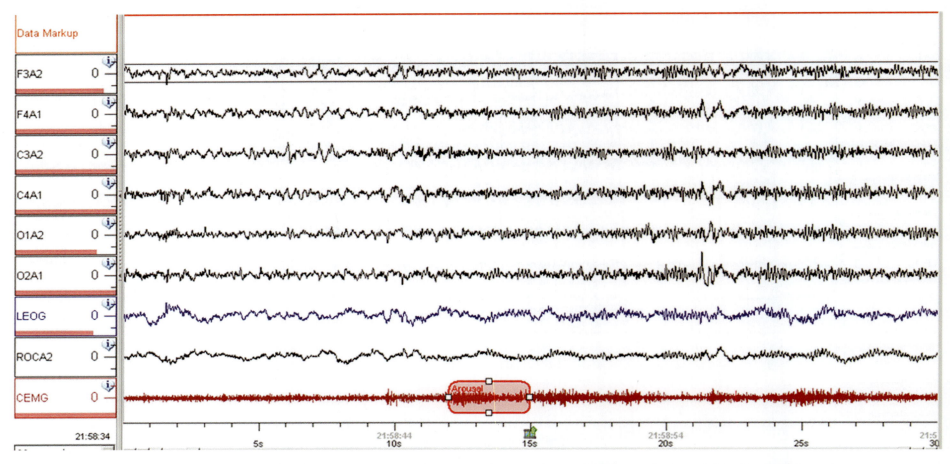

Figure 13.14 B   Change of N2-wakefulness: Since the alpha is seen for more than 50% of the epoch, this epoch will be scored as wake.

Scoring of Data in Adults

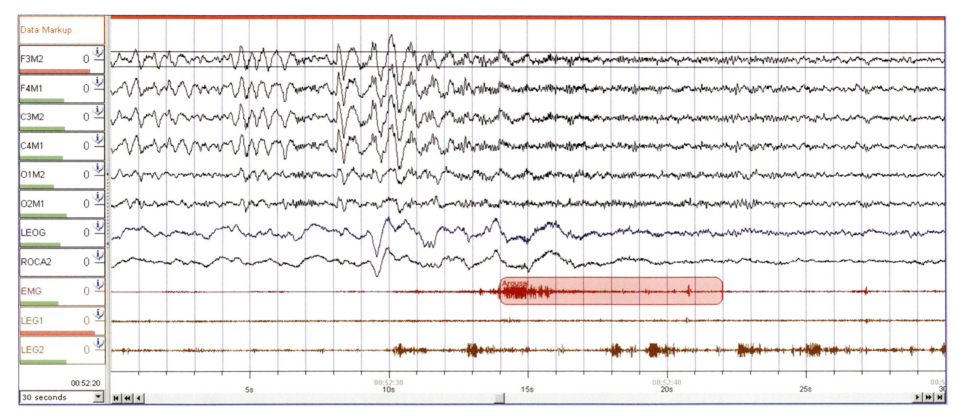

Figure 13.15 A–C Arousal in the later half of N2: A: This epoch shows arousal in second half. However, after arousal, waveforms suggestive of N1 are seen in last part. Since largest part of the epoch is covered by waveforms suggestive of N2, this will be scored as N2.

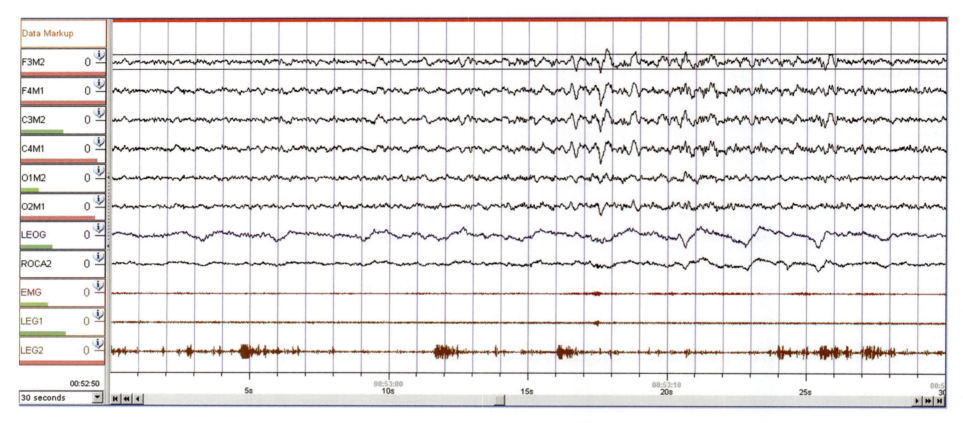

Figure 13.15 B This epoch does not show K complexes or spindles, hence will be scored as N1.

Scoring of Data in Adults

299

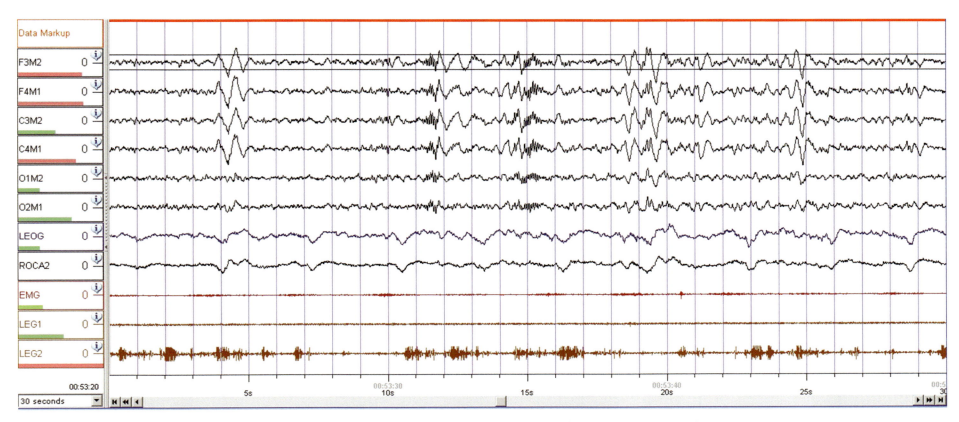

Figure 13.15C This epoch shows K complex in the first half, hence, it will be scored as N2.

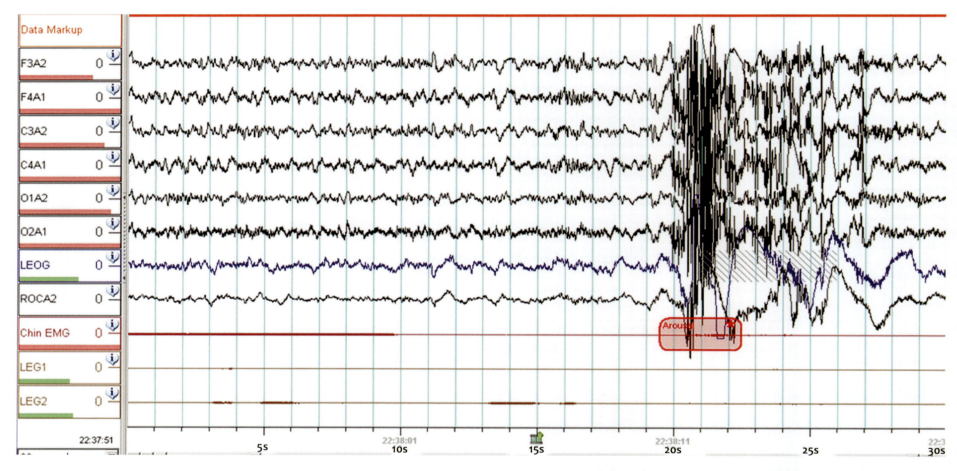

Figure 13.16 Body movement during N2: In this epoch body movement is followed by alpha but low voltage mixed frequency activity is present in most of the epoch, and preceding epoch was N2, hence this will be scored as N2.

Scoring of Data in Adults   301

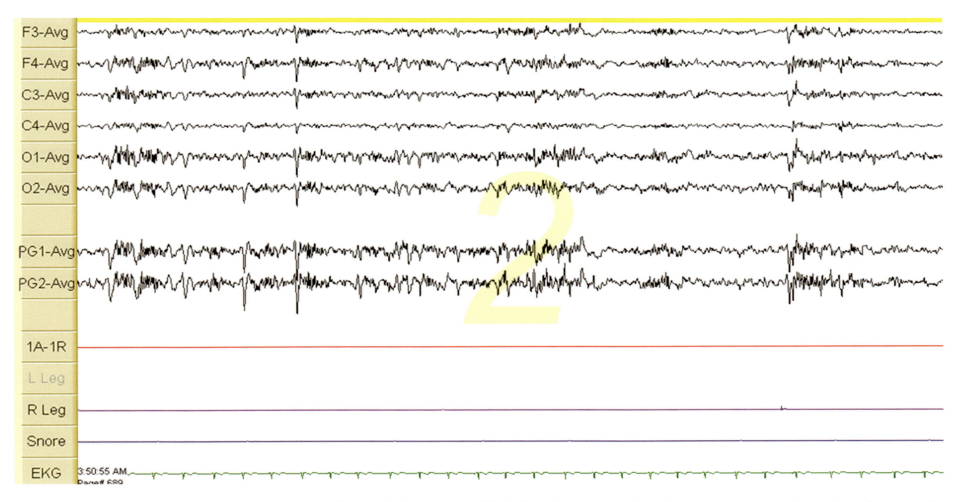

Figure 13.17 A–C   Major body movement during N2: If a major body movement (defined as obscuring EEG beyond recognition in more than 50% epoch as in 13.17 B2) follows an epoch of undisputed N2 epoch (13.17 B1) and also follows an epoch of undisputed N2 (13.17 B3), epoch of movement will be scored as N2.

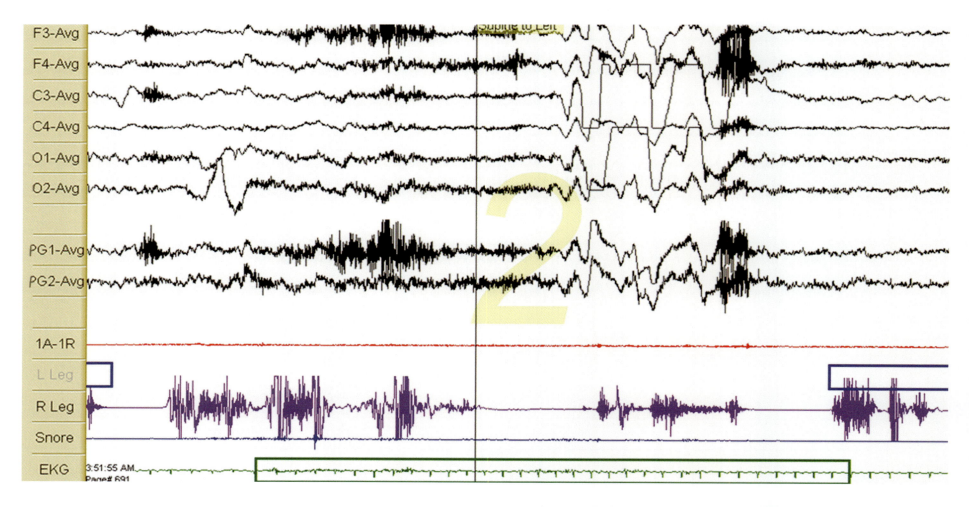

Figure 13.17 B Major body movement during N2: If a major body movement (defined as obscuring EEG beyond recognition in more than 50% epoch as in 13.17 B) follows an epoch of undisputed N2 epoch (13.17 A) and also follows an epoch of undisputed N2 (13.17 C), epoch of movement will be scored as N2.

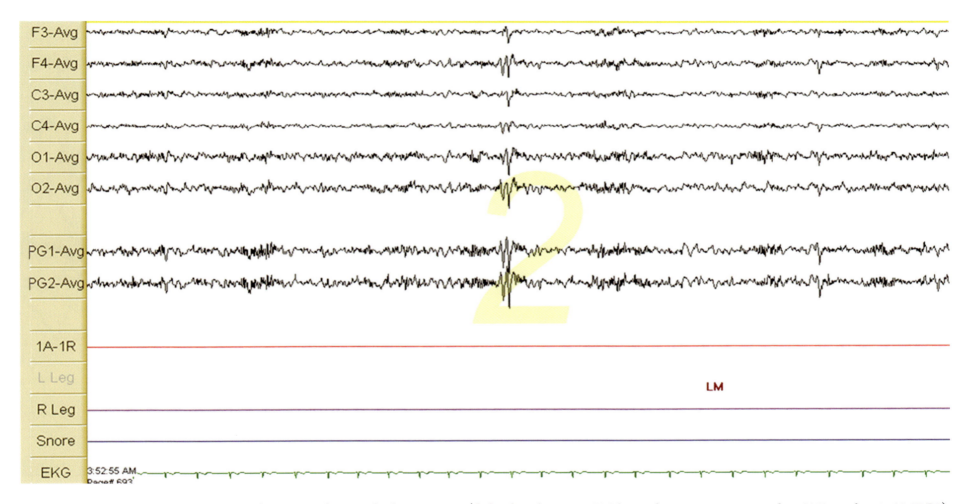

Figure 13.17 C Major body movement during N2: If a major body movement (defined as obscuring EEG beyond recognition in more than 50% epoch as in 13.17 B2) follows an epoch of undisputed N2 epoch (13.17 B1) and also follows an epoch of undisputed N2 (13.17 B3), epoch of movement will be scored as N2.

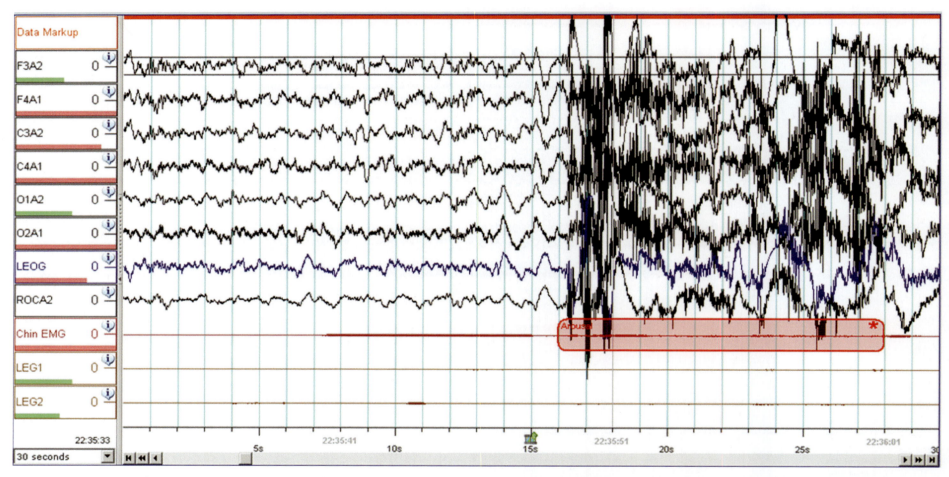

Figure 13.18A–B Slow eye movement in N2 epoch following a major body movement: A: N2 epoch shows a major body movement in second half that continues in first half of B.

Scoring of Data in Adults

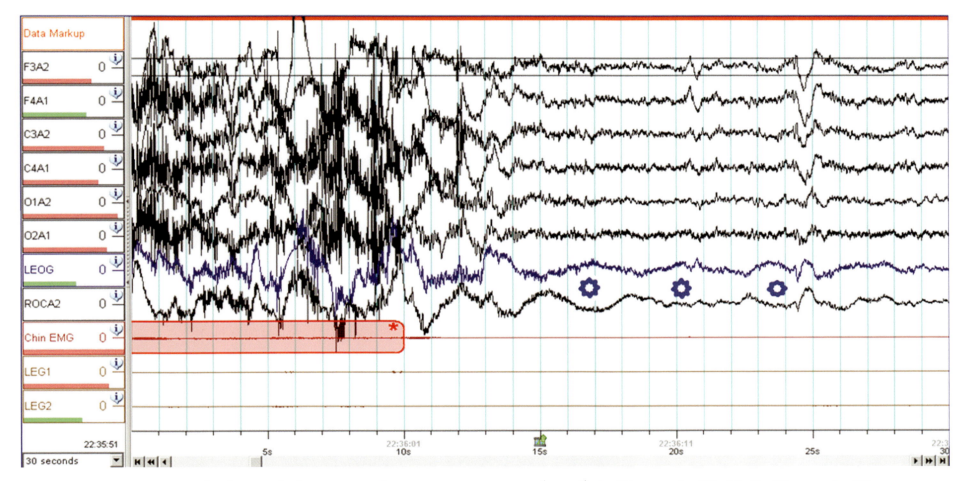

Figure 13.18B After the major body movement, slow eye movements are seen (starred). A will be scored as N2 while B will be scored as N1.

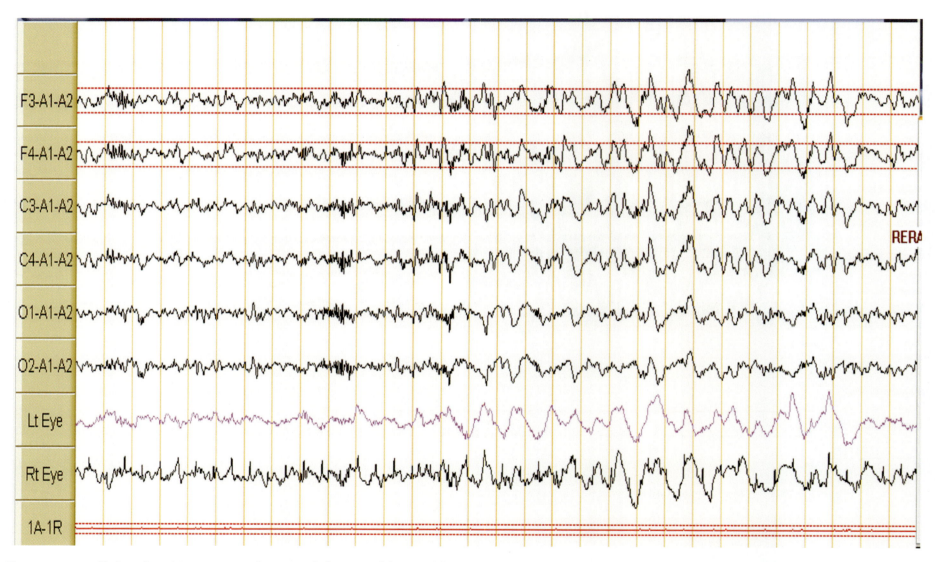

Figure 13.19 Shifting from N2 to N3: N2 sleep may shift to N3. If the epoch has delta waves less than 6 seconds, continue with N2. If the epoch has delta waves for at least 6 seconds, change the scoring to N3. In this epoch delta activity is seen in more than 20% (6 seconds) of epoch, hence, this will be scored as N3.

Scoring of Data in Adults

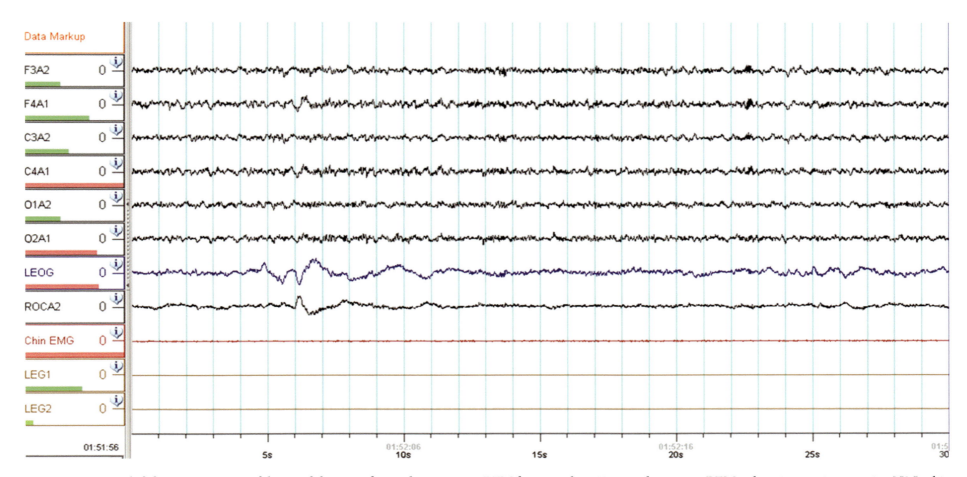

Figure 13.20 A Shift from N2 to REM: If the epoch has waveforms characterizing REM for more than 15 seconds, score as REM, otherwise continue scoring N2 In this epoch see the rapid eye movements and sawtooth waves hence this will be scored as REM. Previous epoch was N2.

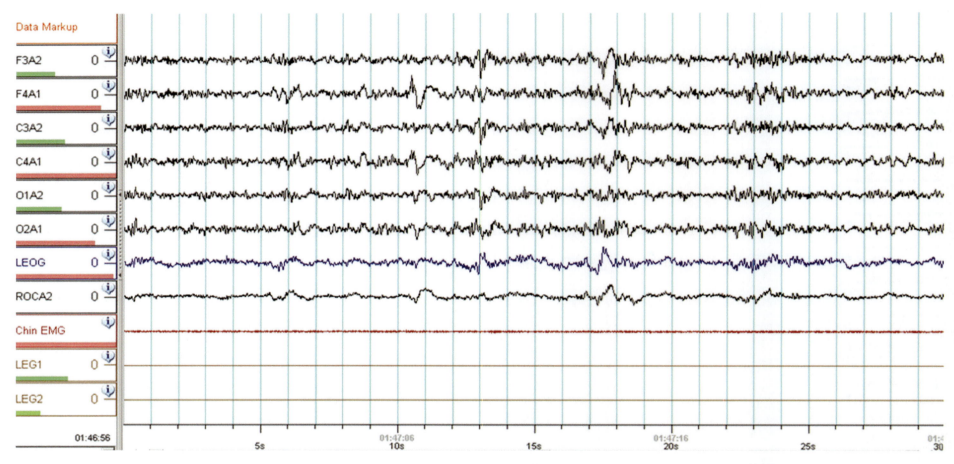

Figure 13.20 B1–B4 Shift from N2 to REM: Sometimes EEG makes it difficult to differentiate between N2 and REM. In such cases, scoring depends upon chin tone. Start scoring REM if an epoch shows lowering of chin tone for more than 15 seconds, does not have K complex or spindle and subsequent epochs can be scored as REM. In this series of epochs also appreciate that all the features of REM (Figure 13.24) do not appear simultaneously. In this series, EEG and EOG changes appeared first and chin tone lowered much later. Since B4 has all the criteria for REM, this will be scored as REM and previous epochs will be scored as N2.

Scoring of Data in Adults

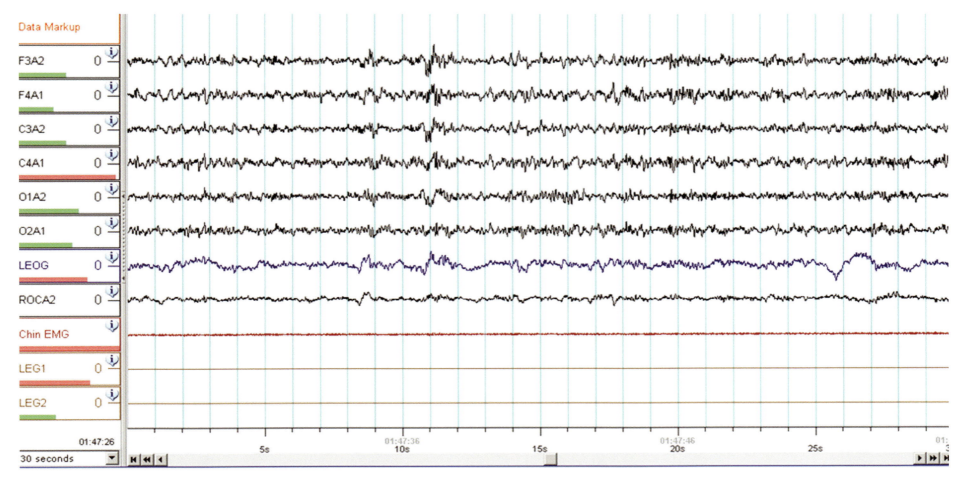

Figure 13.20 B2   Shift from N2 to REM.

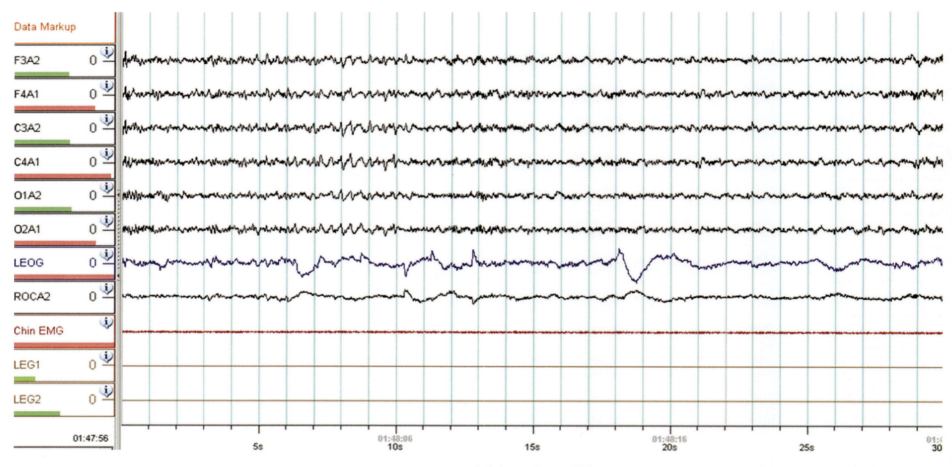

Figure 13.20 B3 Shift from N2 to REM.

Scoring of Data in Adults

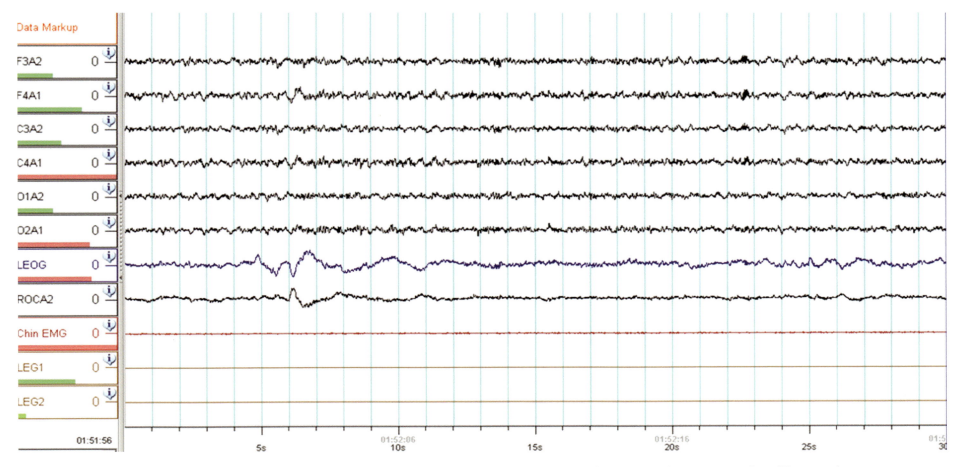

Figure 13.20 B4 Shift from N2 to REM. B4 has all the criteria for REM. This will be scored as REM, and previous epochs will be scored as N2.

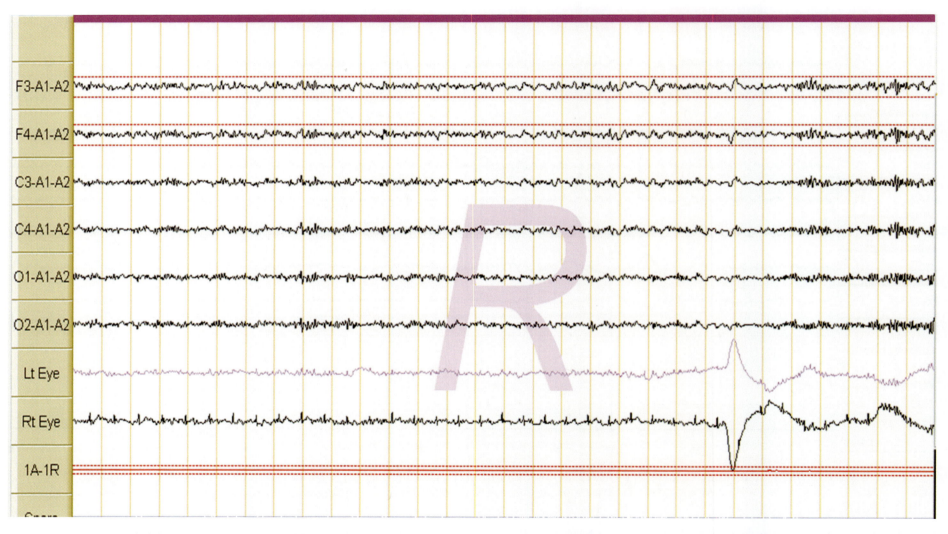

Figure 13.20 C Shift from N2 to REM: Even if the chin tone drops in N2, continue scoring N2 till the last epoch where spindle is seen. Subsequent epoch may be scored as REM, if it has all the characteristics of REM. In this epoch though chin tone is low and rapid eye movement is seen in second half, still this will be scores as N2 as spindles are seen all over the epoch.

Scoring of Data in Adults

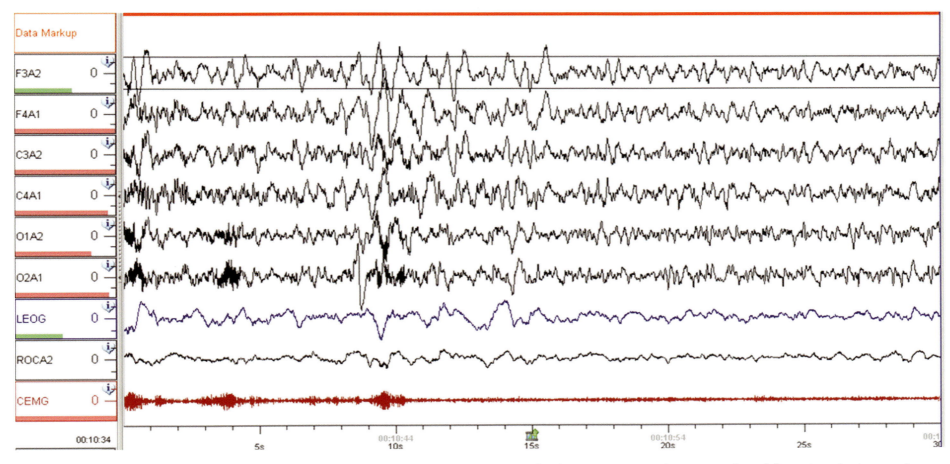

Figure 13.21A Appearance of N3: Score N3, if an epoch has slow wave for at least 6 seconds. Slow wave activity is best seen in frontal derivations, having a peak-to-peak amplitude of 75 µV and frequency of 0.5–2 Hz. These are also known as delta waves. Unlike N2, for scoring of N3, it is immaterial whether delta waves are present in first half or the second half of the epoch (sensitivity 12.5 µV/mm).

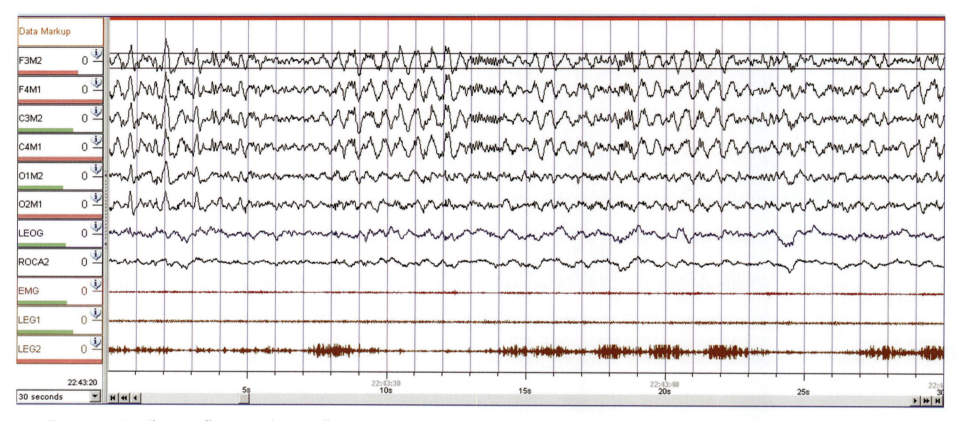

Figure 13.21B Sleep spindles in N3: Sleep spindles may persist along with delta waves. In this case, epoch should be scored as N3 (sensitivity 12.5 μV/mm).

of 0.5–2 Hz and peak-to-peak amplitude of at least 75 μV, is seen in frontal regions and occupies at least 20% of the epoch (Figure 13.21).

Scoring of REM sleep is based upon three characteristics—low chin tone (lower than any other stage of sleep), low-voltage, mixed-frequency activity, and rapid eye movements (Figure 13.24). Sawtooth waves often accompany rapid eye movements and are best seen in central derivations as serrated, triangular waves in the frequency of 2–6 Hz (Figure 13.25). REM is also characterized by transient muscle activity, usually seen in lower limbs, having a duration of less than 250 msec, and superimposed upon lower muscle tone (Figure 13.26). Continue to score REM, even in absence of rapid eye movements (Figure 13.27 A,B). Stop scoring REM, if an epoch of wakefulness appears (Figure 13.34), or epoch shows characteristics of N1 (Figure 13.31), or N2 (Figure 13.32) or N3. If there is a major body movement during REM scoring of epoch, having major body movement and its subsequent epoch depends upon various characteristics (Figures 13.33 and 13.34).

The transition from N2 to REM is especially important and epochs between definite N2 and definite REM may present some difficulty. These epochs may be scored using the following rules:

1. Between these epochs, an epoch that shows a drop of muscle tone to the level of REM in the first half, but in absence of sleep spindles and K-complex, should be scored as REM. Scoring of REM may continue till the epochs showing definite REM.
2. However, if an epoch shows K-complex or spindles in absence of rapid eye movements, but with low chin tone similar to definite REM epoch in the first half of the epoch, it should be scored as N2.
3. An epoch with K-complex or sleep spindles, meeting other criteria for REM and having rapid eye movements should be scored as REM even in presence of spindles/K complexes (Figure 13.20C)

Major body movements are defined as epochs where body movement and muscle artifacts make the EEG incomprehensible to the extent that staging cannot be done. If such an epoch shows alpha waves even for less than 50% epoch, it should be scored as Wake. This epoch should also be scored as Wake if the previous or subsequent epoch can be scored as Wake. In absence of these two features, it should be scored as the sleep stage of the following epoch.

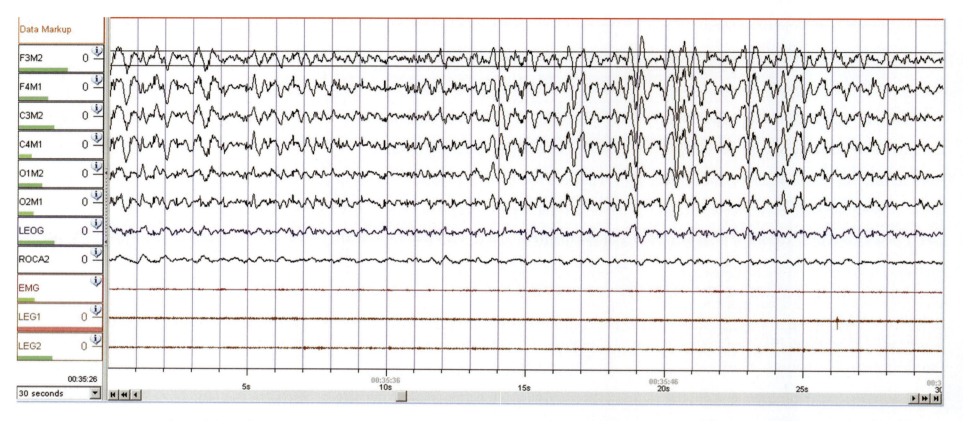

Figure 13.22 K complex as delta: When K complexes have characteristics of delta waves and are present in more than 20% epoch, epoch should be scored as N3.

Scoring of Data in Adults

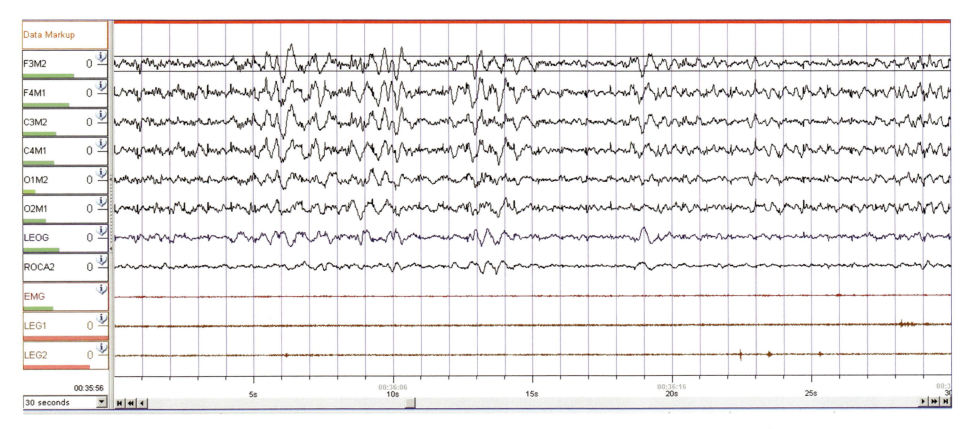

Figure 13.23A–B N3 to REM transition: A: This epoch shows N3 in most of the part, hence it will be scored as N3.

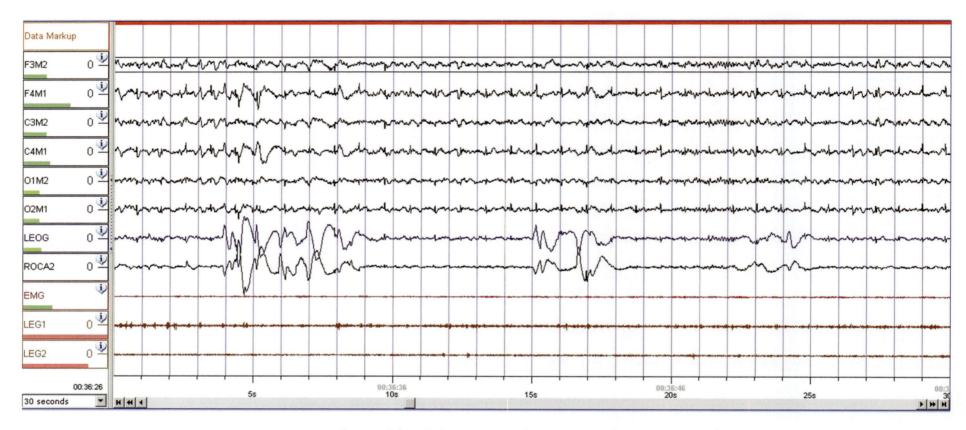

Figure 13.23B This epoch has all characteristics of REM, hence, it will be scored as REM.

Scoring of Data in Adults

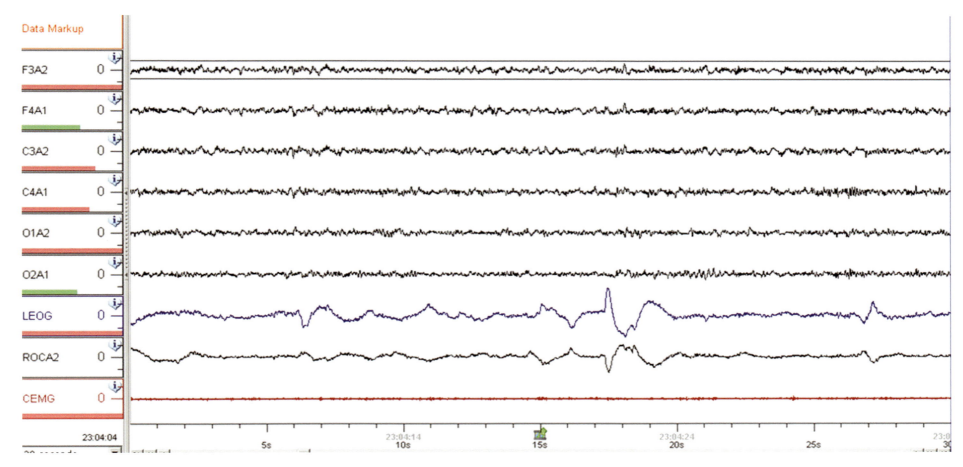

Figure 13.24 REM sleep: Score as REM if the epoch shows all three features of REM-low voltage mixed frequency activity, low chin tone and rapid eye movements (usually lasting less than half a second) for at least 15 seconds.

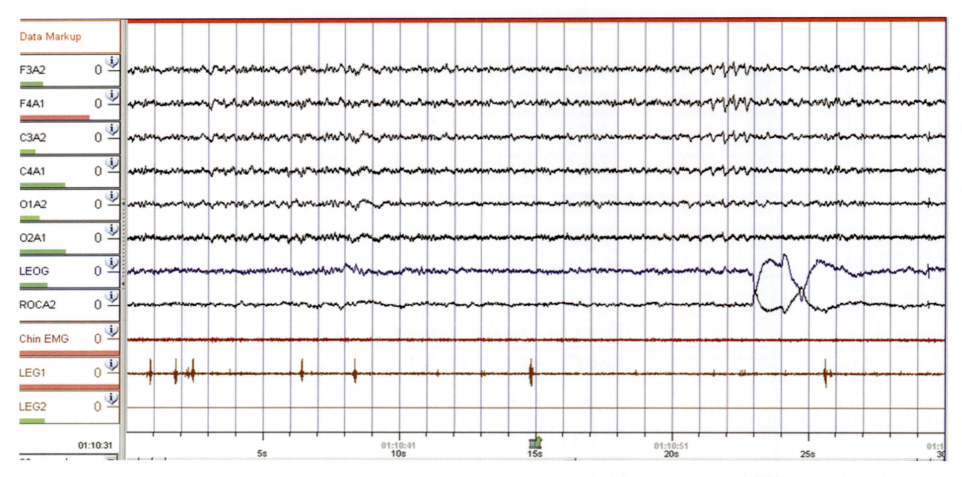

Figure 13.25 Sawtooth waves in REM: Sawtooth waves characterize REM. They are triangular waves with sharp top and appears serrated. They have frequency of 2–6 Hz and maximum amplitude is seen in central derivations. They are usually associated with rapid eye movements.

Scoring of Data in Adults

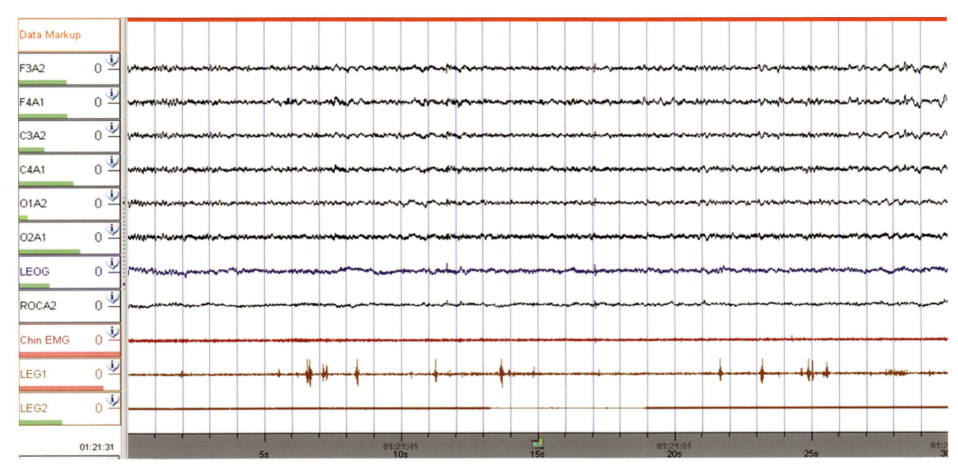

Figure 13.26 Transient muscle activity in REM: REM is characterized by transient muscle activity, which is less than 0.25 seconds in duration along with low EMG tone. This may be seen in any of the derivations (limb, chin, EOG, EEG) depending upon which muscle is involved. In this epoch transient activity is seen in Leg 1.

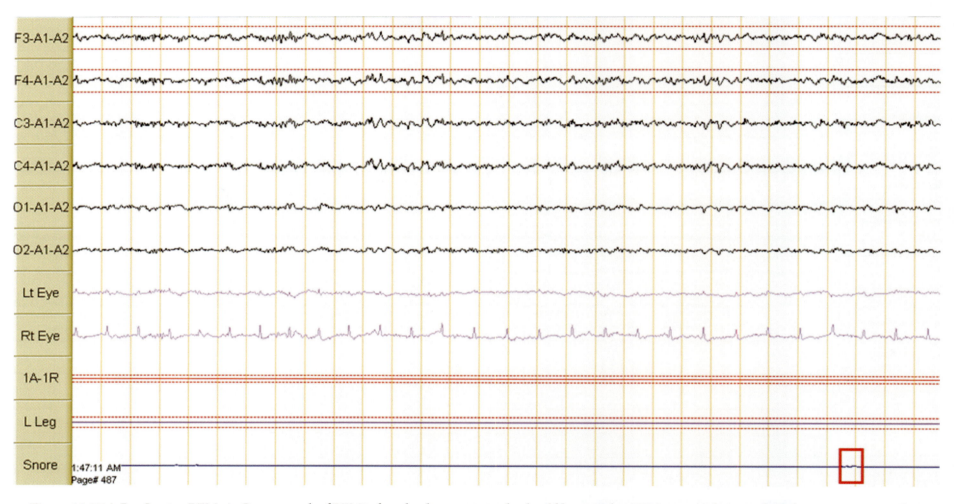

Figure 13.27 A-B   Scoring REM: A: Once a epoch of REM is found, subsequent epochs should be scored as REM, even in absence of rapid eye movements unless interrupted by any other sleep stage. However, the epochs must have EEG and EMG characteristic of REM as described in Figure 13.24.

Scoring of Data in Adults

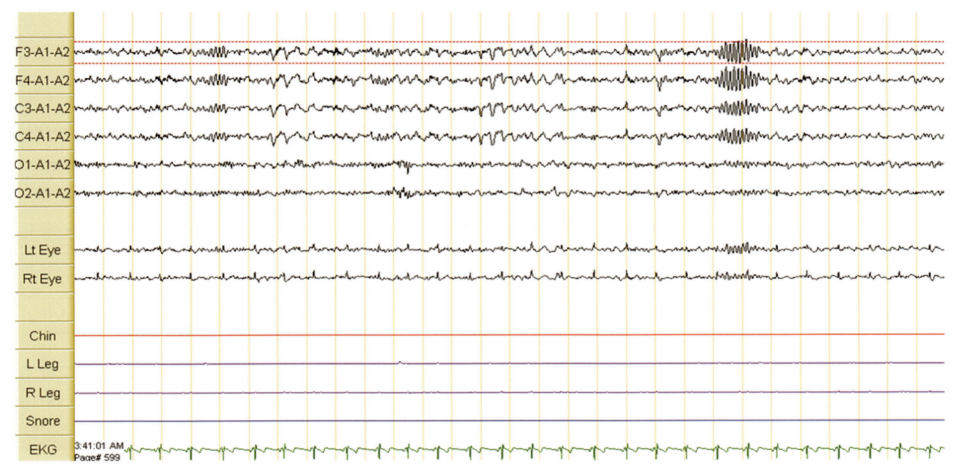

Figure 13.27 B In between epochs of REM, epochs of N2 may appear. If an epoch has spindle of K complex even if preceding and following epochs can be scored as undisputed REM, score as N2 if sleep spindles or K complex is seen, score as N2. In this epoch spindle is seen in first as well as second half.

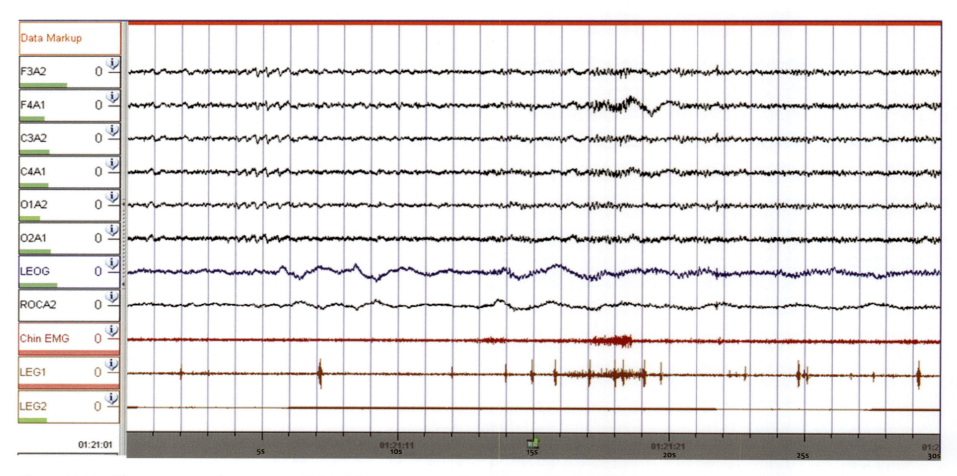

Figure 13.28 Chin tone transiently increasing in REM: If chin tone increases during REM epoch change of scoring depends upon the duration of high tone. If it is less than half the epoch continues scoring REM, however, if it is seen in more than half the epoch, change the score to N1, if other characteristics of N1 are seen or to wake if all characteristics of wakefulness are seen. In this epoch, chin tone increased for more than a second, but lowers down, and is followed by slow eye movements; hence subsequent epoch will be scored as N1. However, in this epoch, REM characteristics are present in more than 50% of epoch, hence, it will be scored as REM.

Scoring of Data in Adults

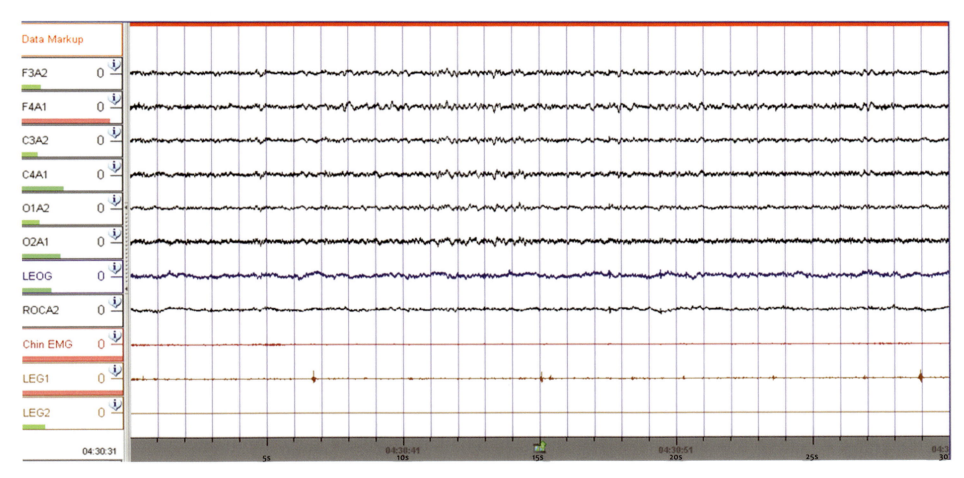

Figure 13.29A-F Arousal in REM: Arousal in REM is scored if the EEG changes suggest arousal (Figure 13.14) along with increased chin tone for at least for 1 seconds. Change to W if EEG and Chin EMG suggestive of arousal persist for more than half the epoch. A: Epoch showing REM.

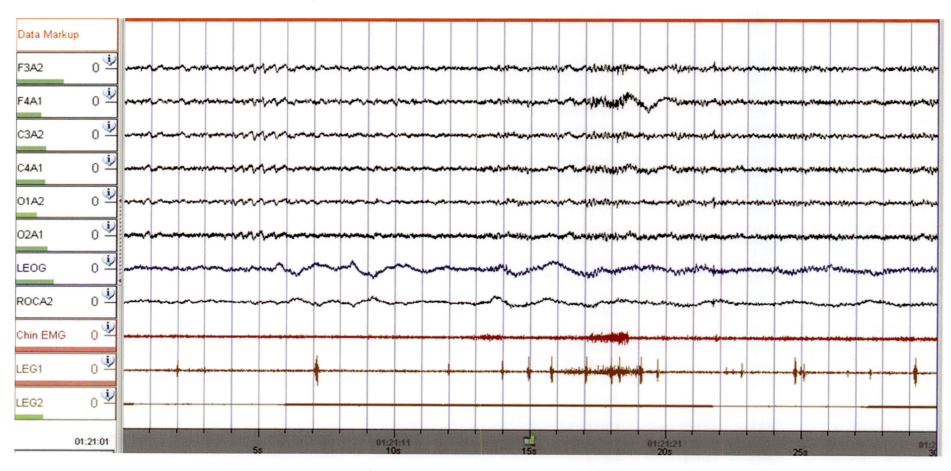

Figure 13.29B Scored as N1 SEM are seen and tone is increased for most of the epoch.

Scoring of Data in Adults

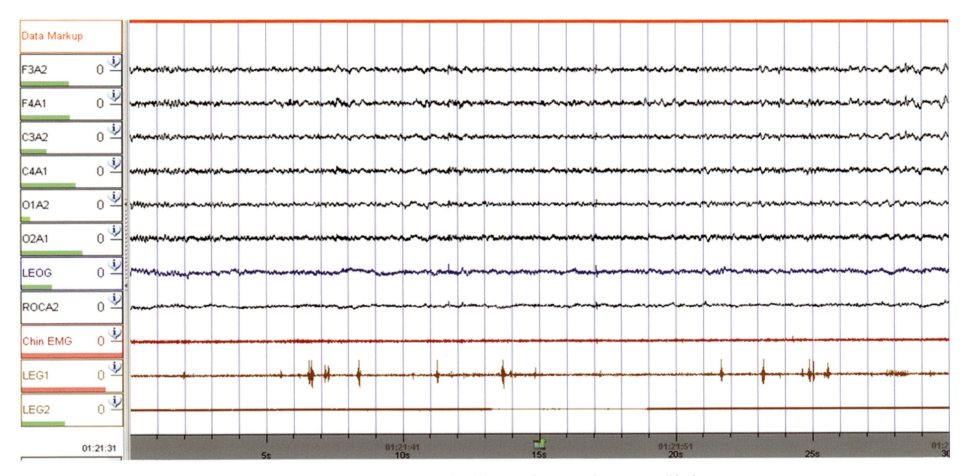

Figure 13.29C Following epoch will be scored as N1 as chin tone is still high.

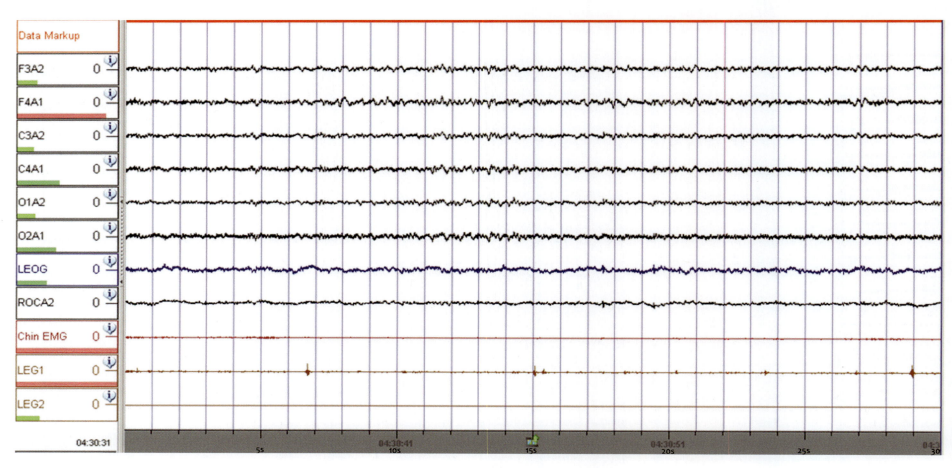

Figure 13.29D Scores as REM as chin tone drops.

Scoring of Data in Adults

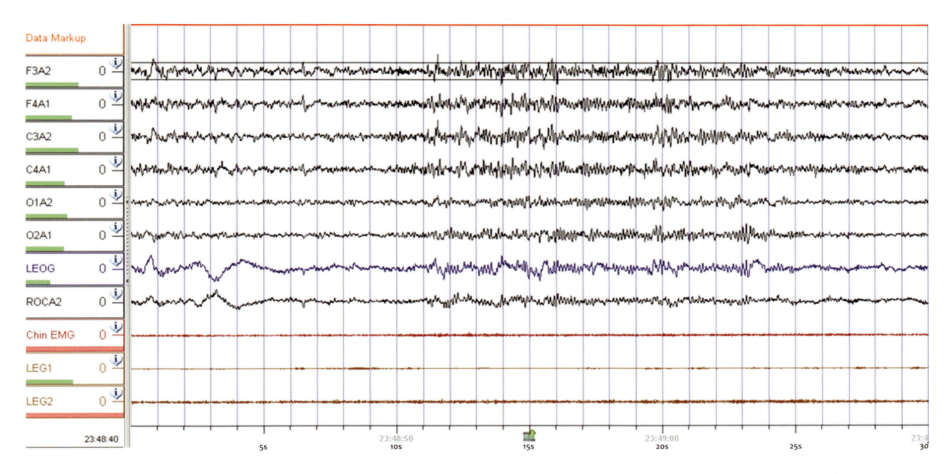

Figure 13.29E This epoch will be scored as wake as chin tone is high in more than 50% epoch and alpha waves also appeared.

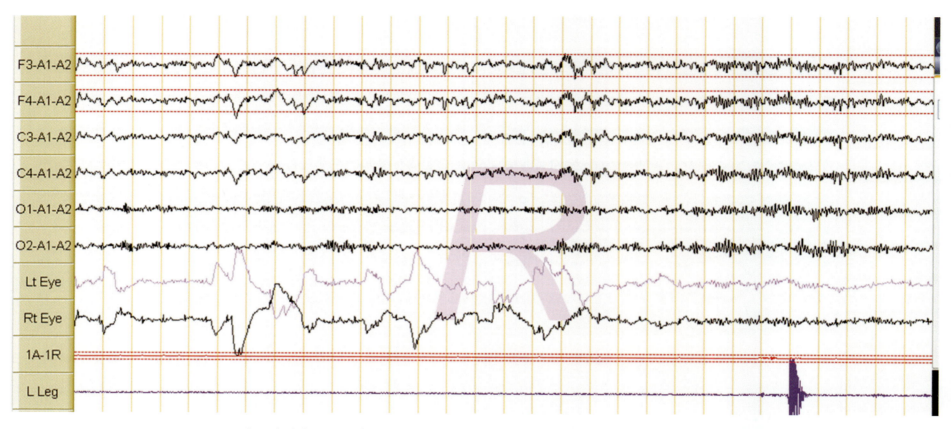

Figure 13.29F Though alpha is seen during REM, but chin tone does not increase and hence, arousal will not be scored.

Scoring of Data in Adults

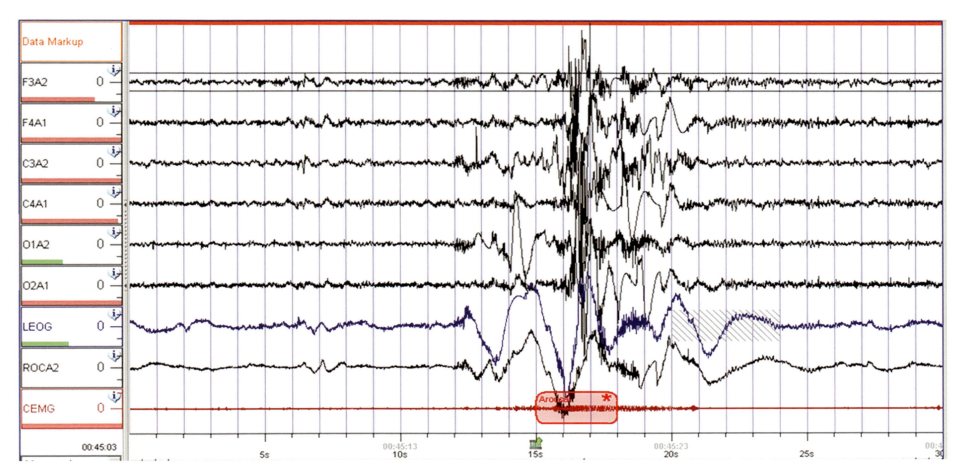

Figure 13.30A-B Arousal in REM: If following arousal, chin tone again goes low and slow eye movement are not seen, continue to score REM A: Lowering of chin tone after arousal and SEM not seen hence scored as REM.

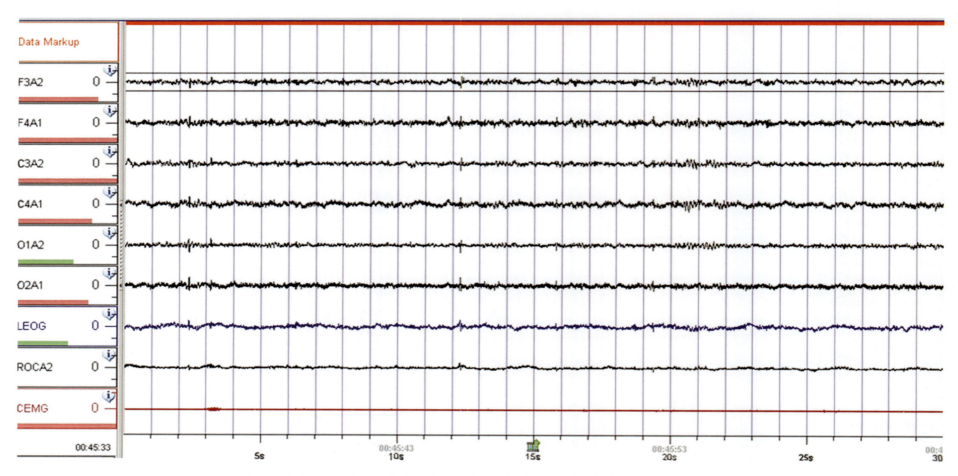

Figure 13.30B This epoch show lowered tone in absence of SEM hence scored REM.

Scoring of Data in Adults

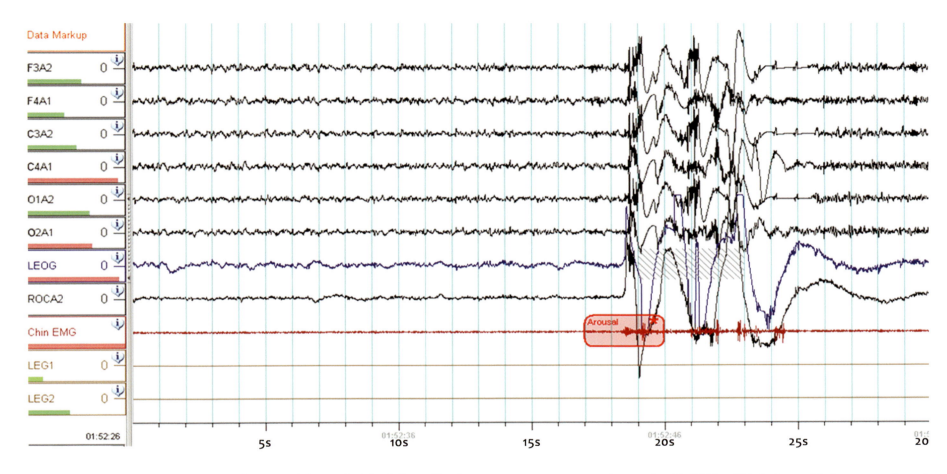

Figure 13.31A-B REM to N1: If an arousal during REM is followed by slow eye movements, score the epoch with SEM as N1. A: Arousal in later part of epoch.

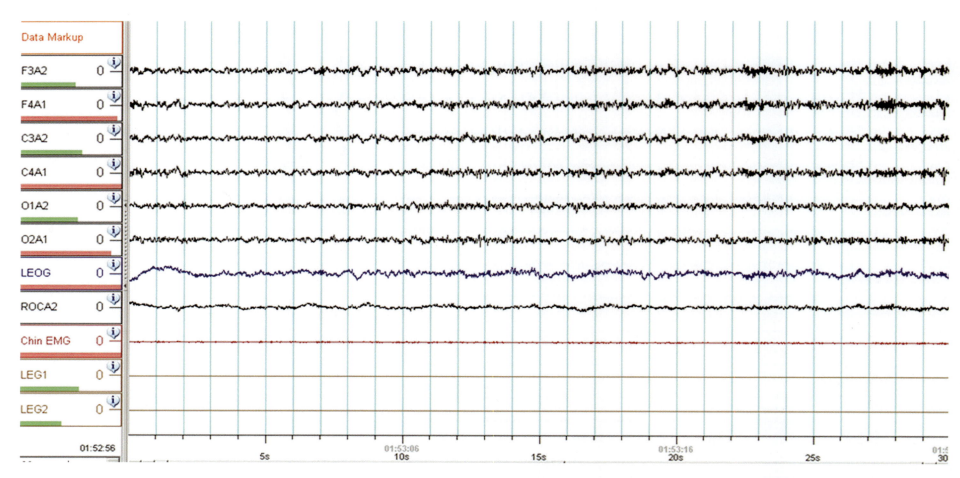

Figure 13.31B Slow eye movements in the initial parts of the epoch. Continue to score N1 till another sleep stage (N2 or REM) appears.

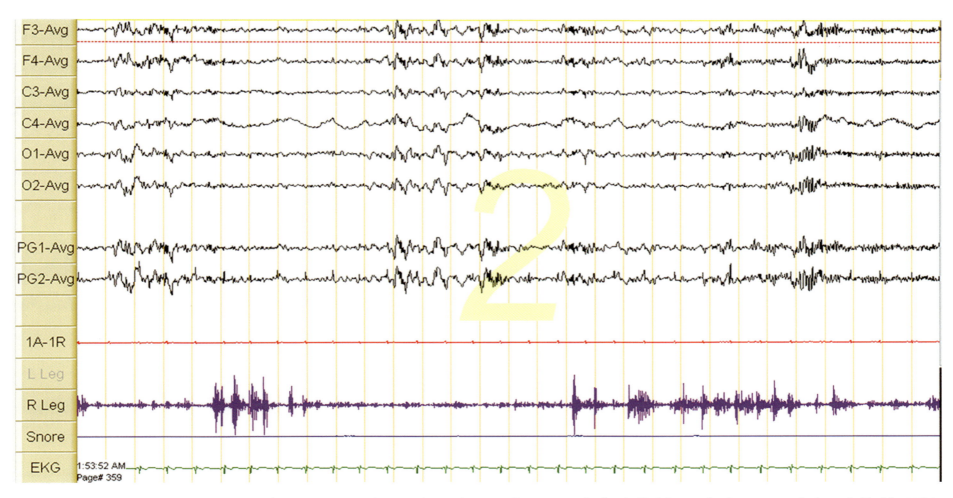

Figure 13.32 REM to N2: Change scoring from REM to N2, if K complex or sleep spindle is seen in the first half of the epoch. If it is seen in the latter half of the REM epoch, score present epoch as REM and subsequent epoch as N2 even in presence of low chin tone.

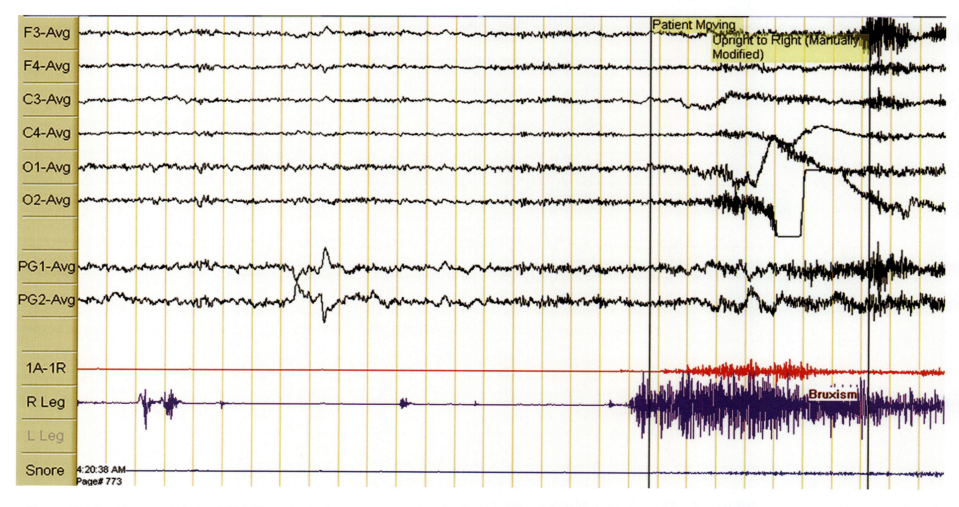

Figure 13.33 Movement during REM: If a major body movement makes the EEG and RMG findings indiscernible in an epoch, but previous and subsequent epochs are REM, score present epoch as REM. This epoch shows a minor movement and REM occupies more than 50% of epoch, hence scored as REM.

Scoring of Data in Adults

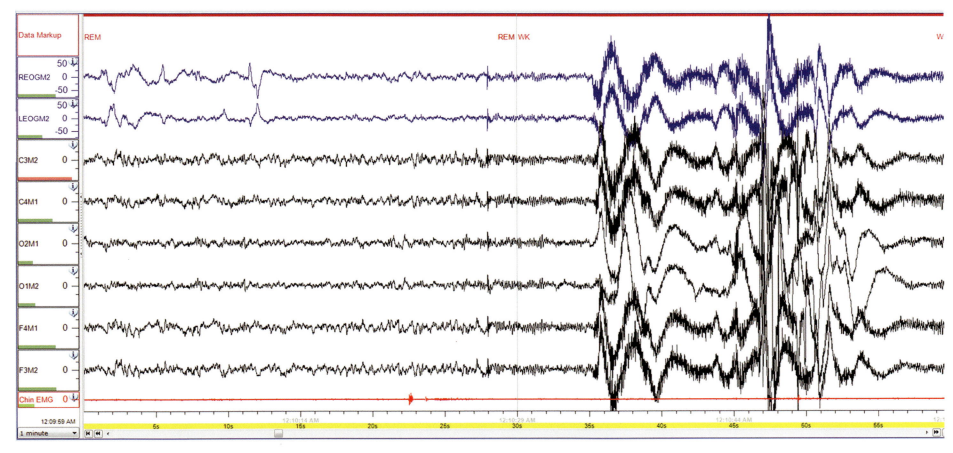

Figure 13.34 Major movement during REM: If an epoch of major body movement during REM is followed by slow eye movements in an epoch, both epochs (with movement and SEM) should be scored as N1.

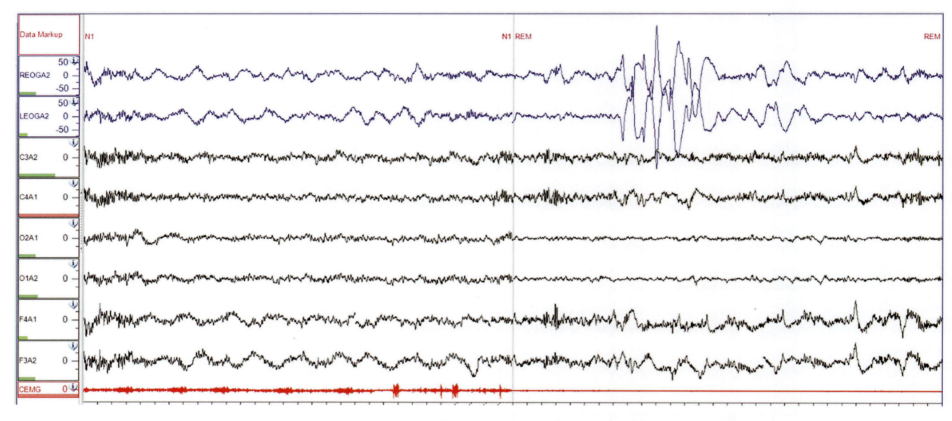

Figure 13.35 N1 to REM: If an epoch following N1, shows characteristics of REM, score as REM.

Scoring of Data in Adults

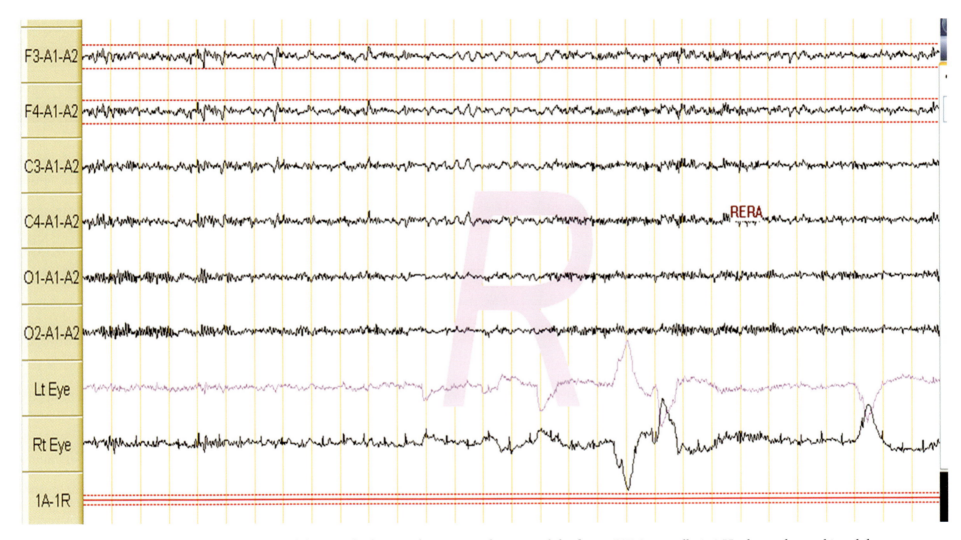

Figure 13.36 Alpha during REM: Alpha may also be seen during REM, however, alpha during REM is usually 1–2 Hz slower than waking alpha.

13.2 SCORING OF RESPIRATORY DATA

Respiratory data is gathered to look for abnormalities of breathing during sleep. These include sleep apnea—obstructive, mixed, and central or hypopnea. In addition, central apnea and central hypopneas may be associated with Cheyne-Stokes breathing (CSB), which needs to be assessed. In addition, periodic respiration and sleep-related hypoventilation need to be assessed. Snoring is an important parameter associated with obstructive sleep apnea and data must provide this information.

As we have already discussed, thermistors or thermocouples are used to score apnea and nasal cannula for the hypopnea along with oxygen saturation. Figure 13.37A depicts normal respiration during sleep and Figure 13.37B shows benign snoring. Though the respiratory effort can be best assessed using esophageal manometry, it is not routinely used due to the invasive nature of the sensor. Hence, thoracoabdominal RIP belts are used to measure the respiratory efforts.

Hypopnea is defined as at least 30% reduction in the amplitude of pressure transducer (cannula) lasting for at least 10 seconds and desaturation of 3% or 4% from the baseline, depending on the definition of desaturation (Figure 13.38A). Duration of event is measured from the lowest point preceding the breath, this clearly reduces in amplitude to the first breath having a baseline amplitude.

If the hypopnea is associated with snoring, flattening of inspiratory signal, and paradoxical movement in the chest and abdominal belts (resulting in a reduction of signal amplitude in effort sum), it is termed as obstructive hypopnea (Figure 13.38 B). However, any hypopnea without these characteristics is termed as central hypopnea.

If there is a malfunction of cannula signals, this information may be obtained from alternate signals, that is, RIP belts from thorax and abdomen, RIP sum or RIP flow to score hypopnea (Figure 13.38C).

Apnea is diagnosed when the thermal sensor signals drop by at least 90% of the baseline value for at least 10 seconds (Figures 13.39–13.41). Note that hypoxemia, though usually accompany apnea, they are not required to score an apnea. If there is a malfunction of cannula signals, this information may be obtained from alternate signals, that is, RIP belts from thorax and abdomen, RIP sum or RIP flow to score hypopnea (Figure 13.38C).

Depending upon the respiratory effort, apnea is divided into obstructive, mixed and central apnea (Figures 13.41–13.43). During obstructive apnea, respiratory flow ceases while effort continues. Central apnea is characterized by cessation of

Scoring of Data in Adults 341

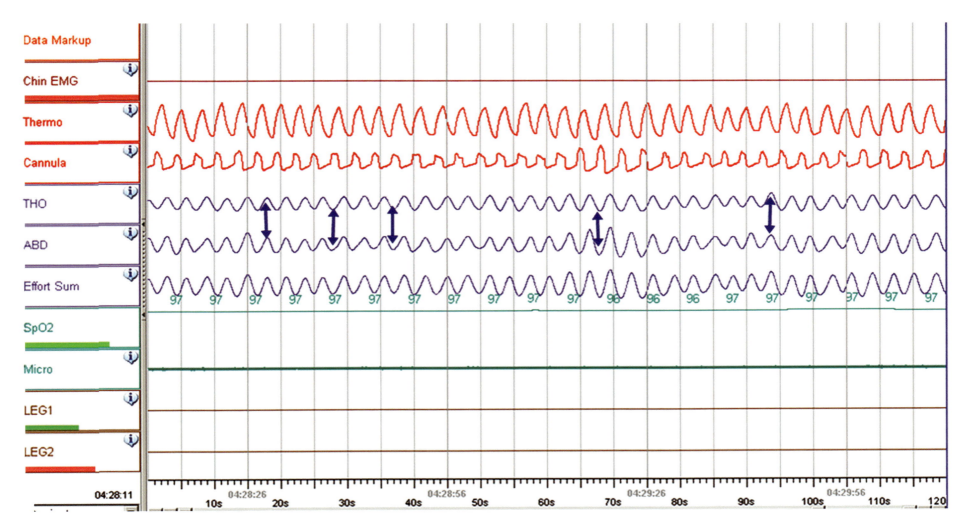

Figure 13.37A Normal Respiration during sleep: 2 min epoch showing normal respiration. See that flow in the thermistor and cannula is regular, rhythmic, and has equal amplitude across the epoch. Chest and abdomen RIP belts show in phase movement and effort sum (RIP-Sum) does not show any change in amplitude. Oxygen saturation is maintained throughout the epoch and microphone does not show any signal.

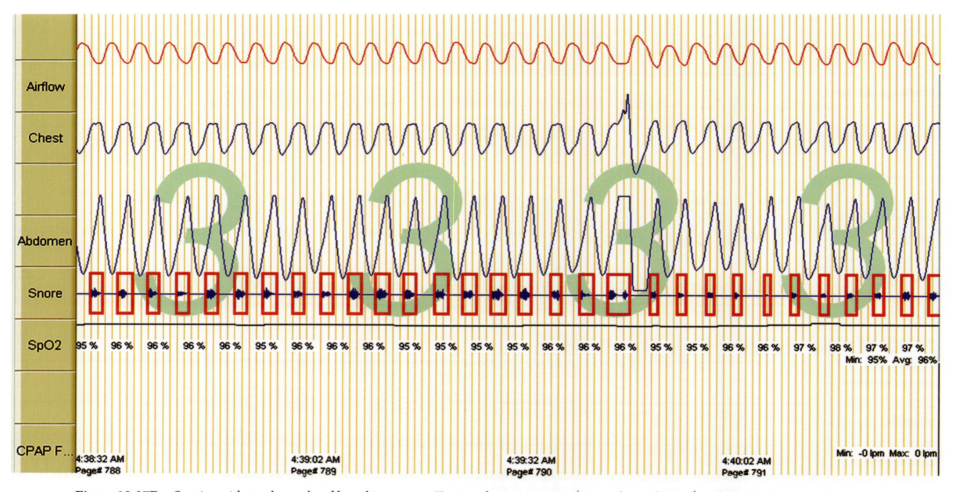

Figure 13.37B Snoring without sleep related breathing events: Tracing showing snoring (microphone channel) without the hypopnea or apnea.

Scoring of Data in Adults

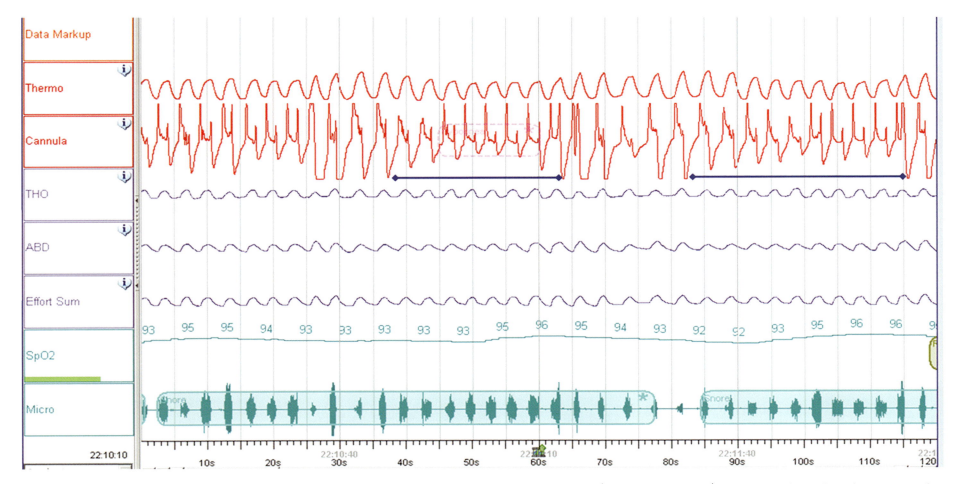

Figure 13.38A Hypopnea is characterized by at least 30% reduction in the amplitude of nasal airflow (pressure transducer) with retained signals in thermistor and regular breathing effort (chest and abdominal belts) and desaturation in oximetery channel (3% or 4% depending upon the criteria followed). Duration must be at least 10 seconds (Blue lines). In this epoch regular snoring is also visible.

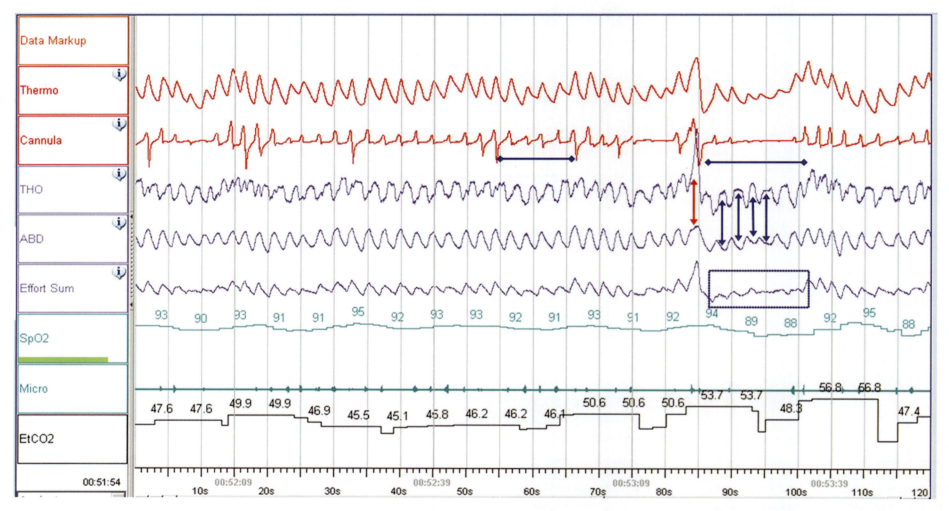

Figure 13.38B Hypopnea is characterized by at least 30% reduction in the amplitude of nasal airflow (pressure transducer) with retained signals in thermistor and regular breathing effort (chest and abdominal belts) and desaturation in oximetery channel (3% or 4% depending upon the criteria followed). Duration must be at least 10 seconds (horizontal blue arrows). In this epoch regular snoring is also visible. Also see in-phase breathing (red arrows)) during normal breathing and paradoxical breathing (blue vertical arrows) and reduction in effort-sum (box).

Scoring of Data in Adults

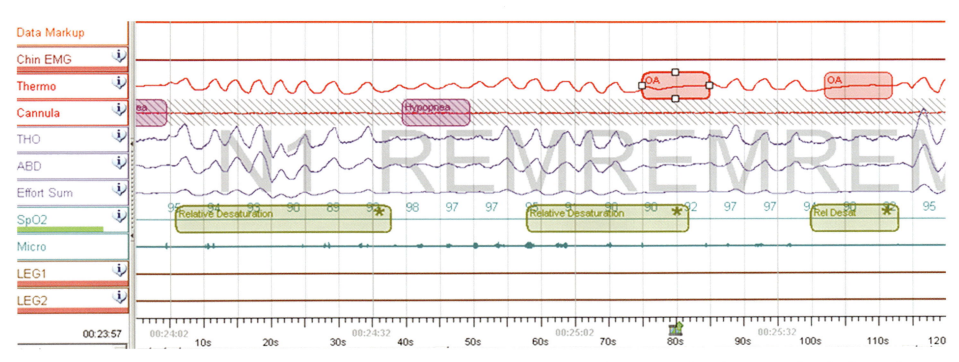

Figure 13.38C Hypopnea and apnea during pressure transducer malfunction: In case of malfunction of pressure transducer, hypopnea and may be diagnosed using signals from RIP belts. During Hypopnea sum of Chest and abdominal belts show paradoxical movement followed by desaturation with reduction in effort-sum signals, whereas in apnea, signals from thermistor as well as effort-sum are flat.

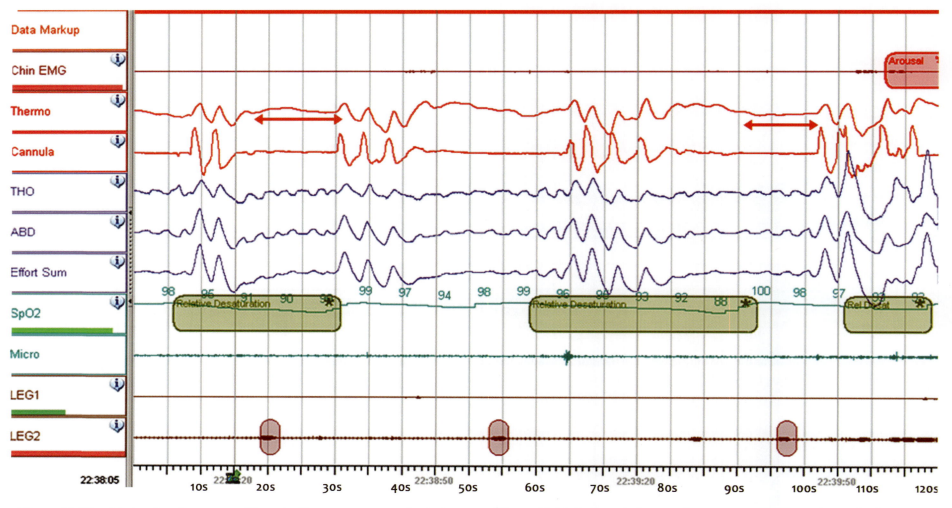

Figure 13.39 Obstructive sleep apnea: If a part of hypopnea meets criteria for apnea (Figure 13.39: Red arrows), score the event as apnea. Note that in the latter part thermistor signals showing flattening, hence this will be scored as apnea.

Scoring of Data in Adults

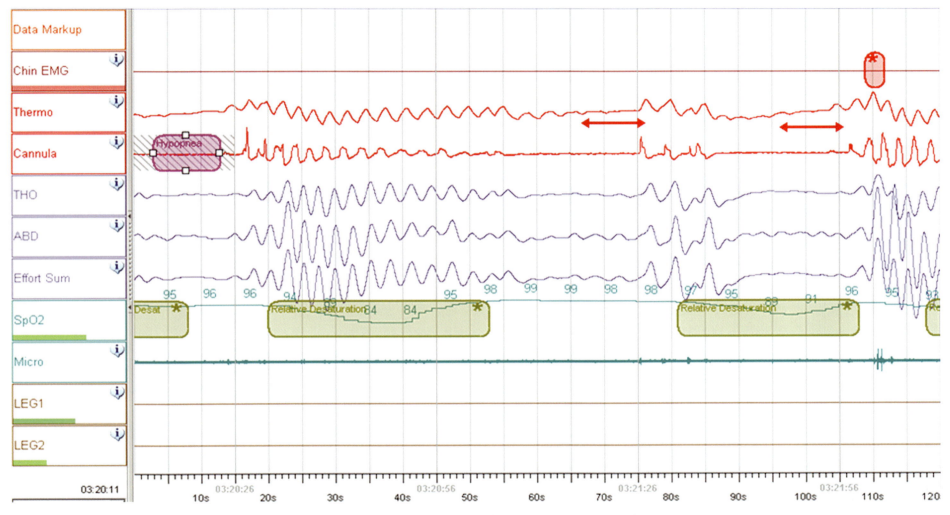

Figure 13.40 Obstructive sleep apnea: If a part of hypopnea meets criteria for apnea (Figure 13.39: Red arrows), score the event as apnea. Note that in the latter part of thermistor signals showing flattening, hence this will be scored as apnea.

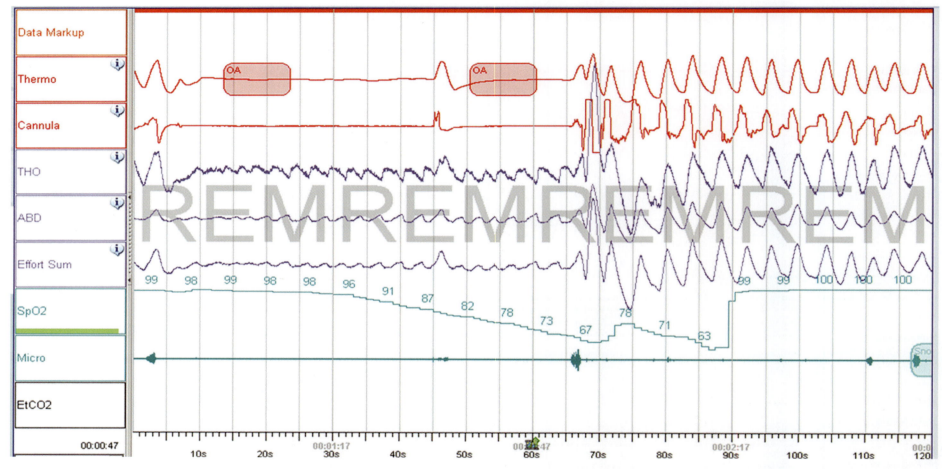

Figure 13.41 Obstructive sleep apnea: Obstructive apnea is characterized by absence of signals (for at least 10 seconds) in thermistor and pressure transducer. There is usually reduction of signals in RIP belts and effort-sum followed by hyperpnoea (belts) and subsequent desaturation. However, desaturation is not necessary to diagnose OSA. In this epoch you can appreciate paradoxical breathing (movement of chest and abdomen in opposite phase) during apnea.

Scoring of Data in Adults

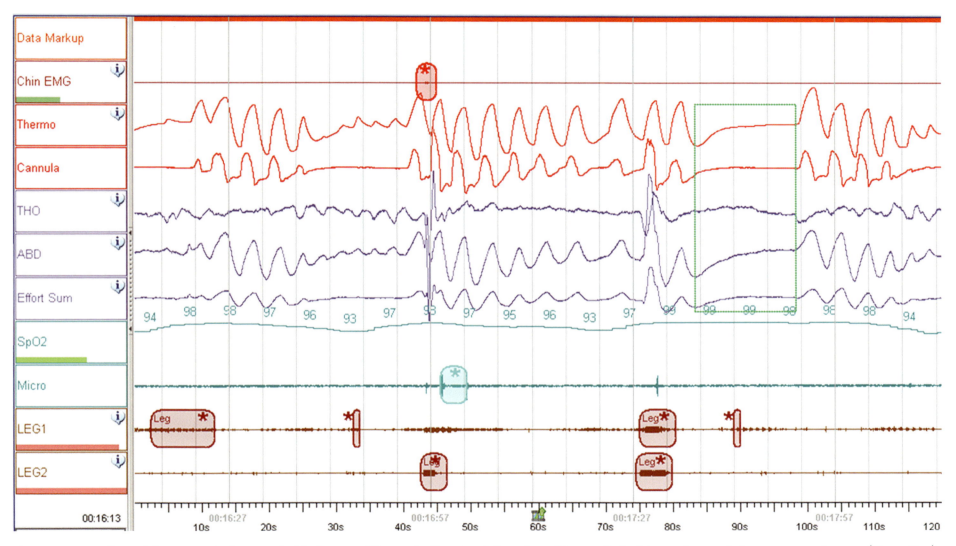

Figure 13.42 Central sleep apnea is diagnosed with absence of flow and effort signals for at least 10 seconds. In this epoch central sleep apnea can be seen (Green Box).

respiratory flow as well as respiratory efforts, and mixed apnea (Figure 13.43) is scored when the initial portion of an event shows morphology of central apnea and the latter part that of obstructive apnea. During titration study, apnea and hypopnea are scored using signals from the PAP sensor.

Respiratory effort related arousal (RERA) is diagnosed when a progressively increased breathing effort is noticed for 10 or more seconds and culminates in an arousal but does not meet the criteria for hypopnea or apnea (Figure 13.44 A–E).

Cheyne-Stokes breathing is scored when 3 or more central sleep apnea or hypopneas are separated by crescendo-decrescendo breathing and the cycle length is at least 40 seconds. In such cases, central AHI should be at least 5 along and the crescendo-decrescendo breathing pattern must be seen for at least 2 hours during the study (Figure 13.45).

Sleep-related hypoventilation is scored when there is at least 10 mmHg increase in $PaCO_2$ during sleep (compared to wake supine value) to at least 50 mmHg for 10 or more minutes or $PaCO_2$ increases to 55 mmHg or more for at least ten minutes during sleep. During a diagnostic study, arterial PCO_2 or transcutaneous PCO_2 or end-tidal CO_2 ($EtCO_2$) may be used for this purpose, however, end tidal CO_2 value is not reliable during PAP titration because of dilution from the external air that is delivered through PAP. However, transcutaneous CO_2 or $EtCO_2$ does not represent arterial CO_2. Transcutaneous CO_2 values usually lag behind arterial values by at least 2 min and have to be routinely calibrated with $PaCO_2$. Similarly, nasal secretions, nasal obstruction, supplemental oxygen and mouth breathing can provide spurious values of $EtCO_2$.

13.3 LEG MOVEMENTS

Leg movement is defined as an increment of at least 8 µV in leg EMG signals from the resting signals for the duration of 0.5 to 10 seconds. The point where the voltage increases at least 8 µV is considered as the starting point of the leg movement and terminates where this EMG remains increases by at least 2 µV for at least half a second (Figure 13.46).

For scoring periodic limb movement series (PLMS), at least four consecutive movements are required separated by at least 5 seconds (considering the time of onset) and maximally by 90 seconds (considering the time of onset). However, movement in different legs separated by less than 5 seconds (considering the time of onset) is considered as one movement (Figures 13.47 and 13.48).

If a leg movement occurs within 0.5 seconds of the onset or offset of a hypopnea or

Scoring of Data in Adults

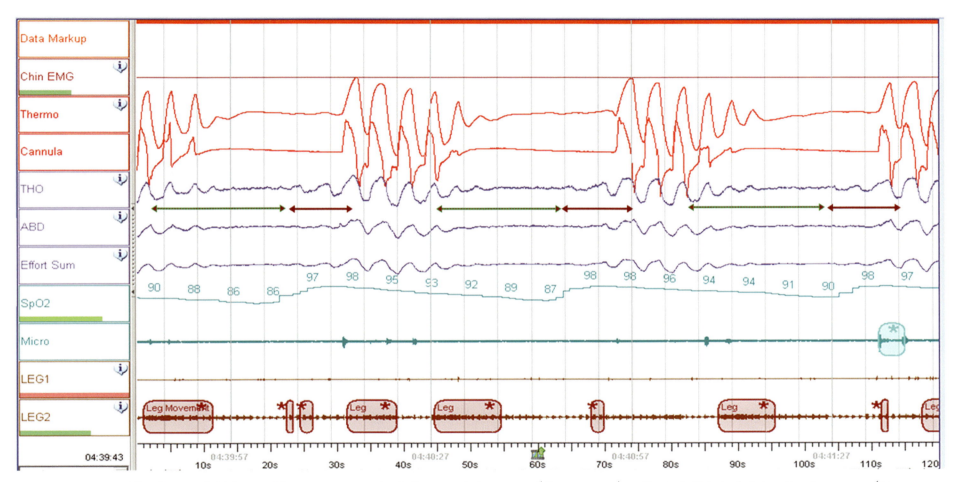

Figure 13.43 Mixed apnea: Initial part of the apnea meets criteria for central sleep apnea (Green arrows) and latter half that of obstructive sleep apnea (Maroon Arrows). In this epoch chest resumes movement before apnea terminates in flow signals.

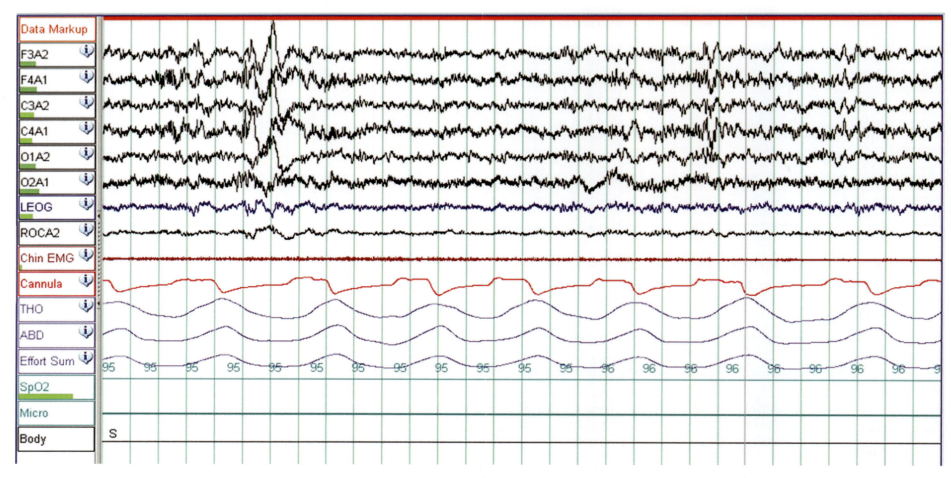

Figure 13.44A-B Respiratory event related arousal: A series of breaths with flattened waveform showing increased efforts to breath for at least 10 seconds and associated with an arousal. These breaths do not meet criteria for hypopnea. A: All the breaths in the nasal cannula have nearly equal amplitude.

Scoring of Data in Adults

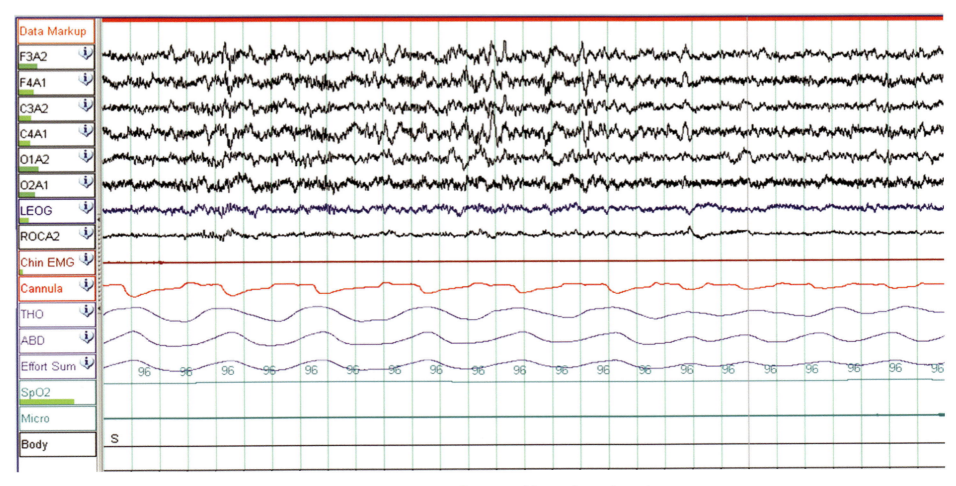

Figure 13.44B Progressive flattening of the nasal cannula tracing.

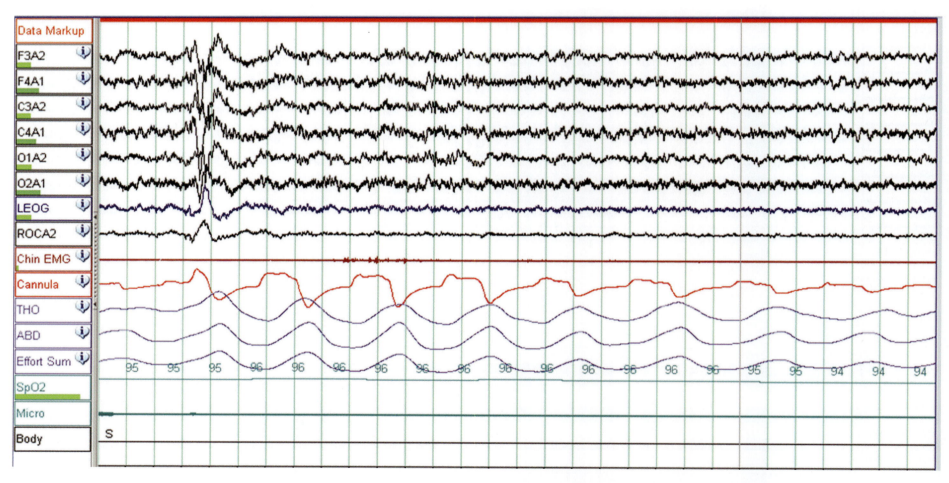

Figure 13.44C Normal amplitude of respiration appears with an arousal, later part of epoch shows progressive flattening.

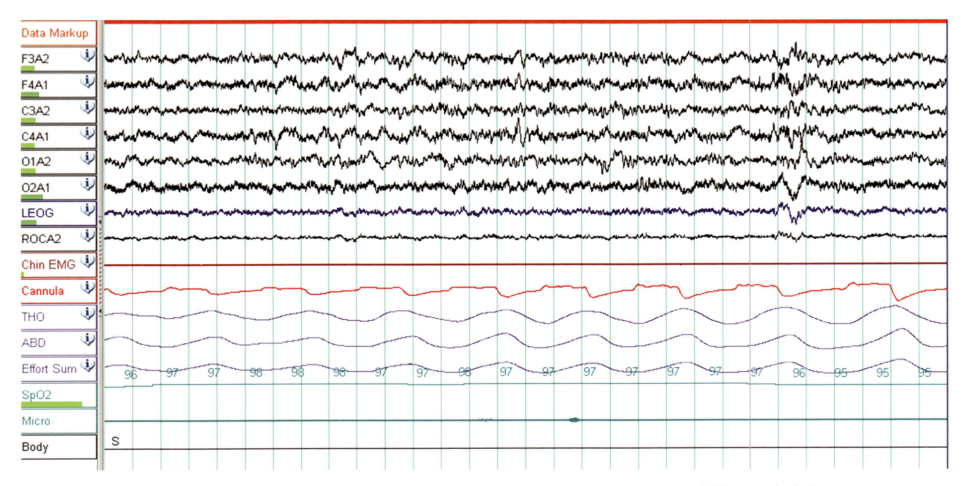

Figure 13.44D Normal amplitude appears with an arousal. See in all epochs, oxygen desaturation does not fulfill criteria for the hypopnea.

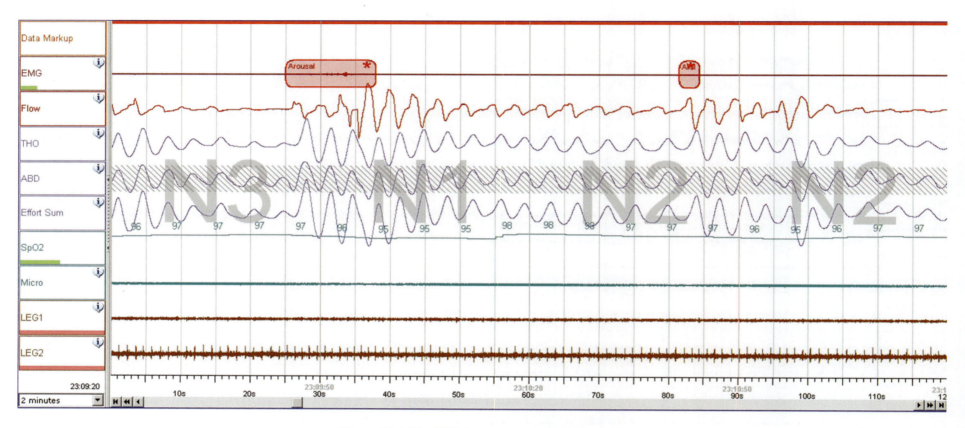

Figure 13.44E RERA as appears in 120 min epoch.

Scoring of Data in Adults

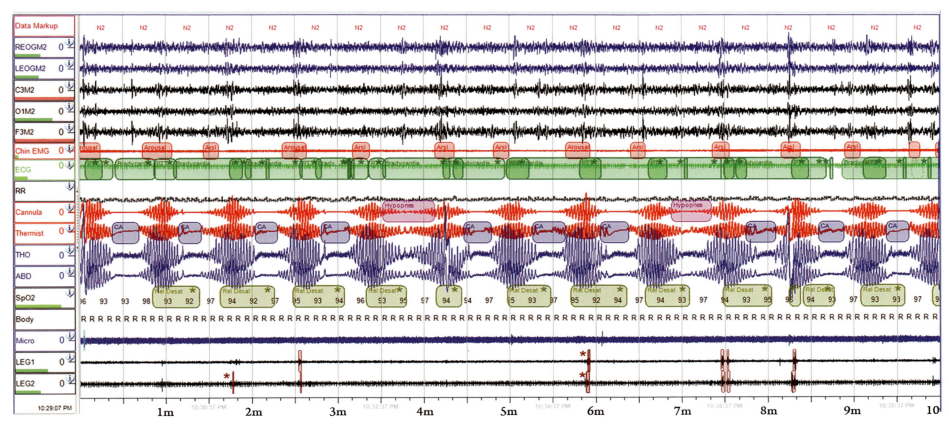

Figure 13.45A Cheyne–Stokes breathing: A: Cheyne–Stokes breathing is characterized by three or more central apnea/hypopnea that are separated by crescendo-decrescendo pattern of breathing and this cycle lasts at least 40 seconds. In addition central apnea/hypopnea index should be at least 5. In addition, crescendo-decrescendo breathing should be present at least 2 hours of total sleep time.

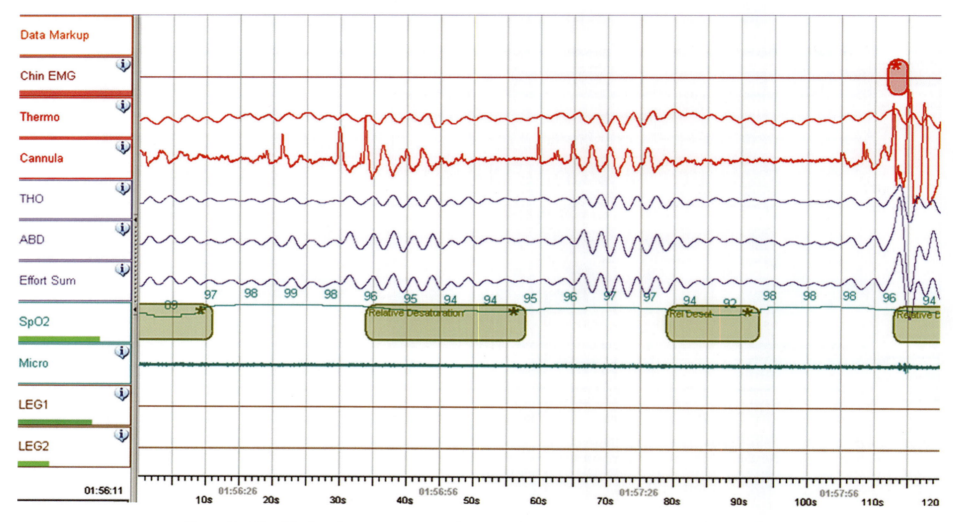

Figure 13.45B This breathing pattern seen in obstructive sleep apnea should not be confused with CSB.

Scoring of Data in Adults

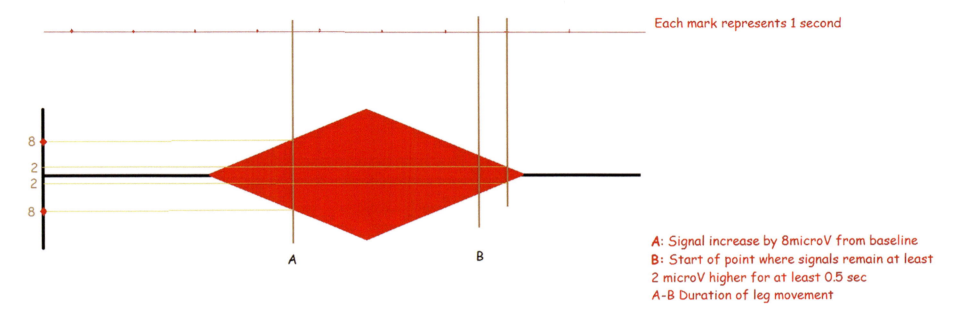

Figure 13.46 Leg movement: To be considered as movement, leg EMG must rise at least 8 µV above the baseline EEG and should last between 0.5 to 10 seconds. End point of leg movement is marked where EMG remains elevated by at least 2 µV for at least half a second.

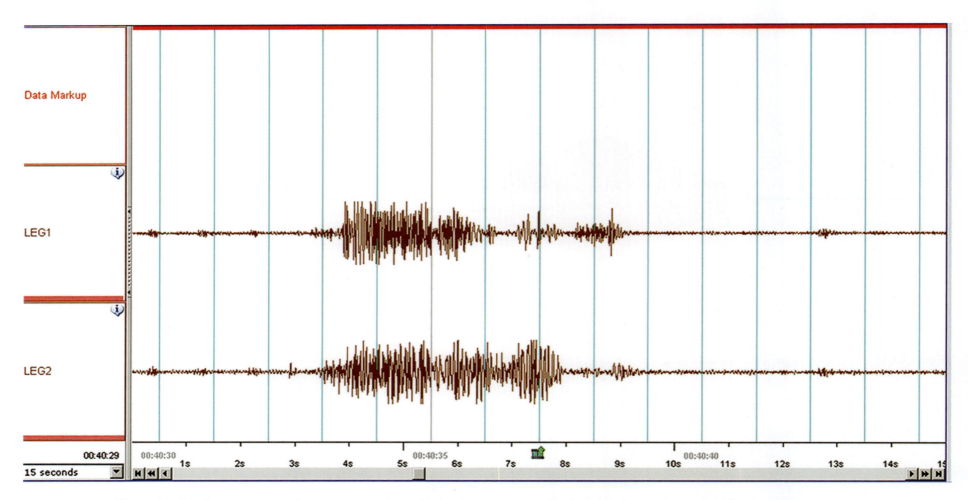

Figure 13.47 Leg movement: Leg movements in two limbs that are separated by less than 5 seconds are considered as one movement.

Scoring of Data in Adults

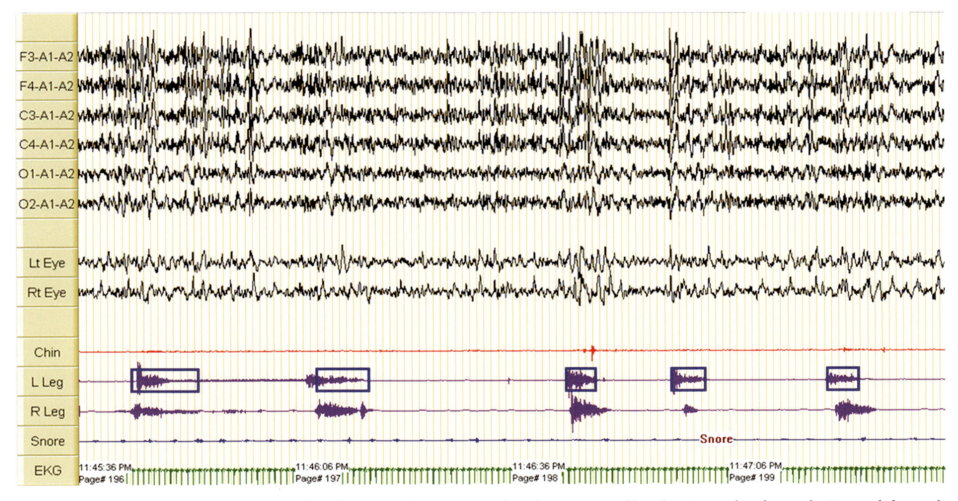

Figure 13.48A Periodic limb movement series: At least four or more leg movements where they are separated by at least 5 seconds and maxim by 90 seconds from each other are termed as leg movement series.

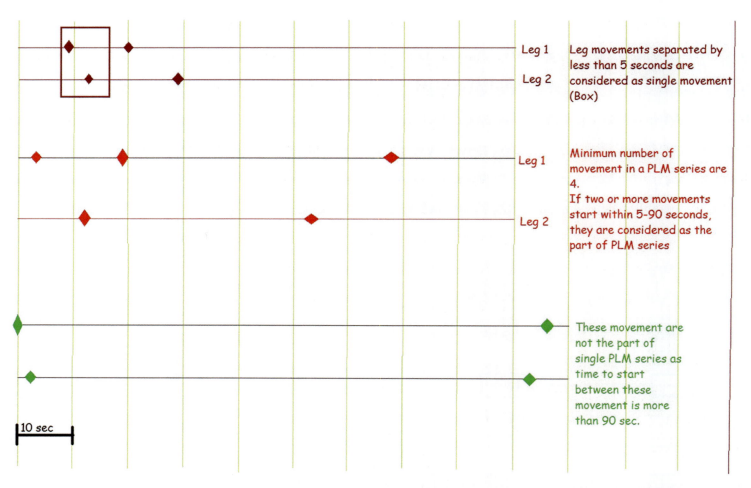

Figure 13.48B Periodic limb movement series: At least four or more leg movements where they are separated by at least 5 seconds and maxim by 90 seconds from each other are termed as leg movement series.

apnea or RERA or any other sleep disordered breathing, it is not scored as LM (Figure 13.49). Similarly, if an arousal (start or termination) is within 0.5 seconds of the start of a leg movement, they are considered associated with each other (Figure 13.50).

13.4 BRUXISM

It is characterized by brief (at least three sequential signals of 0.25–2 seconds duration) or sustained (at least 2 seconds) elevation of chin EMG, where chin EMG amplitude at least doubles the background activity (Figures 13.51 and 13.52). To be considered different, two bruxism signals must be separated by at least 3 seconds of baseline activity in chin EMG. In addition, if audio signals provide concomitant evidence of at least 2 episodes of teeth grinding (but not with seizure) along with polysomnography during the study night, the reliability of scoring improves. For better signals, EMG electrodes may be placed over masseter muscles.

Rhythmic masticatory movement activity (RMMAI) is common during sleep but without audio signals of teeth grinding.

13.5 REM SLEEP BEHAVIOR DISORDER

REM sleep behavioral disorder can be diagnosed when at least half of the REM epoch shows activity in chin EMG that is higher than the minimum activity seen during REM epoch in the same study (Figure 13.54), or when excessive transient muscle activity is seen (ETMA). ETMA is scored when any 5 of the mini-epochs (created by dividing a 30 seconds REM epoch into 10 equal epochs) show phasic muscle bursts lasting 100 m seconds to 5 seconds and at least four times higher than background EMG in chin EMG or leg EMG (Figure 13.55).

13.6 RHYTHMIC MOVEMENT DISORDER

RMD can be diagnosed when the EMG activity increases at least twice the background activity, movements occur at the frequency of 0.5–2 Hz, and at least one cluster of four movements is seen.

364 Clinical Atlas of Polysomnography

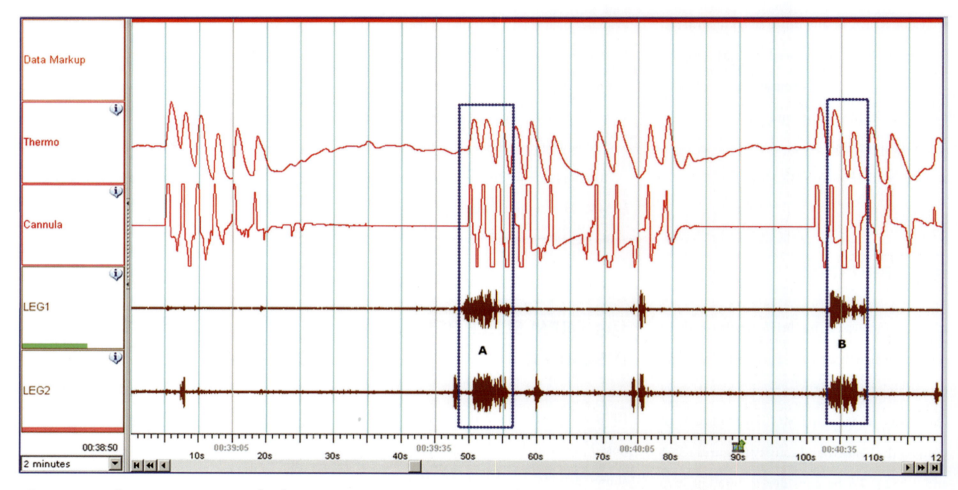

Figure 13.49 Leg movements associated with respiratory events: Leg movement preceding or following a sleep related breathing disorder event by 0.5 seconds is not considered as leg movement (A). However, if the start of leg movement is beyond 0.5 seconds, it is cored as leg movement (B).

Scoring of Data in Adults

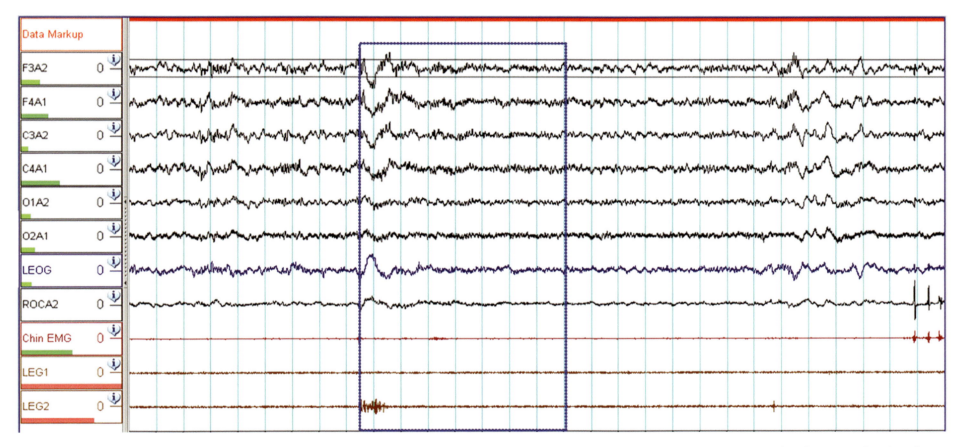

Figure 13.50 Leg movement related arousals: If a leg movement and an arousal is separated by less than half a second, they are considered associated with each other (box).

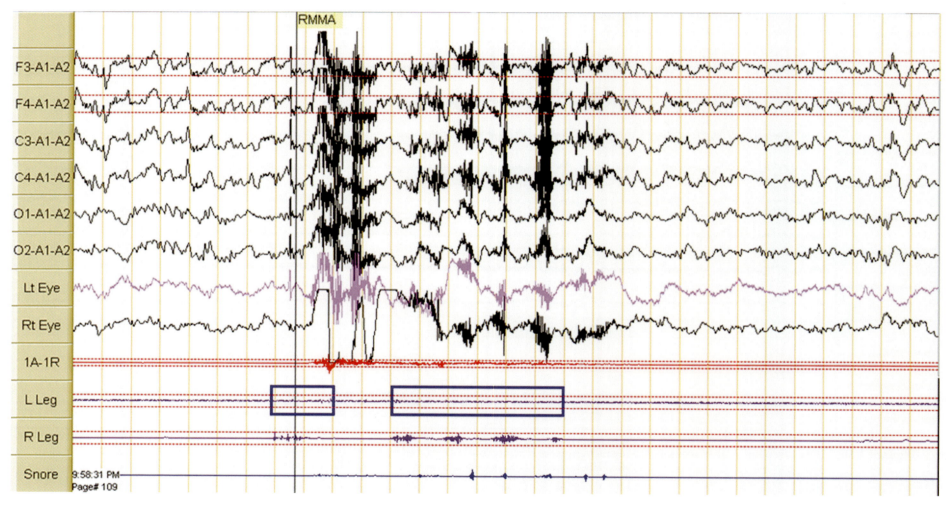

Figure 13.51 Bruxism: 2 seconds elevation in chin EMG suggesting sustained activity during bruxism.

Scoring of Data in Adults

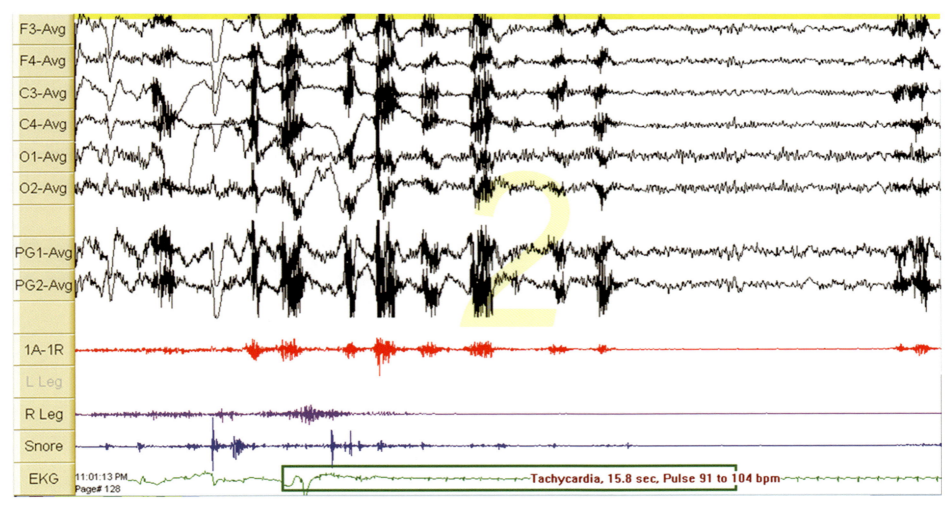

Figure 13.52 Bruxism: Phasic elevation in chin EMG suggesting bruxism. See the teeth grinding in microphone channel.

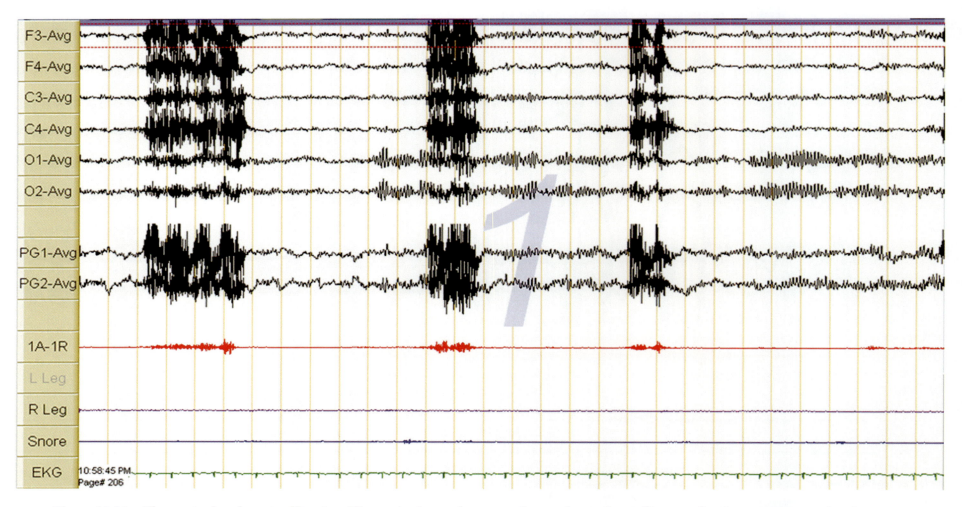

Figure 13.53 Three episodes of sustained bruxism: Three episodes are three seconds apart, hence they will be considered as separate episodes of Brixusm.

Scoring of Data in Adults

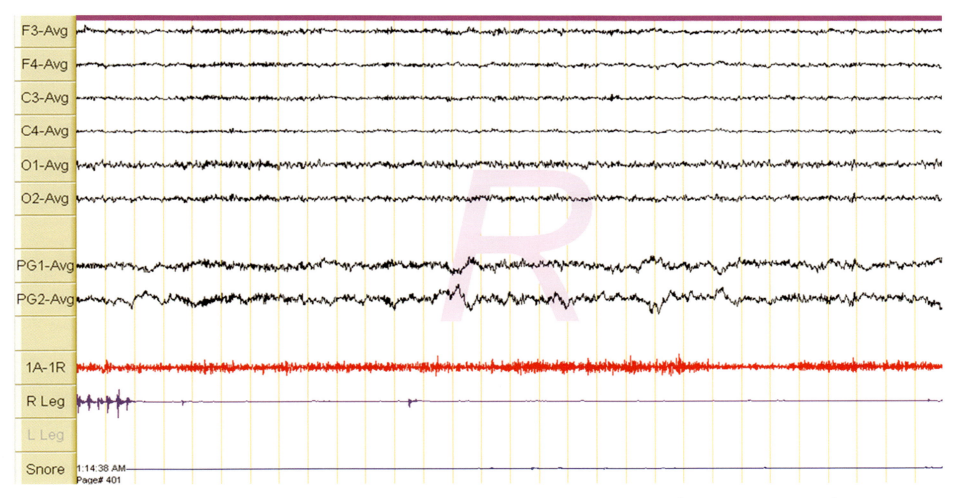

Figure 13.54A REM sleep without Atonia. Sustained muscle activity is seen in this epoch throughout (chin EMG: Channel 1A-1R).

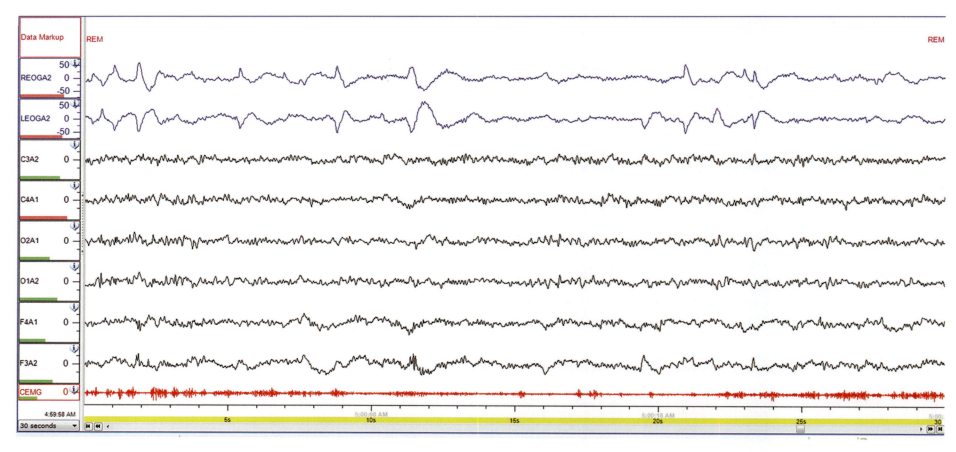

Figure 13.54B　REM sleep without atonia: Most of epoch shows loss of muscle atonia in chin EMG during REM.

Scoring of Data in Adults

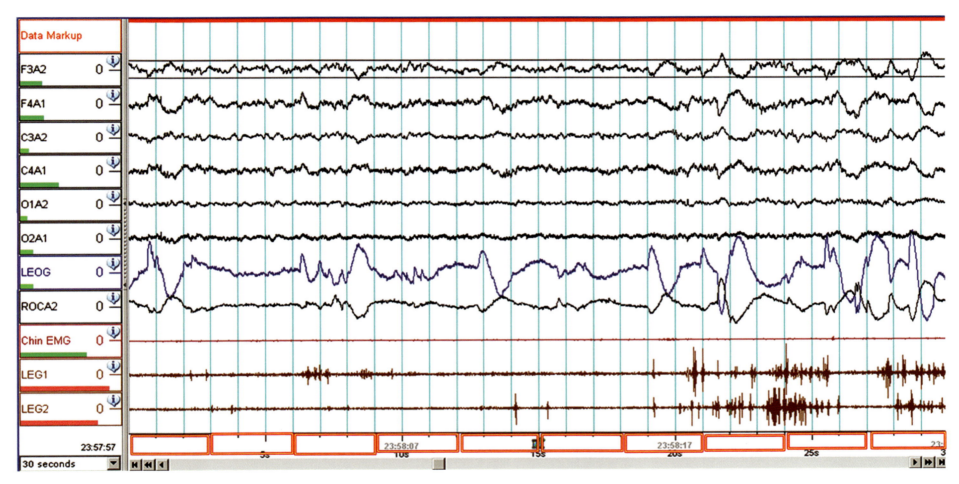

Figure 13.55 Excessive fragmentary transient muscle activity also suggests REMBD. It is scored by diving a 30 seconds epoch into 10 mini-epochs of 3 seconds duration each.

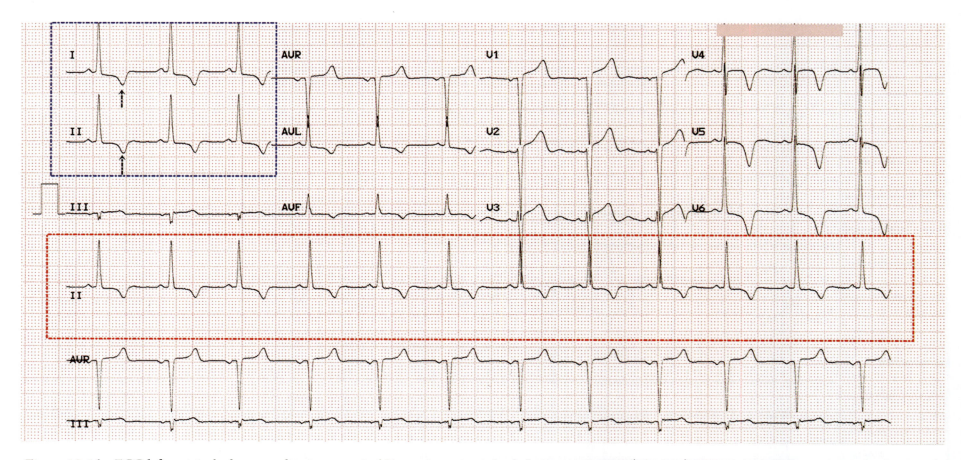

Figure 13.56 ECG-left ventricular hypertrophy: Asymmetrical T wave inversion in both derivations I and II (Blue box). Also see in the long run in Lead II (Red Box). In such cases, Physician's opinion must be sough as this may be seen in a number of non-life threatening conditions, e.g., Left ventricular hypertrophy as well as potentially life threatening conditions, e.g., Myocardial infarction. However, in myocardial infarction, inverted T waves are symmetrical.

Scoring of Data in Adults

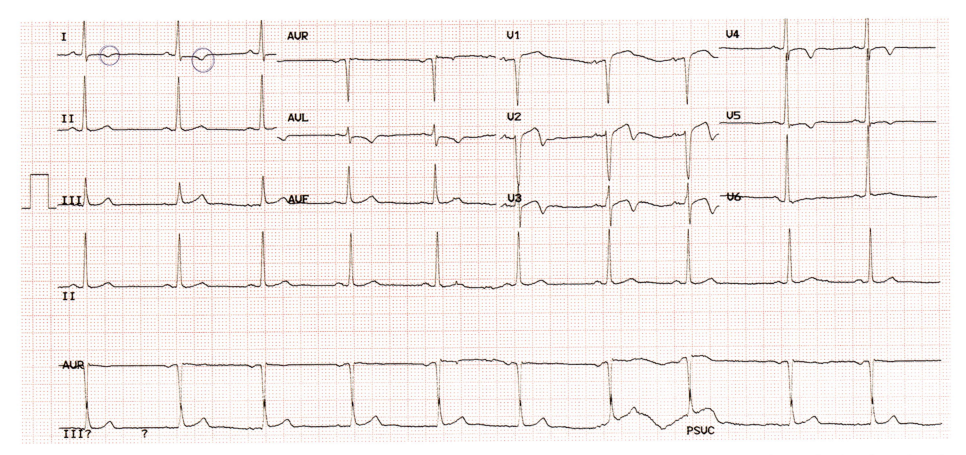

Figure 13.57 T wave inversion: Symmetrical T wave inversion is seen in lead I. in such cases opinion from a physician should be sought immediately and complete 12 derivations ECG must be obtained.

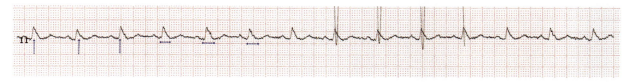

Figure 13.58 Wide QRS complex: This ECG shows wide QRS complex (arrows) (duration >120 msec; line segments) suggestive of bundle branch block.

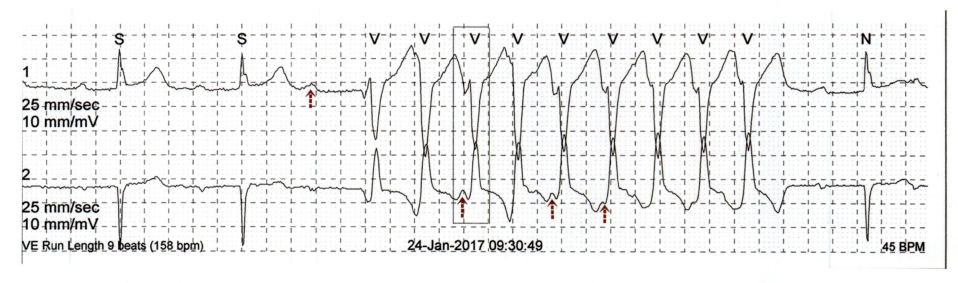

Figure 13.59 This ECG shows short lasting Ventricular Tachycardia. In this ECG, P wave (red arrows) is independent of QRS complex even during tachycardia. Such events are pathological and should be immediately reported to attending Physician.

Scoring of Data in Adults

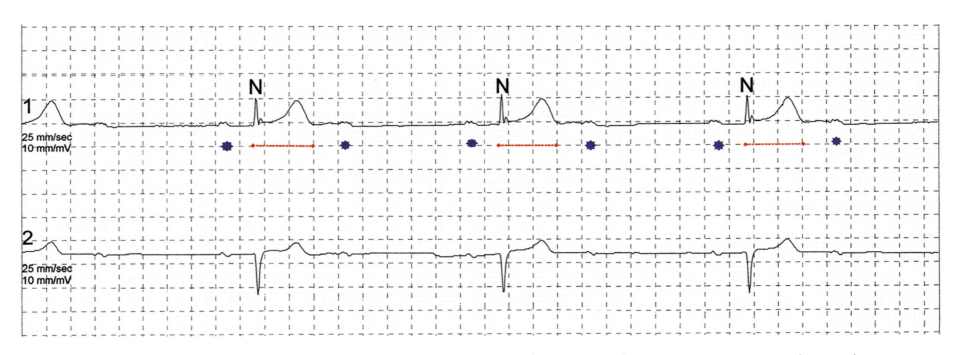

Figure 13.60 2:1 heart block: This ECG shows 2:1 AV block. QRS complex (redline segment) appears after every two waves (blue stars).

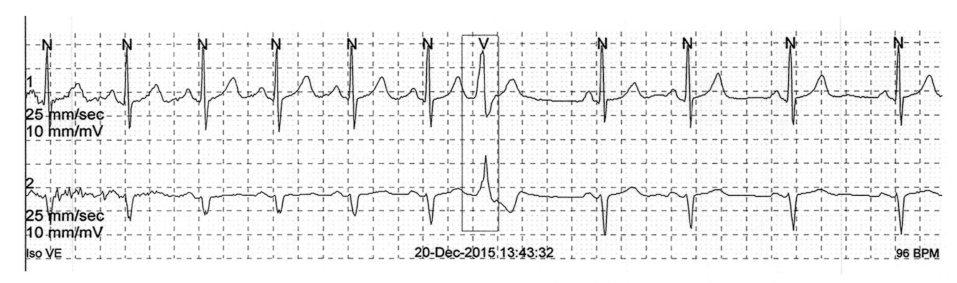

Figure 13.61 VPC: This ECG shows ventricular premature beat (marked in box as V) in between normal beats (Marked as N). Look that VPC lacks P wave, hence, ventricular in origin.

Scoring of Data in Adults

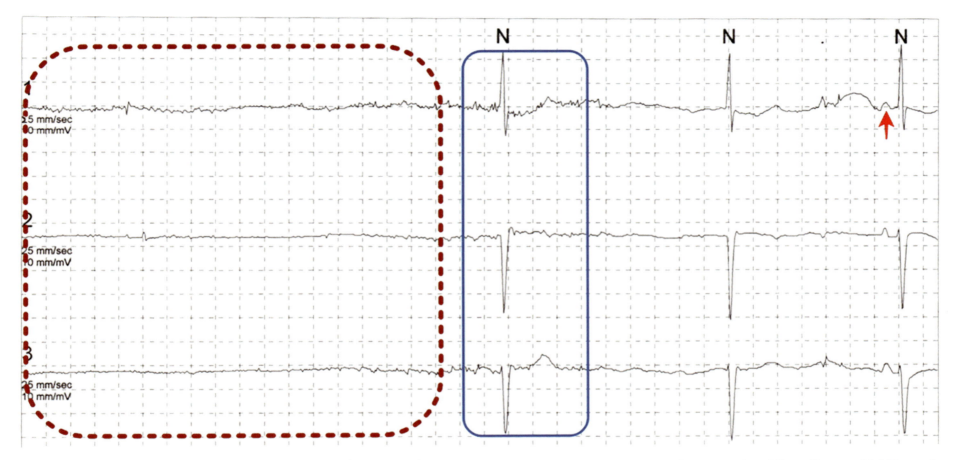

Figure 13.62 Asystole: This ECG shows cardiac asystole (Maroon box). The baseline is flickering giving appearance of AF. Asystole is followed by one QRS-T complex (Blue box), which is functional in origin and P wave is absent. Last beat is QRS complex is preceded by P wave (red arrow).

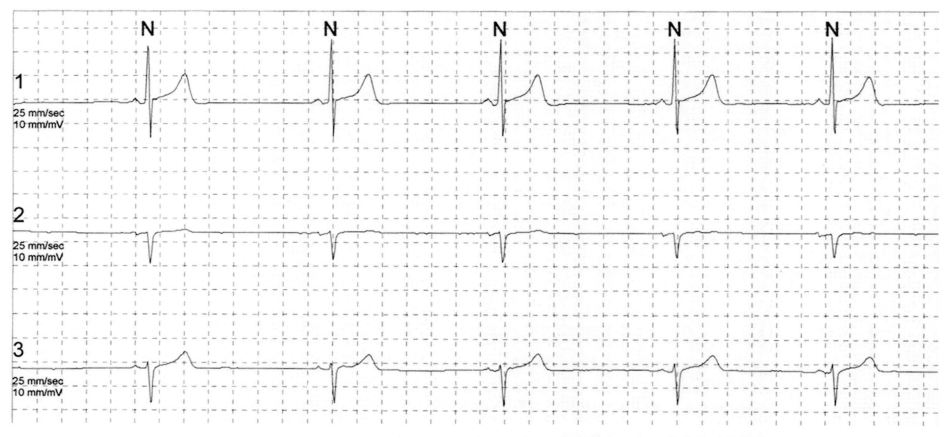

Figure 13.63 Bradycardia: This ECG shows sinus bradycardia with PQRST waves but they are widely placed. One big square of ECG represents 0.2 seconds duration. Heart rate is calculated by dividing 60 with the RR interval. In this case RR interval is 1.4 seconds. Hence, heart rate is 42/min..

Scoring of Data in Adults

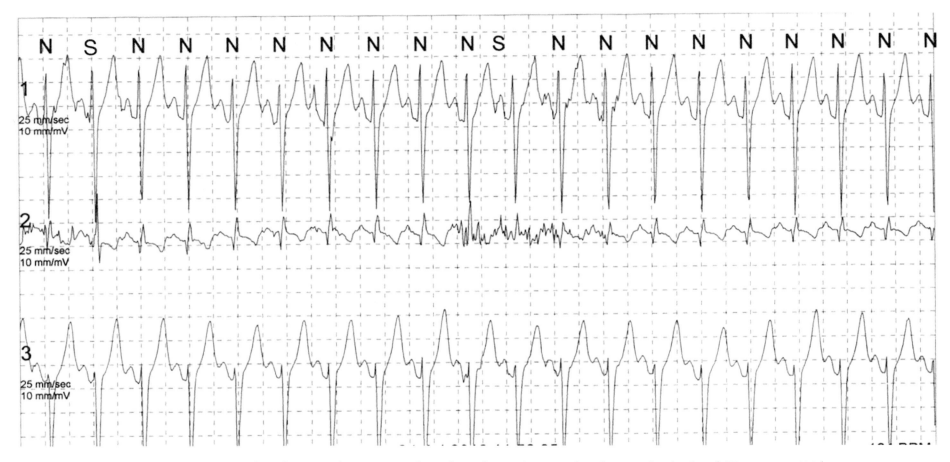

Figure 13.64 Tachycardia: This ECG shows sinus tachycardia with PQRST waves but they are closely placed. Heart rate is 150/min.

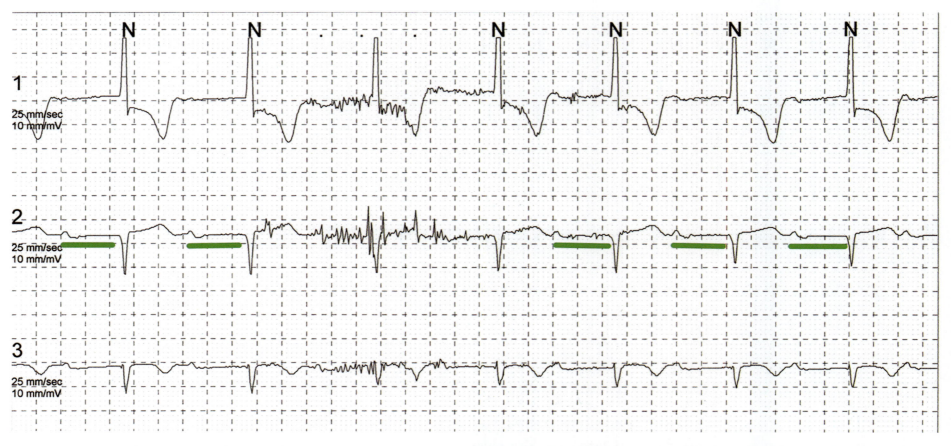

Figure 13.65 Grade 1 AV block: This ECG shows prolonged PR interval. If it is longer than 0.2 seconds, it is considered as AV block. This ECG shows PR interval of nearly 0.4 seconds which is equal duration throughout the strip and regular QRS complex following each P wave. Hence, this is first degree AV block.

Scoring of Data in Adults

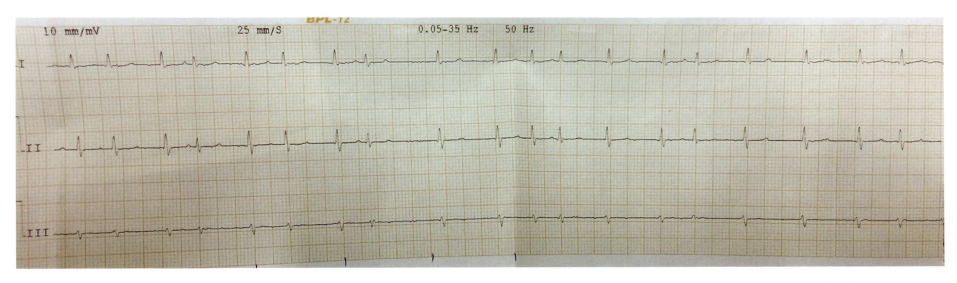

Figure 13.66 Atrial fibrillation: In this ECG P wave can not be seen and heart rhythm is irregularly irregular. This is suggestive of atrial fibrillation.

13.7 SCORING OF EKG DATA

Scoring of EKG data is depicted in Figures 13.56–13.66.

FURTHER READING

1. Berry, R. B., Brooks, R., Gamaldo, C. E., Harding, S. M., Lloyd, R. M., Marcus, C. L., et al., (2017). The AASM manual for scoring of sleep and associated events: rules, terminology and technical specifications. Version 2.4. www.aasmnet.org. Darian, Illinois: American Academy of Sleep Medicine.

REVIEW QUESTIONS

1. An epoch is scored the sleep wake stage that occupy at least following percent of the epoch:
   A. 20%
   B. 50%
   C. 70%
   D. 90%

2. Wakefulness is scored when:
   A. Eyes show slow movement
   B. EKG shows normal rhythm
   C. EEG shows alpha rhythm in posterior leads
   D. Respiration flow channels show flattening

3. N1 is scored when:
   A. Low amplitude mixed frequency activity appears for more than 15 seconds
   B. Eye blinks are seen
   C. Chin tone is lowest
   D. Limb EMG shows no activity

4. In patients who do not produce alpha, N1 is scored when:
   A. K-complexes appear
   B. Delta activity appear
   C. Eye blinks are absent
   D. EEG activity is at least <1 Hz slower than waking and theta appear

5. When N2 is interrupted with arousal, subsequent epochs will be scored:
   A. N2 even when spindle or K-complexes
   B. N1 till spindles or K-complexes appear in the epoch
   C. Wake till N2 appears
   D. N3 even when epoch does not show delta waves

6. When an arousal interrupts REM sleep and subsequent epoch shows low voltage mixed frequency activity with chin EMG equal to REM will be scored as:
   A. N2
   B. REM
   C. N1
   D. N3

7. If an epoch subsequent to N3 epoch do not show delta waves, but is followed by N3 epoch should be scored as:
   A. N2, unless it meets criteria for REM or wake
   B. N2, irrespective of waveforms
   C. N3, irrespective of waveforms
   D. Wake, irrespective of waveforms

8. If an epoch shows spindle in second half and preceded by N1 epoch, will be scored as:
   A. Wake
   B. N2
   C. N1
   D. REM

9. Following may be seen during N3 sleep EXCEPT:
   A. Delta waves > 6 seconds
   B. Spindles
   C. K-complexes meeting criteria for delta waves
   D. Sawtooth waves

10. If an epoch of N2 is followed by an epoch with K-complex in first half but has low voltage mixed frequency activity in most of the epoch along with chin EMG to REM level, it will be scored as:
    A. REM
    B. N2
    C. N1
    D. Wake

11. If an epoch of REM is followed by an epoch that has high chin EMG and low voltage mixed frequency activity for more than 15 seconds and is followed by an epoch with K-complex in first half, epoch with high chin EMG will be scored as:
    A. N2
    B. REM
    C. Wake
    D. N1

12. An epoch subsequent to an epoch of REM with an arousal in second half, but without eye movements will be scored as:
    A. N1
    B. REM
    C. N2
    D. Wake

13. An epoch of major body movement after a REM epoch will be scored as:
    A. REM if subsequent epoch shows slow eye movement
    B. N1 irrespective of other waveforms in subsequent epoch
    C. N1 if slow eye movement appears in subsequent epoch

D. Wake, irrespective of waveform in subsequent epoch

14. If an epoch has REM waveforms in majority of epoch but has K-complex, will be scored as:
    A. N2
    B. REM
    C. N1
    D. N3

15. Mixed apnea is scored when:
    A. Respiratory flow and effort show flattening
    B. First half of the event shows CSA followed by OSA
    C. First half of the event shows OSA followed by CSA
    D. Respiratory flow shows flattening with regular efforts

16. For Cheyne-Stokes breathing following is required:
    A. Crescendo-decrescendo breathing
    B. Crescendo-decrescendo breathing with intervening central sleep apnea
    C. Crescendo-decrescendo breathing with intervening central sleep apnea with central AHI at least five
    D. Crescendo-decrescendo breathing with intervening central sleep apnea with central AHI at least five and this pattern is seen for at least 2 hours

17. Two leg movements are scored as one movement when:
    A. They occur less than 5 seconds apart
    B. They occur more than 5 seconds apart
    C. Their starting points are separated by less then 5 seconds
    D. Terminal point of first is separated by less then 5 seconds from the starting point of other

18. Leg movements are considered as a part of PLM series when:
    A. Their starting points are separated by minimum 5 seconds and maximally by 90 seconds
    B. Their terminal points are separated by minimum 5 seconds and maximally by 90 seconds
    C. Their starting points are separated by minimum 10 seconds and maximally by 90 seconds
    D. Their terminal points are separated by minimum 10 seconds and maximally by 90 seconds

19. A major body movement is scored when:
    A. EMG artifacts are seen in the epoch
    B. EMG artifacts are seen in the epoch to such an extent that staging is not possible

C. EMG artifacts are seen in two consecutive epochs

D. EMG artifacts are seen along with change in body position signals

20. In REM sleep behavior disorder patient, scoring of REM is dependent upon:

A. EKG

B. EMG

C. EOG

D. EEG

ANSWER KEY

1. A 2. C 3. A 4. D 5. B 6. C

7. A 8. C 9. D 10. A 11. D 12. B

13. C 14. B 15. B 16. D 17. C 18. A

19. B 20. D

14

USE OF VIDEO POLYSOMNOGRAPHY

LEARNING OBJECTIVES

After reading this chapter, the reader should be able to:

1. Discuss the importance of audio-video signals during a sleep study.

CONTENTS

Information Provided by Video 387

Further Reading 388

Review Questions 388

Answer Key 389

A synchronized video recording can provide invaluable information to differentiate between parasomnia and seizures. Therefore, we stress that video recording should always be available with the PSG data.

Any movement made by the patient causes disturbance in the ECG. In such cases, absence of a video would make accurate diagnosis difficult. This is especially true for cases of RLS, sleep-related rhythmic movement disorders, sleep-walking, nocturnal seizures, and REM sleep behavior disorder.

For optimum results, it is essential to have an infrared camera that can capture pictures in the dark and also has a zoom facility. Besides, it should be possible to be maneuver the camera through a console in the technician's room. A good camera would be able to capture even the most trivial movement, for example, lip smacking during sleep. Thus, information captured on the video can help to arrive at a more accurate diagnosis.

Video can be used to detect localized movement that may be suggestive of REM sleep behavioral disorder, for example, hand movement along with chin muscle atonia. Video recording may also be used to show the movement to the bed partners and to analyze whether the movement observed in the sleep laboratory is similar to what happens at home ; in other words, to understand if it is stereotyped. Stereotyped movements usually favor seizures, especially if associated with EEG changes. These movements may also be observed during sleep-related rhythmic movement disorder, however, in that case, they are not associated with the EEG changes seen during seizures.

Without a video, it may be difficult to differentiate between the "turning in bed" and having an abnormal movement. Sleep-related rhythmic movement disorder, seizures, parasomnia and turning in bed, especially if brief, produce similar kinds of movement artifacts.

Audio forms an important part of the sleep study. Teeth grinding and sleep talking are two activities that produce similar kind of signals in the microphone channel. However, the microphone only provides visual signals. In such cases, having an external microphone may be helpful to differentiate between these conditions.

Different characteristics of the audio signals have been used in recent years to detect snoring and sleep apnea, however, this technique is not recommended by the AASM scoring manual.

FURTHER READING

1. Roebuck, A., Monasterio, V., Gederi, E., et al. (2014). A review of signals used in sleep analysis. *Physiological measurement*. 35(1), R1–57. doi:10.1088/0967–3334/35/1/R1.

REVIEW QUESTIONS

1. Synchronized video recording during sleep study helps in recognizing:
   A. Sleep apnea
   B. Seizure activity

C. Insomnia

D. Hypersomnia

2. Requirement of light for video recording during sleep study is overcome by:

   A. Keeping the lights turned on
   B. Use of ultraviolet waves during data recording
   C. Intermittent use of flash during recording
   D. Use of infrared waves during recording

3. Video camera in the sleep laboratory must have following functions:

   A. Facility for audio recording
   B. Facility for the flash light
   C. Facility for the zooming and maneuvering
   D. Facility for daytime recording

4. Video camera should be placed:

   A. Close to the head of patient
   B. In such position that whole body can be seen
   C. Focused on the legs of the patient
   D. Focused on the trunk of the patient

5. Video recording is not essential during:

   A. Suspected parasomnia
   B. Suspected seizure disorder
   C. Suspected Sleep related movement disorder
   D. Multiple sleep latency test

ANSWER KEY

1. B   2. D   3. C   4. B   5. D

15

USE OF SLEEP HISTOGRAM

LEARNING OBJECTIVES

After reading this chapter, the reader should be able to:

1. Comment on sleep and sleep disorders looking at histogram and hypnogram.

CONTENTS

Interpretation of Histograms 391

Review Questions 403

Answer Key 403

INTERPRETATION OF HISTOGRAMS

The graphical display of the sleep architecture, or sleep stages, along a time axis depicting, is called a hypnogram. However, the graphical display of the sleep stages, sleep-related events and other measured sleep parameters along a time axis,

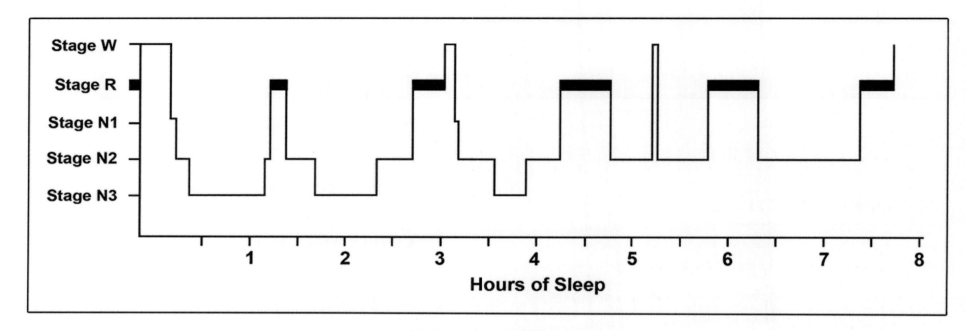

Figure 15.1 Normal hypnogram.

Use of Sleep Histogram 393

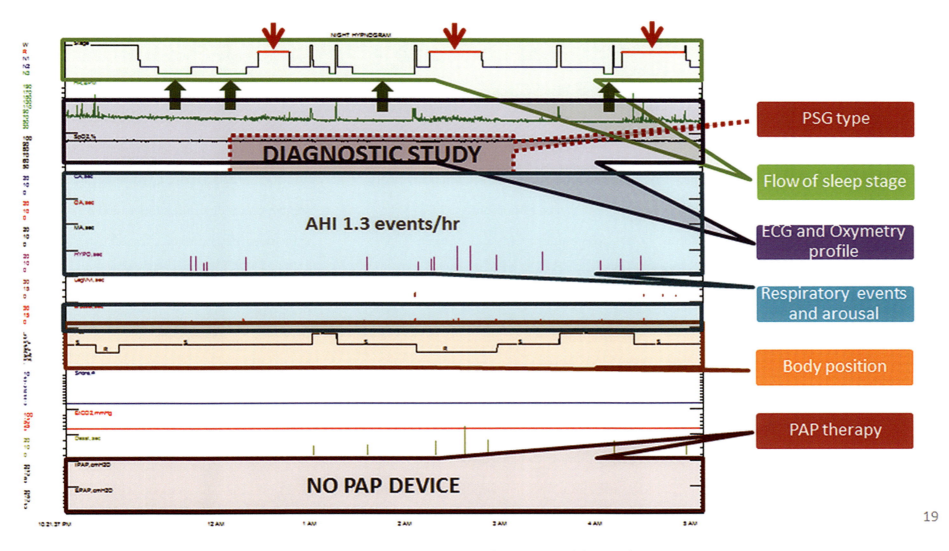

Figure 15.2 Histogram showing normal sleep study.

is called histogram. Figures 15.1 and 15.2 shows hypnogram and histogram, respectively.

A histogram is the end result of a scored sleep study. It provides a wealth of information and enables the health care provider to access the major findings of a sleep study on a single page.

A histogram graphically depicts the progression of sleep stagees, and provides information on factors such as continuity of sleep, latency of various sleep stages, major awakenings, respiratory events, heart rate, and response to positive airway pressure therapy.

We propose a systematic approach to histogram interpretation that includes:

A. PSG type: The histogram tells us if the study is diagnostic, therapeutic, or split-night study (Figure 15.3).
B. The flow of sleep stages: The hypnogram shows if the patient has progressed into all sleep stages in a normal pattern. It also shows the effect of PAP therapy on sleep architecture.
C. ECG and oximetry profile. This shows the effect of respiratory events on heart rate variability (bradycardia and tachycardia) and oxygen saturation.
D. Respiratory events and arousals. REM-related OSA can be easily detected in the histogram (Figure 15.4).
E. Body position.
F. PAP therapy.

Figure 15.5 demonstrates that there is no slow wave (N3) in the diagnostic part due to severe OSA. During the therapeutic part, sleep progresses into N3 after eliminating the obstructive events, indicating a good subjective response to CPAP therapy. The ECG variability line is thicker during the diagnostic part due to the alternating bradycardia and tachycardia, secondary to the obstructive events. The signal is more stable in the therapeutic part. The obstructive events (apneas and hypopneas) are represented by red and pink lines. The length of the lines represents the duration of the events. During an REM sleep, the apnea and hypopnea duration is longer and is associated with more desaturation. Desaturation is eliminated at a pressure of 15 cm H_2O; however, the patient continues to snore associated with arousals, indicating RERA. Upon increasing the CPAP pressure, snoring is eliminated. However, during REM sleep, hypopneas reappears again; hence, pressure is increased to eliminate OSA during REM sleep.

Figure 15.6 shows short REM latency with SOREM sleep. Causes of SOREM include, narcolepsy, sudden withdrawal of REM suppressants, major depression, and sleep deprivation.

Use of Sleep Histogram

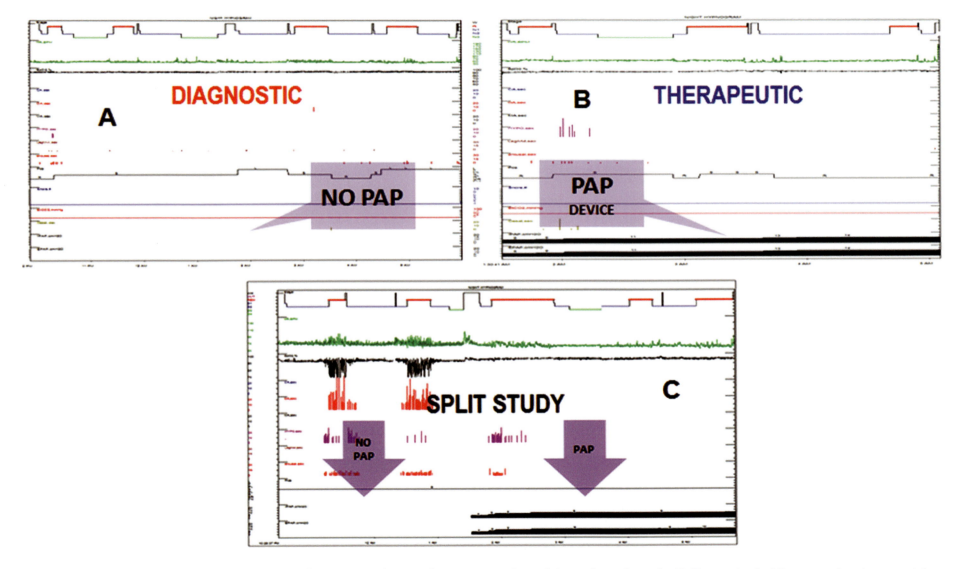

Figure 15.3 Three histograms showing: A: A diagnostic study; B: A therapeutic study; and C: A split-night study: Difference in the Histogram showing overnight diagnostic study, overnight PAP titration and split-night study.

396 Clinical Atlas of Polysomnography

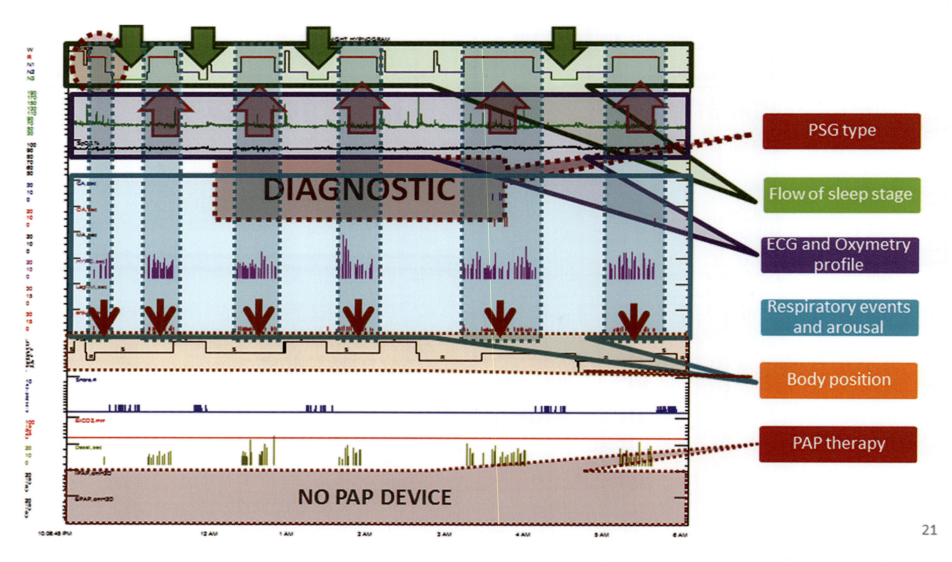

Figure 15.4 Histogram showing REM related OSA: Normal pattern of the sleep architecture as can be seen more N3 in the first half of the night. Frequent dips in saturation that appear as darker oxygen saturation line due to the intermittent desaturation. Hypopneas are restricted to REM sleep only and occur even in the lateral position during REM.

Use of Sleep Histogram

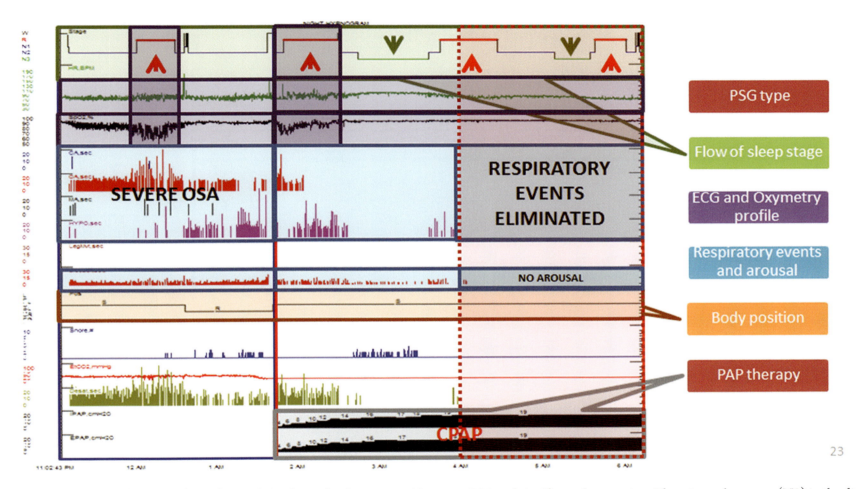

Figure 15.5 Histogram of a split night study: A split-night study of a patient with severe OSA and significant desaturation. There is no slow wave (N3) in the diagnostic part due to severe OSA. During the therapeutic part, sleep progressed into N3 after eliminating the obstructive events indicating a good subjective response to CPAP therapy. ECG variability line is thicker during the diagnostic part due to alternating bradycardia and tachycardia secondary to the obstructive events. The signal is more stable in the therapeutic part. The obstructive events (apneas and hypopneas) are represented by red and purple lines. The length of lines represents the duration of the events. During REM sleep, apnea and hypopnea duration is longer and is associated with more desaturation. Desaturation was eliminated at a pressure of 15 cm H_2O; however, the patient continues to snore associated with arousals indicating RERA. Upon increasing CPAP pressure, snoring was eliminated. However, during REM sleep, hypopneas reappeared again; hence, pressure was increased to eliminate OSA during REM sleep.

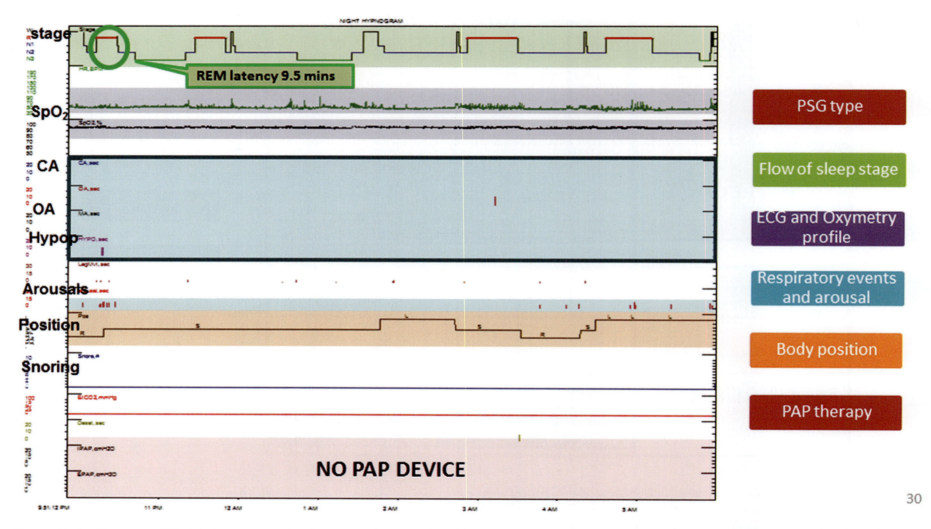

Figure 15.6 Sleep onset REM: A diagnostic study of a patient who was referred for sleep study to rule out OSA. However, the study as can be seen showed sleep onset REM (SOREM) with a REM sleep latency of 9.5 min. Causes of SOREM include, narcolepsy, sudden withdrawal of REM suppressants, major depression, and sleep deprivation.

Use of Sleep Histogram

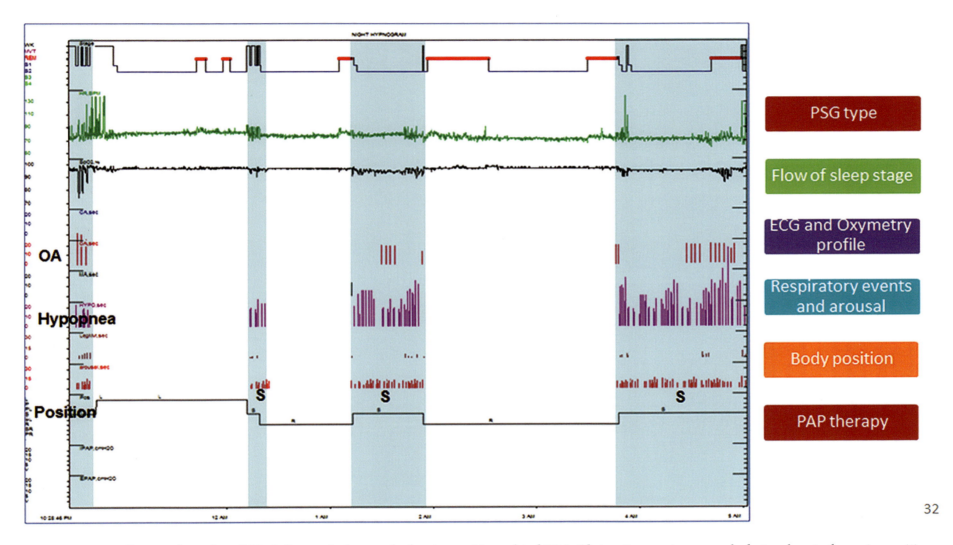

Figure 15.7 Position dependent OSA: A diagnostic sleep study showing position-related OSA. Obstructive events occur only during sleep in the supine position. Obstructive apneas and hypopnea appear only in the supine position and are associated with desaturation.

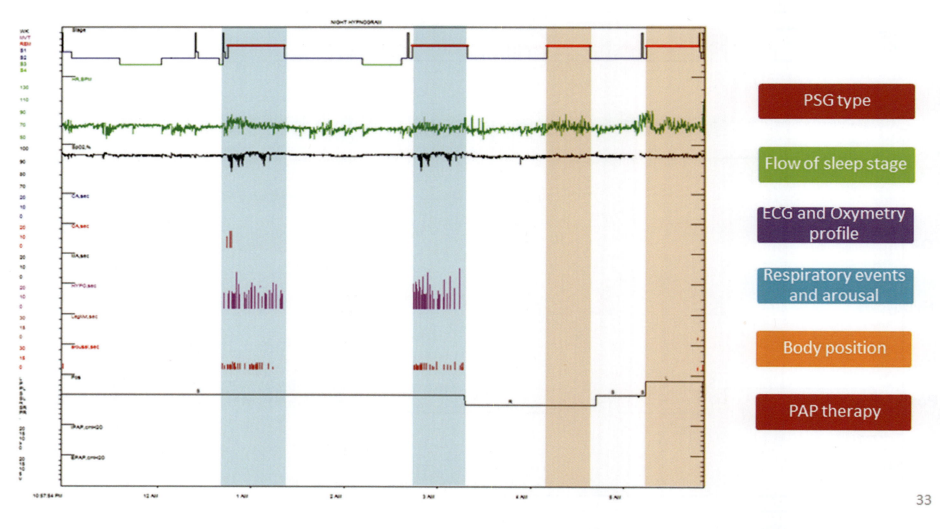

Figure 15.8 Position and sleep stage dependent OSA: A diagnostic study showing position and REM-related OSA. Obstructive apneas and hypopneas occurred during REM sleep in the supine position and are associated with desaturation. During REM sleep in the lateral position, there were no obstructive events or desaturation.

Use of Sleep Histogram

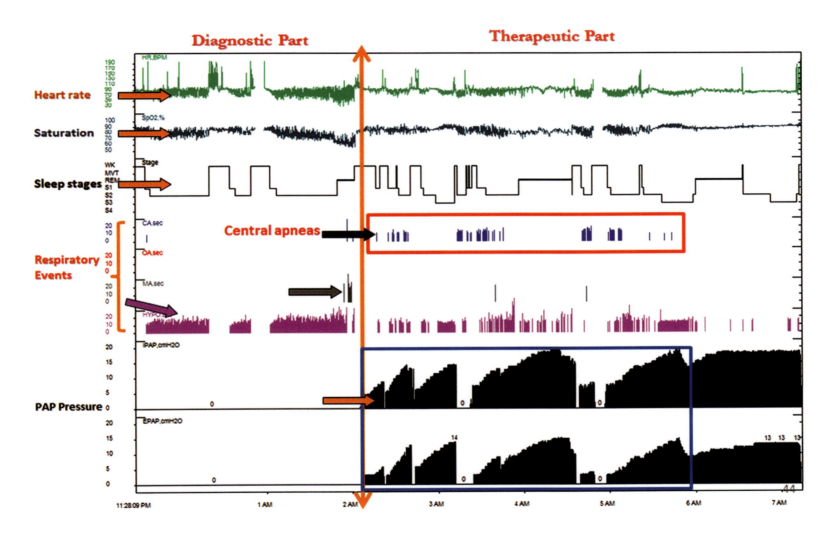

Figure 15.9 Split-night study with treatment emergent CSA (complex sleep apnea): A split-night study of a patient with severe OSA. It shows incorrect rapid CPAP pressure build up. The rapid increase in pressure increases the chance in developing central apneas. This patient developed central apneas (CPAP emergent apneas) associated with desaturation.

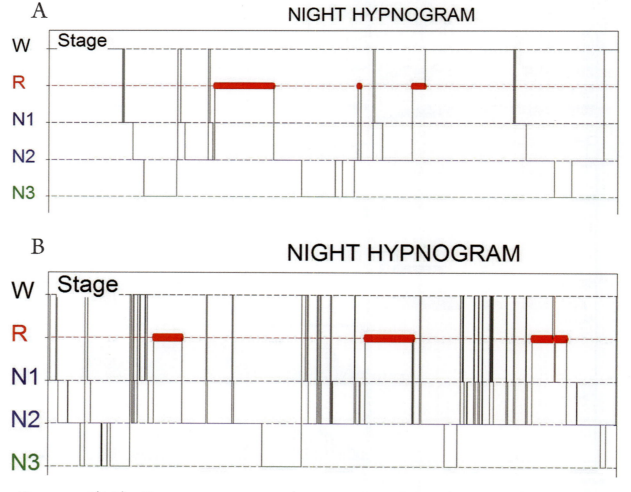

Figure 15.10(A,B) Hypnogram in a patient with insomnia: A: Hypnogram shows prolonged sleep latency and middle insomnia; B: Multiple awakenings in patient with insomnia.

Histograms can help in detecting position-related and sleep stage related OSA (Figures 15.7 and 15.8).

Additionally, histograms provide a clear picture of PAP titration pattern and patient's response to therapy. Figure 15.9 shows incorrect rapid CPAP pressure build up. The rapid increase in pressure increases the chance in developing central apneas. This patient developed central apneas (CPAP emergent central apneas) associated with desaturation.

Hypnogram should also be seen while preparing report. Insomnia is characterized by delayed sleep onset or poor maintenance of sleep (Figure 15.10).

REVIEW QUESTIONS

1. Histogram is important as it:
   A. Provides a comprehensive view of the data
   B. Helps in assessing the severity of sleep disordered breathing
   C. Helps in diagnosing hypersomnia
   D. Provides the guidance to the management

2. Sleep stage related sleep apnea can best be viewed in:
   A. Hypnogram
   B. Histogram
   C. Tabular depiction of data
   D. Raw data

3. To understand the effect of body position on various parameters:
   A. Look at the raw data
   B. Look at the video data
   C. Look at the tabulated data
   D. Look at the histogram

4. Differential diagnosis for insomnia in hypnogram is:
   A. Reverse first night effect
   B. Second night effect
   C. First night effect
   D. Reverse second night effect

5. To explain to the patient regarding the illness, especially sleep disordered breathing use:
   A. Raw data
   B. Tabulated data
   C. Hypnogram
   D. Histogram

ANSWER KEY

1. A   2. B   3. D   4. C   5. D

16
POLYSOMNOGRAPHY IN CHILDREN: SCORING RULES

LEARNING OBJECTIVES

After reading this chapter, the reader should be able to:

1. Discuss changes in EEG seen during sleep among children.
2. Correlate EEG changes with the age among children.
3. Discuss the rules for respiratory scoring among children.

CONTENTS

16.1 Respiratory Events 408

Further Reading 415

Review Questions 415

Answer Key 416

Children are different from adults in terms of neuronal development. Myelination of brain completes by adolescence, hence, their polysomnographic recording is not similar to that of adults. For this reason, different rules are followed for scoring of sleep data among children. These rules apply to children aging from 2 months post-term to adolescents.

Vertex sharp waves, similar to adults in morphology, first appear at the age of 16 months post-term. Sleep spindles may be seen at the age of 4–6 weeks post-term and are usually seen in post-term 2–3 months old infants, but are often asynchronous over hemispheres. K complexes usually develop at the age of post-term 4–6 months. Slow waves usually appear by the age of post term 2 months and are well-developed at the post-term age of 4–5 months.

N1, N2, and N3 can be diagnosed at the age of post-term 4 months. Non-EEG signals, for example, irregular respiration, atonia, rapid eye movements and transient muscle activity may help to differentiate between REM and NREM sleep. Similarly, posterior dominant activity slowly develops, 3.5–4.5 Hz at the age of 3–4 months post-term, 5–6 Hz by 5–6 months and 7.5–9.5 Hz by 3 years of age.

Due to their small face, chin EMG electrodes in children are placed at a distance of 1 cm and EOG electrodes at a distance of 0.5 cm.

Among children, sleep stages are divided as adults, that is, awake, N1, N2, N3 and REM in addition to N (NREM) stage. N is scored when all epochs of NREM sleep do not contain characteristic waveforms, that is, K-complex, spindle, or slow waves. However, epochs that contain K-complex are scored as N2 and epochs that show greater than 20% of slow-wave activity are scored as N3. The rest of the epochs are scored as N.

Wakefulness is scored when more than 50% of the epoch contains age-appropriate posterior dominant rhythm (PDR) (Table 16.1). PDR is usually intermixed with posterior slow waves of youth (2.5–4.5 Hz bilaterally present slow waves that block with eye opening and disappear with sleep seen between 8–14 years of age) or random occipital slowing (<100 µV, 2.5–4.5 Hz activity lasting less than 3 seconds seen between 1–15 years age). However, if alpha reactive to eye opening or age-appropriate PDR is not seen, eye blinks, reading eye movements and high chin tone can be used to score wakefulness.

N1 is scored when PDR is attenuated by low-amplitude-mixed-frequency activity in more than half of the epoch. However, where PDR is not seen, any of the following is sufficient to score N1: 4–7 Hz activity, which represents 1–2 Hz slowing compared to wake activity, slow eye movements, vertex sharp waves, rhythmic anterior theta (5–7 Hz theta predominantly seen in frontal and/or central regions, appears at 5 years

Table 16.1 Development of EEG Waveforms Across Age in Children

	<3–4 months	3–4 months	5–6 months			36 months		
Posterior Dominant Rhythm during wake	Slow irregular potential changes	Irregular 50–100 μV, 3.5–4.5 Hz activity reactive to eye opening	50–110 μV, 5–6 Hz activity			>8 Hz activity		
Drowsiness		8–36		6–8 months	8–36 months		>3 years	
				Diffuse, high amplitude 3–5 Hz and slower by 1–2 Hz than waking background activity	Diffuse or burst activity 75–200 μV, 3–4 Hz over occipital regions and > 200 μV, 4–6 Hz theta in frontal and/or central regions		1–2 Hz slowing of PDR or PDR become low voltage mixed frequency activity	
N1				6–8 months		16 months		5 years-Adults
				Paroxysmal run or burst bilaterally synchronized activity 75–350 μV, 3.5–4.5 Hz waves maximum over frontal and or central derivations—hypnogogic hypersynchrony		Appearance of vertex sharp waves similar to that of adults in morphology		Rhythmic anterior theta 5–7 Hz theta in frontal derivations
N2	8–9 weeks		5–6 months					
	Well-developed sleep spindles		Well developed K-complexes					
N3		3–4 months						
		Slow wave activity 0.5–2 Hz 100–400 μV in frontal derivations						
REM			5 months		9 months		1–5 years	5 years-onwards
			4-5 Hz with sawtooth waves		4–6 Hz		5-7 Hz theta	Similar to adults

of age and continues till adulthood), diffuse or occipital predominant 3–5 Hz activity, or hypnagogic hypersynchrony (Figure 16.1). N2, N3, and REM are scored as in adults though waves gradually develop with age. In infants < 3 months post-term age, REM appears at sleep onset rather than N1.

16.1 RESPIRATORY EVENTS

Obstructive apnea is scored as an apnea that meets the duration of at least two normal breaths (duration calculated by the duration of two normal breaths) with more than 90% fall in signal amplitude for the duration of the event; it is associated with continued respiratory efforts during the entire period of the event. Starting and terminal points of the event are scored as in adults.

Mixed apnea is scored when the initial portion of the event is characterized by a loss of respiratory effort but it is resumed before the end of the event.

Central apnea is diagnosed when the absence of airflow is associated with an absent respiratory effort of at least 20 seconds duration. If the duration of the event is less, then it should last the duration of at least two normal breaths, and should be associated with either an arousal or an awakening, or at least 3% oxygen desaturation to be scored as central apnea.

Classification of apneas among children into subtypes should be avoided in the absence of quantitative assessment of respiratory efforts through either esophageal balloon or calibrated respiratory inductance plethysmography.

Hypopnea is cored when there is at least 30% fall in amplitude of airflow signals for the duration of at least two breaths and is associated with either an arousal (Figure 16.2A–C) or at least 3% oxygen desaturation (Figure 16.3).

Respiratory effort related arousal (RERA) is scored when there is an increased respiratory effort or there is flattening of the nasal pressure waveform for the duration of at least two normal breaths or snoring, or an increase in $EtCO_2$ or $TransCO_2$ from the pre-event baseline. If the esophageal manometry is used, it should show progressively increased respiratory effort for the duration of at least two normal breaths and is associated with snoring, noisy breathing, increased respiratory efforts, or increased $EtCO_2$ or $TransCO_2$.

Polysomnography in Children: Scoring Rules

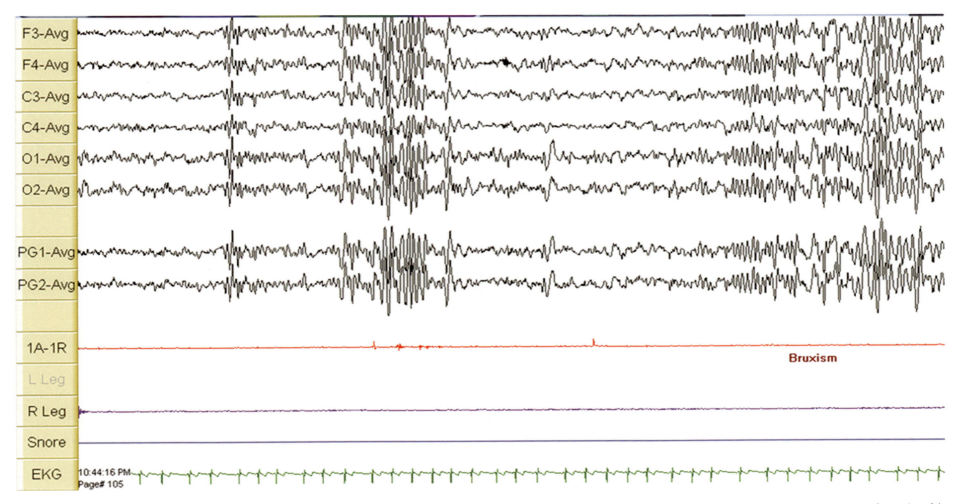

Figure 16.1A Hypnogogic hypersynchrony is paroxysmal EEG activity of 75–350 µV, 3–4.5 Hz in frequency bisynchronous activity that is usually seen in frontal and/or central derivations (30 sec epoch).

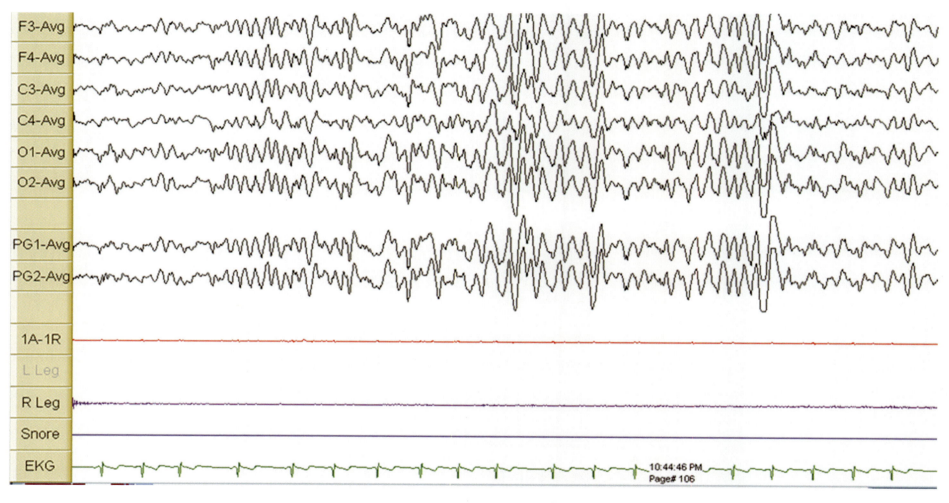

Figure 16.1B Hypnogogic hypersynchrony is paroxysmal EEG activity of 75–350 µV, 3–4.5 Hz in frequency bisynchronous activity that is usually seen in frontal and/or central derivations (15 sec epoch).

Polysomnography in Children: Scoring Rules

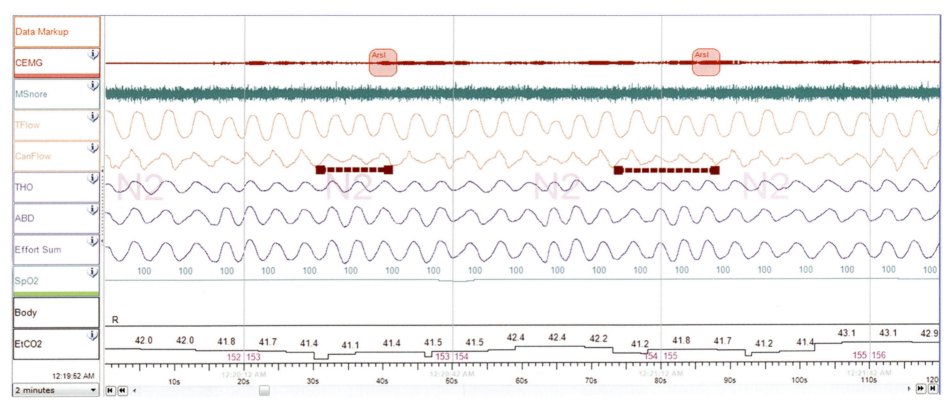

Figure 16.2A Hypopnea among children can be diagnosed when with 30% reduction in amplitude of airflow for the duration of at least two normal breaths is associated with an arousal.

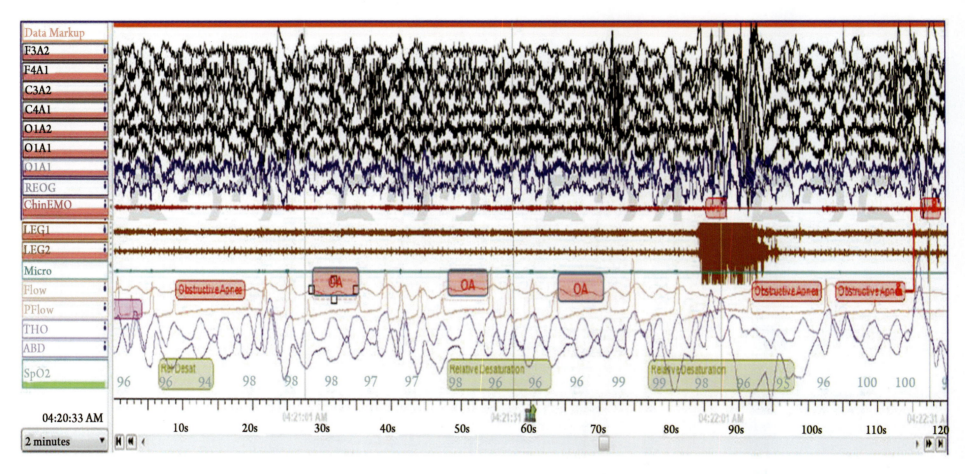

Figure 16.2B Hypopnea among children can be diagnosed when with 30% reduction in amplitude of airflow is associated with an arousal.

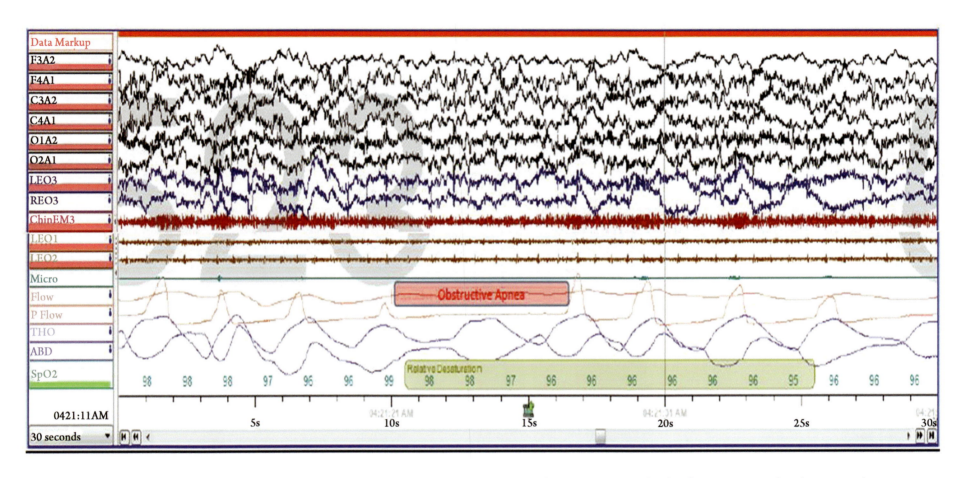

Figure 16.2C Hypopnea among children can be diagnosed when with 30% reduction in amplitude of airflow is associated with an arousal.

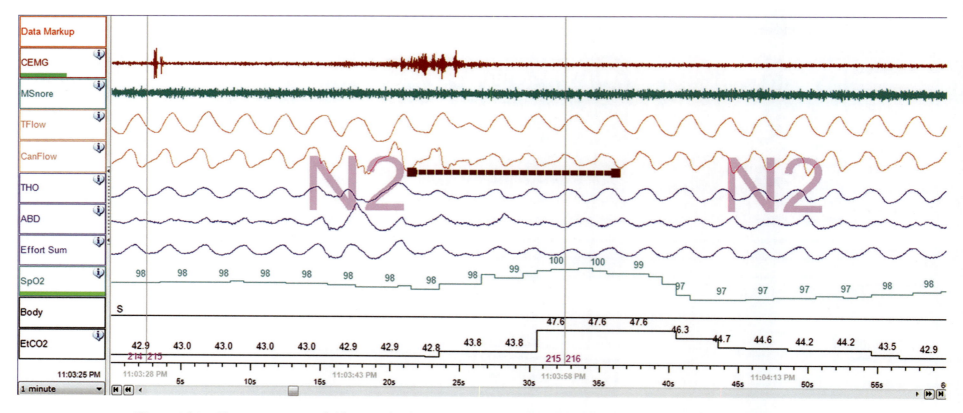

Figure 16.3 Hypopnea among children can be diagnosed with 30% reduction in amplitude and at least 3% oxygen desaturation.

FURTHER READING

1. Berry, R. B., Brooks, R., Gamaldo, C. E., Harding, S. M., Lloyd, R. M., Marcus, C. L., et al., (2017). The AASM manual for scoring of sleep and associated events: rules, terminology and technical specifications. Version 2.4. www.aasmnet.org. Darian, Illinois: American Academy of Sleep Medicine.

REVIEW QUESTIONS

1. Duration of a respiratory event to be scores as obstructive apnea among children is:
   A. 5 seconds
   B. 10 seconds
   C. Duration equal to 2 normal breaths
   D. Duration equal to 4 normal breaths

2. Scoring of central apnea is different among children as all EXCEPT:
   A. Duration is longer than required in adults
   B. If duration is less than 20 seconds, then associated desaturation or arousal is required
   C. May be scored based upon the change in heart rate associated with event, if the duration is short
   D. Presence of hypercapnea is required

3. In case oro-nasal thermal sensor goes dysfunctional, alternate acceptable sensor to measure airflow among children is:
   A. Oximetry
   B. End tidal CO_2
   C. Piezoelectric effort belts
   D. Heart rate

4. Periodic breathing is scored among children when:
   A. >2 central apneas each event lasting at least 2 seconds and interevent period is >20 seconds
   B. >3 central apneas each event lasting at least 5 seconds and interevent period is <20 seconds
   C. >3 central apneas each event lasting at least 3 seconds and interevent period is ≤20 seconds
   D. >5 central apneas each event lasting at least 10 seconds and inter-event period is <40 seconds

5. Which of the following is important for assessing the normalcy of EEG among infants:
   A. Conceptional age
   B. Gestational age
   C. Chronological age
   D. Legal age

6. Differentiation between NREM sleep stages is difficult:

   A. Till conceptional age of 52 weeks

   B. Till chronological age of 58 weeks

   C. Till conceptional age of 48 weeks

   D. Till chronological age of 37 weeks

7. Trace alternance is seen among:

   A. Elderly

   B. Adults

   C. Children

   D. Infants

8. Rapid eye movements are required to score REM among:

   A. Children

   B. Infants

   C. Adults

   D. Elderly

ANSWER KEY

1. C 2. D 3. B 4. C 5. A 6. C

7. D 8. B

17
TEST PROTOCOLS

LEARNING OBJECTIVES

After reading this chapter, the reader should be able to perform sleep studies as per protocols to ensure recording of good quality data.

CONTENTS

17.1	Protocol for the Diagnostic Sleep Study	418
17.2	Protocol for the Manual Titration with PAP	418
17.3	Protocol for the Multiple Sleep Latency Test (MSLT)	419
17.4	Protocol for the Maintenance of Wakefulness Test (MWT)	420
Further Reading		421
Review Questions		421
Answer Key		422

With the help of polysomnography, we perform a number of tests. Tests may be performed at night, for example, diagnostic polysomnography and manual titration with PAP. In some cases of severe OSA, split-night polysomnography is done where first half of the night is dedicated to the diagnostic study and the next half of the night is dedicated to the manual titration with PAP. In addition, in certain circumstances, to assess the somnolence, multiple sleep latency tests (MSLTs) or maintenance of wakefulness test (MWT) is performed during the day. In this chapter, we will present the protocols for these tests.

17.1 PROTOCOL FOR THE DIAGNOSTIC SLEEP STUDY

1. The patient should be given time to get acclimatized to the room. Hence, it is better that he/she is called at least 6–8 hours before his/her usual bed time.
2. Request the patient to fill the details in the "Pre-sleep questionnaire."
3. Make sure that the scalp and face of the patient are free of any grease/oil.
4. The patient's clothes must be comfortable and should not be synthetic.
5. The temperature of the room should be kept at around 22°C.
6. The patient should have taken all the prescribed medications
7. Start the preparation at least 40–60 min before the patient's usual bedtime, as preparation may take around 20–30 min. Any deviation from this will cause spuriously long or short sleep onset latency. This may be checked by looking at the data from the patient's sleep diary
8. Apply all the channels on the patient's body, as discussed in Chapter 5.
9. Fill all the details as requested by the software.
10. Check for the impedance in the electrical channels and try to keep it below 5 Kilo Ohms.
11. Do the electrical calibration and follow it by bio-calibration. Make sure that all the channels are working properly.
12. If any channel is malfunctioning, try to rectify it before the start of the study.
13. After calibration, let the patient sleep. Reassure him/her that you are watching his/her output and you are available in case of any emergency.
14. The patient should be woken up at his/her usual wake time in the morning.
15. In the morning patient should be given the post-diagnostic study questionnaire.

17.2 PROTOCOL FOR THE MANUAL TITRATION WITH PAP

1. It should be done during the night after the diagnostic polysomnography.

2. The patient should be informed about the results of the diagnostic study and should be given time to acclimatize with the mask.

3. For acclimatization, show the patient the PAP and mask during the day. Explain how it works and give him/her the mask during the day before the titration study.

4. Ask him/her to hold the mask in front of the nose and breathe through it. For this, the patient should be allowed to hold the mask in hand (this should not be strapped at once). Request him/her to practice it by keeping it in position for longer periods during the day.

5. Follow the steps 2–12 as mentioned in the "diagnostic study" section above. However, during titration, nasal cannula and thermistor should not be placed as it may lead to leakage. Instead, flow signals are acquired from the mask.

6. Raise the pressure of PAP, as discussed in Chapter 20.

7. The patient should be requested to fill the details in the "post-CPAP night questionnaire."

17.3 PROTOCOL FOR THE MULTIPLE SLEEP LATENCY TEST (MSLT)

1. This should be done after the diagnostic study. During the diagnostic study, sleep apnea and any other cause that may lead to sleep disruption should be ruled out.

2. Total sleep time during the night should be at least 6 hours.

3. Get the sleep log for at least one week prior to the MSLT.

4. Stimulant medications or medications that suppress REM should be withdrawn at least two weeks ahead of the test. Physician consultation should be done to avoid the influence of hypnotic medication on the day of the test. Vigorous physical activity, exposure to bright light, smoking, caffeine should be avoided on the day of the test as they may be counterproductive to the sleep. Any other measure to promote wakefulness should be terminated at least 15 min prior to every nap opportunity. Drug screening may be performed in selected cases.

5. The test is started in the morning at least two hours after waking up after a light breakfast.

6. Five opportunities for sleep are provided, 2 hours apart during the day.

7. Sleep lab should be sufficiently dark and quit to promote the sleep. The temperature of the lab should be optimal.
8. Make sure that the scalp and face of the patient are free of any grease/oil.
9. The patient's clothes must be comfortable and should not be synthetic.
10. Make sure the patient's sleep is not disrupted because of any other reason, for example, going to the washroom.
11. Apply the EEG electrodes, EOG electrodes, chin EMG, and EKG for the test.
12. Perform bio-calibration to assess alpha waves in the EEG, eye movement, and chin EMG. Preferably, this should be done in every nap.
13. Instruct the patient "Please close your eyes and try to fall asleep" before the nap and request him/her to lie down and be comfortable in bed.
14. Recording should be done for 20 min. If the patient does not fall asleep during 20 min, record sleep latency as 20 min.
15. If the patient falls asleep during this 20 min, from the first epoch-defining sleep, another 15 min should be provided for REM sleep to appear. For example, if the epoch-defining sleep is seen at the 10th min, the test should be terminated at the 25th min (10 min + 15 min) instead of the 20th min.
16. After each nap, the patient should be asked and information should be filled in the "MSLT questionnaire."
17. Light lunch should be provided during noon.
18. Report "start and stop time," "sleep onset latency from lights out," for each recording, "mean sleep latency," and "number of sleep onset REM (SOREM) out of five naps."

SOREM is defined as REM appearing within 15 min of the sleep onset during a recording.

17.4 PROTOCOL FOR THE MAINTENANCE OF WAKEFULNESS TEST (MWT)

1. The test should be started at least 2–3 hours after the night nap.
2. Diagnostic polysomnography in the preceding night and sleep logs are not mandatory before MWT. They may be conducted using clinical judgment.
3. Light breakfast should be provided before the first recording.
4. Sleep lab should be sufficiently dark and quit to promote sleep. The temperature of the lab should be optimal.
5. Make sure that the scalp and face of the patient are free of any grease/oil.

6. The patient's clothes must be comfortable and should not be synthetic.
7. Make sure the patient's sleep is not disrupted because of any other reason, for example, going to the washroom.
8. The patient is not allowed to take any stimulant unless approved by the sleep physician. They are instructed to avoid using any other measure to stay awake, e.g., singing, etc. A drug screen may be performed in selected cases.
9. The patient should be made to sit in a reclining position in bed.
10. Apply the EEG electrodes, EOG electrodes, chin EMG, and EKG for the test.
11. Perform bio-calibration to assess alpha waves in the EEG, eye movement, and chin EMG. Preferably, this should be done before every recording.
12. Instruct the patient "Please sit still and look in front of you. Try to stay awake as long as possible," before starting the recording.
13. Four recordings are done, each lasting 40 min (if no sleep occurs) and two hours apart.
14. A recording may be terminated earlier than 40 min if three consecutive epochs of N1 sleep are seen or one epoch of any other stage of sleep.
15. Light lunch should be provided during noon.
16. Report "start and stop time," "sleep onset latency from lights out," for each recording, and "mean sleep latency," "total sleep time," and "sleep stages recorded during each trial."

FURTHER READING

1. Littner, M. R., Kushida, C., Wise, M., Davila, D. G., Morgenthaler, T., Lee-Chiong, T., et al., (2005). Standards of Practice Committee of the American Academy of Sleep Medicine. Practice parameters for clinical use of the multiple sleep latency test and the maintenance of wakefulness test. *Sleep 28*(1), 113–121.

REVIEW QUESTIONS

1. Subjective assessment of sleep is important after diagnostic polysomnography as it:
   A. Gives the idea of underlying illness
   B. Adds to the scoring of data
   C. Provides the clue to the missed pathology, if the illness is mild
   D. Has been shown to improve over time

2. Titration of PAP should optimally be manual because of all EXCEPT:

   A. It provides lower pressure

   B. Leak if large, may be managed for optimal pressure

   C. Helps in removing all sleep related breathing event

   D. Gives an idea about mask tolerance

3. MSLT should be started:

   A. Earliest in the morning

   B. After two hours of waking up

   C. Late in the noon

   D. Late in the evening

4. Best measure to rule out the excessive daytime sleepiness is:

   A. MSLT

   B. Daytime diagnostic sleep study

   C. Nighttime diagnostic sleep study

   D. MWT

5. To overcome the issue of mask intolerance:

   A. Always use full face mask

   B. Always use nasal mask

   C. Leave the choice to the patient

   D. Always use nasal pillows

ANSWER KEY

1. C   2. A   3. B   4. D   5. C

18

DOCUMENTATION

LEARNING OBJECTIVES

After reading this chapter, the reader should be able to:

1. Recognize the importance of proper documentation during sleep study monitoring.
2. Enumerate names of records to be maintained in a sleep laboratory.
3. Design their own document for recording the information.
4. Discuss the importance of maintaining these records.

CONTENTS

Various Documents and Questionnaires 425

Review Questions 430

Answer Key 430

One of the essential parts of every sleep study is the proper documentation by sleep technologists. A sleep technologist's responsibility is to maintain a good quality sleep study that will assist the treating physician to reach accurate diagnosis. Included in the technologist's documentation are all the events especially unusual sleep-related events that were noted during the PSG, as well as, any medication taken by or given to the patient during the time of the study. Table 18.1 demonstrates examples of subjective and objective events that need documentation during PSG.

Here, we are presenting with some questionnaires that will help you in your clinical practice. As the name suggests, these are to be distributed to the patient. After getting them filled, please supply them to the Sleep Physician. If you have any significant observation, please mention it in the provided space (Tables 18.2–18.6).

Table 18.1 Examples of the Subjective and Objective Events that Need Documentation During PSG

Subjective	Objective
Patient's perception regarding sleep quality during the study.	Sleep-wake patterns of patient's sleep.
Any unusual complaints, i.e., a headache, pain, nightmares, etc.	Unusual events, i.e., short REM sleep latencies, leg movements, hypopneas, apneas, sleep walking, parasomnias, sleep talking, etc.
The subjective feeling when a PAP device titration was implemented.	Artifacts
Patient's perception of PAP device therapy.	Response to PAP device titration

Documentation

Table 18.2 Pre-Sleep Questionnaire

	Name:	**Gender: M/F**
	Age: ……… years	UHID: …………
	Date of testing: …………	
1.	Do you take any medication regularly?	Yes/No
2.	If yes, please mention their names below: ……………………………………………………… ………………………………………………………	
3.	Have you taken them today?	Yes/No
4.	If not, why? A. Suggested by Sleep Physician B. Not available with me	
5.	Have you taken a nap today?	Yes/No
6.	If yes: How many hours? At what time?	………… Hours Between ………… and …………
7.	Do you regularly take alcohol?	Yes/No
8.	If yes, when did you last take it?	…………………….
9.	Do you smoke/chew tobacco?	Yes/No
10.	If yes, when did you last take it?	…………………….
11.	Have you prepared yourself for the sleep study as per the instructions are given in flyer?	Yes/No
12.	If no, why? ……………………………………………………… ………………………………………………………	
13.	Are you feeling comfortable here?	Yes/No
14.	If no, why? ………………………………………………………	
15.	How was your sleep during past week?	Good Bad Too bad
If you think that you have any concerns regarding the test, please feel free to discuss it with Mr./Ms.……………… who is a Sleep technician.		

Table 18.3 Post-Diagnostic Night Questionnaire

	Name:	**Gender: M/F**
	Age: … … years	**UHID:** … … … …
	Date of testing: … … … …	
1.	How was your sleep last night?	Better than usual As usual Worse Too bad
2.	If it was not good, what was the reason? … … … …	
3.	Have you had the problem that you have come for, last night as well?	Yes/No
4.	At what time did you go to bed?	… … … PM
5.	How much time did it take to fall asleep?	… … … min
6.	Did you wake up at night?	Yes/No
7.	If yes, how many times?	… … …
8.	Were you able to fall asleep easily after awakenings?	Yes/No
9.	If not, why? … … … … … … … … … … … … … … … …	
10.	Overall, how do you rate your experience in a sleep lab on a scale of 0–10? 0 being worst and 10 being satisfying	… … … … … …
If you think that you have any concerns regarding the test, please feel free to discuss it with Mr./Ms. … … … … … … … who is a Sleep technician.		

Documentation

Table 18.4 Post CPAP Night Questionnaire

	Name:	**Gender: M/F**
	Age: …… years	UHID: …………
	Date of testing: …………	
1.	How was your sleep last night?	Better than usual As usual Worse Too bad
2.	If it was not good, what was the reason? ………………………………………………… …………………………………………………	
3.	How did you feel with the PAP at night? ………………………………………………… …………………………………………………	
4.	How are feeling today?	Better than usual As usual Worse Too bad
5.	How much time did it take to fall asleep?	……… min
6.	Did you wake up at night?	Yes/No
7.	If yes, how many times?	………
8.	Were you able to fall asleep easily after awakenings?	Yes/No
9.	If not, why? ………………………………………………… …………………………………………………	
10.	Overall, how do you rate your experience with PAP on a scale of 0–10? 0 being worst and 10 being satisfying	………………
11.	Would you consider using CPAP at home?	Yes/No
12.	If not, why? ………………………………………………… …………………………………………………	
If you think that you have any concerns regarding the test, please feel free to discuss it with Mr./Ms.…………… who is a sleep technician.		

Table 18.5 MSLT Questionnaire

	Name:	Gender: M/F				
	Age: …………years	UHID: ………				
	Date of testing: ………					
	Item	Nap 1	Nap 2	Nap 3	Nap 4	Nap 5
1.	At what time was the test started?	…………	…………	…………	…………	…………
2.	At what time did the test finish?	……	……	……	……	……
3.	Did you fall asleep during this test?	Yes/No	Yes/No	Yes/No	Yes/No	Yes/No
4.	If yes, how much time did it take to fall asleep?	………min	………min	………min	………min	………min
5.	Did you dream during the nap?	Yes/No	Yes/No	Yes/No	Yes/No	Yes/No
6.	Did you fall asleep between two tests?	Yes/No	Yes/No	Yes/No	Yes/No	Yes/No
7.	If yes, for how long?	………min	………min	………min	………min	………min
8.	Did you dream during that nap?	Yes/No	Yes/No	Yes/No	Yes/No	Yes/No

Documentation

Table 18.6 PAP Questionnaire (For Technician)

			Please fill the details after every half an hour				
	Name:	**Gender: M/F**					
	Age: …… years	UHID:… … … …					
	Date of testing: … … …						
Time	PAP pressure (cm H_2O)		Leak (L/min)	Apneas/Hypopneas observed (Yes/ No)	Oxygen rate	Sleep stage	Body position
	EPAP	IPAP					

REVIEW QUESTIONS

1. Documentation in the sleep laboratory is important as it:
   A. helps the scorer in improving the scoring
   B. provides additional information for the patient
   C. provides additional information that helps in scoring of data and management of patient
   D. is done customarily, so it should be done

2. Information after the PAP night will help the physician:
   A. to manage the pressure of PAP
   B. to address PAP compliance issues
   C. to diagnose the problem
   D. to include the patient in management

3. MSLT questionnaire provides:
   A. objective record of sleep
   B. subjective record of REM sleep
   C. objective record of REM sleep
   D. subjective record of sleep

4. Post diagnostic night questionnaire helps in:
   A. analyzing the subjective sleep quality
   B. analyzing the objective sleep quality
   C. analyzing the subjective sleep quantity
   D. analyzing the objective sleep quantity

5. PAP questionnaire for technician helps the:
   A. technician to increase the pressure after the test
   B. sleep physician to decide the pressure after the test
   C. technician to address the pressure and leak issues during test
   D. sleep physician to address the pressure and leak issues during test

ANSWER KEY

1. C   2. B   3. B   4. A   5. C

19

TROUBLESHOOTING

LEARNING OBJECTIVES

After reading this chapter, the reader should be able to:

1. Identify and troubleshoot common problems encountered during recording of polysomnography.

CONTENTS

Steps at Troubleshooting 431

Review Questions 436

Answer Key 437

Polysomnography equipment consists of a software and hardware. Unlike in any other equipment, in this case, different manufacturers may develop hardware and software. In this case, it becomes prudent that the hardware can support the platform that you are loading in the computer system to run other software (e.g., Windows/Linux/Mac-OS). Similarly, the platform (depends upon which version of the platform you are using) of the computer system must also support the software provided by the polysomnography manufacturer. If any of these parts are incompatible, the system will not work properly. This should be taken care of by the sleep technologist at the time of purchasing the instrument. At times, the antivirus software of the computer system can also interfere with the functioning of the polysomnography software. In such cases, the PSG software does not open, or takes a long time while opening or may automatically shut down. In such cases, a trial may be given to turn off the antivirus software.

Sensors placed on the body are connected to the head-box; head-box is connected through cable to the base system, which, in turn, feeds the signals to the computer system. If a sensor is not generating signals (improper placement, high impedance, battery dead if it requires battery power for signal generation), or if any of the connections is loose or compromised in any other way (short-circuit, generation of local electrical potential), an interference with either signal acquisition or signal quality will appear. These communication cables must be checked time-to-time.

Troubleshooting of the artifacts has been discussed in Chapter 12. Similarly, all manufacturers provide the common trouble-shoots in their manual. The reader is advised to consult those manuals for issues that may be specific to a particular machine. In this chapter, we will discuss other issues related to in-lab attended polysomnography.

1. Some of the EEG derivations are showing dangling waves.

    This may be seen when the EEG electrodes are in poor contact with the skin. To resolve this issue, look if:

- All derivations with poor waveform have a common reference
- All derivations showing poor waveform are from one side of the scalp (right or left)
- Derivations with poor waveform don't have anything in common.

If the derivations with poor waveform have common reference (M1 or M2), check if the reference electrode is properly placed and its impedance is within normal limits. If there is a problem, fix it.

If derivations with poor waveform belong to one side of scalp, see if their contact with skin is improper. It may happen when patient takes a turn in the bed during sleep. Fix them. If these derivations belong to the side of scalp/head that is resting on pillow, it may be possible that the patient is

Troubleshooting

Figure 19.1 Software and hardware required in a polysomnography laboratory.

sweating from that side. Adjust the environmental temperature, if necessary.

If derivations don't have anything in common, see if the scalp has been cleaned properly before hooking up; if in doubt, remove electrode, clean the scalp, and place the electrode again. With repeated use, electrodes also loose the conductive material and impedance goes high. If cleaning of scalp does not work, change the electrodes.

2. Waveform is attenuated in one of the EOG channels.

    It usually occurs when the impedance of the electrode is high or the eye is not able to generate the electrical potential. To sort out this issue,

    - Ask if the patient has an artificial eye on that side (a rare possibility).

- See if the filter and sensitivity setting of that channel is optimum.
- Remove the electrode and place it again after cleaning the underlying area.
- If this does not work, consider changing the electrode with a new one.

3. Signals from chin EMG are distorted. Common issues that lead to distorted chin EMG signals include high impedance between the skin and electrode. This is especially common among patients who wear a beard. If the patient allows, an area large enough to place the EMG electrodes may be cleaned using curved scissors, and electrodes may be placed after that.

4. Flow signals are not apparent during diagnostic study.

   Check whether signals are missing from thermistor or pressure transducer or both.

   If they are missing from both sensors, make sure that:
   - Both sensors are connected to the right slots in the head box.
   - Patient is a nose breather and nasal airway is patent.
   - If the patient is nose breather, sensors are placed correctly.
   - If sensors are correctly placed but signals generation require some electrical current, make sure the battery is working.
   - If this is ok, look for the sensitivity and filter setting in the software. If it is not set as depicted in the AASM manual, change the settings.

5. Flow signals are not appearing during the titration study.

   During the titration study, flow signals are generated from the PAP device. If they are not apparent, make sure that:
   - The PAP device is working.
   - Mask fitting is adequate and leakage is within the prescribed limits.
   - The PAP device is connected with the system, as depicted in the manual of PSG.
   - If this is ok, look for the sensitivity and filter setting in the software. If it is not set as depicted in the AASM manual, change the settings.

6. Signals from chest and/or abdominal belts are not appearing.

   Please check if:
   - Belts are loose or too tight. Ideally, an un-stretched RIP belt should

cover around two-third of the chest circumference.
- If the belts require battery to generate signals, it is not discharged. If there is a problem, fix it.
- Jack from the Z-RIP-module is placed properly in the head box. If not, fix it.
- Sensitivity and filter setting of the waveform in the software is correct. If not fix it.

7. $EtCO_2$ signals not coming up.

    Please check if:
    - Cannula is properly placed. If not, fix it.
    - Patient is a nose breather and nasal airway is patent. If the patient is mouth breather, you may switch to a cannula that has a port for oral breathing as well.
    - If you are using $EtCO_2$ signals, the cannula is free of moisture.
    - It is connected to the hardware properly.

8. Signals from the pulse oximeter are intermittent.

    This can happen if there is an interference with the transmission of light waves from one side to the other.

    In such case, please ensure that:
    - Probe is placed correctly. A loose probe may lead to intermittent loss of signals. Fix it.
    - Fingernail is completely clean. If some part of the fingernail is covered with any kind of opaque material, it may lead to intermittent loss of signals.
    - Flow of blood is not obliterated in the arm in which the sensor is placed. Intermittently, it may happen when the patient is using this arm as a pillow.

9. Mask leakage too high.

    Mask leakage is subjected to the poor fitting of the mask. Please ensure that an appropriate mask has been chosen depending upon the shape of the face and other factors.

    Poor mask fitting is seen when:
    - It is tied too tightly on the face. Tight fitting will lose the cushioning effect of the mask and promote leakage.
    - Mask is lying loose on the face.
    - The patient is not able to breathe through the mask as pressure is considered too high.
    - The patient is fiddling with the mask.
    - The patient has taken turn in bed and the mask has been drawn to one side of the face. In such case, place the PAP in the back of the bed and place the hose pipe loose and in such a manner that it

passes over the head. These maneuvers will reduce the dragging of the mask and improve its fitting.

- Cannula/thermistor is placed below the mask.

10. Study not downloading in the computer system.

    Some PSG machines store the data in a base station and transfer it to the computer system when you close the study, for example, Alice 6. This transfer may be hampered if the available disk space is inadequate. In such case, create some space in your computer system and then try to download the data from the base station again.

11. Video recording stops suddenly in between the study.

    This results from poor feed from the video or inadequate disk space in the computer system. Poor feed from the video may be related to interference in the electrical supply to camera or due to interference in the communication cable. At times, software that runs the video may also malfunction. Take the following steps:

    - Ensure that infrared lights are tuned on in the camera. For this, you may have to go to the patient's room. If lights are turned on, electrical supply is OK.
    - Look for the available disc space. If space is low, waveforms may continue to be recorded, but not the video.

REVIEW QUESTIONS

1. If the signals in all EEG channels show electrical artifact, consider:

    A. Checking the reference electrodes
    B. Checking the ground electrode
    C. Checking the filter setting
    D. Checking the individual electrode

2. If the flow signals from PAP show flattening after optimal titration:

    A. Look for the ground electrode
    B. Look for the oxygen saturation
    C. Look for the leakage
    D. Look for the respiratory rate

3. If the signals from effort RIP Chest belt are present intermittently, but from abdomen regularly:

    A. Check the battery of RIP module
    B. Check if the chest belt is loose or twisted
    C. Check if the problem in software
    D. Check the filter settings

4. If one EEG derivation shows dangling waveform:
   A. Lower down the temperature of room
   B. Check the reference electrode
   C. Check the ground electrode
   D. Clean and replace the electrode that shows dangling waveform

5. If signals from the oximeter are appearing intermittently, then all of following are possible EXCEPT:
   A. Patient is sleeping on the arm where oximeter is placed
   B. Patient is not breathing regularly
   C. Oximeter is loose
   D. Oximeter wire is loosely fit with the headbox

ANSWER KEY

1. A   2. C   3. B   4. D   5. B

20
MANUAL TITRATION WITH POSITIVE AIRWAY PRESSURE

LEARNING OBJECTIVES

After reading this chapter, the reader must be able to:

1. Understand the steps of manual PAP titration among adults and children.
2. Select the proper mask for the patient.
3. Discuss and recognize the quality of titration study.

CONTENTS

20.1	Pressure Limits	440
20.2	CPAP Titration for OSA	440
20.3	When to Switch From CPAP to BPAP in Patients with OSA?	441
20.4	BPAP Titration for OSA	441
	Further Reading	444
	Review Questions	444
	Answer Key	445

Continuous positive airway pressure (CPAP) and bi-level positive airway pressure (BPAP) are the gold standard and the most effective treatment for OSA and other sleep-related breathing disorders. Non-invasively, positive airway pressure applied through a mask provides a continuous stream of air pressure through the nose and/or mouth. Positive air pressure prevents airway collapse, allowing the patient to breathe freely during sleep.

The mask provides a bridge between patient and device. It is an important component of a successful PAP therapy.

Usually, titration of the PAP device pressure is done gradually under PSG monitoring. Prior to PAP titration, all patients should receive adequate education about PAP, hands-on demonstration, careful mask fitting, and acclimatization prior to the process. The technician has to size and fit a patient with proper mask size. Some laboratories provide time for mask acclimatization to the patients. After selecting the correct mask, patients are given the mask in the morning before the titration study is planned, and they are asked to keep it on their nose and breathe normally. As they get comfortable with the mask, the mask is fixed with the strap on their hand, for progressively longer periods. Finally, have the patient try CPAP or BPAP for some time while awake, starting at a very low setting (4–5 cm H_2O) and titrate until 10 cm H_2O. This simple maneuver improves the compliance to titration at night.

PAP titration should aim for an AHI less than or equal to 5/hr, SpO2 >88%, respiratory arousal index less than or equal to 5/hr, and to eliminate snoring.

The AASM has published Clinical Guidelines for the Manual Titration of PAP in Patients with OSA. In the coming section, we will summarize those guidelines.

20.1 PRESSURE LIMITS

The recommended minimum starting CPAP should be 4 cm H_2O for pediatric and adult patients. On the other hand, the recommended maximum CPAP should be 15 cm H_2O for patients <12 years, and 20 cm H_2O for patients ≥12 years. PAP should be increased until the following obstructive respiratory events are eliminated (no specific order); apneas, hypopneas, respiratory effort-related arousals (RERAs), snoring, or the recommended maximum CPAP is reached.

20.2 CPAP TITRATION FOR OSA

- CPAP should be increased if at least 1 obstructive apnea is observed for patients <12 years or at least 2 obstructive apneas are observed for patients ≥12 years.

- CPAP should be increased if at least 1 hypopnea is observed for patients <12 years or at least 3 hypopneas are observed for patients ≥12 years.

- CPAP should be increased if at least 3 RERAs are observed for patients <12 years or at least 5 RERAs are observed for patients ≥12 years.
- CPAP should be increased if at least 1 minute of loud or unambiguous snoring is observed for patients <12 years or at least 3 minutes of loud or unambiguous snoring are observed for patients ≥12 years.
- Increase CPAP pressure by at least 1 cm H_2O with an interval no shorter than 5 minutes.
- Make sure that the increment in CPAP pressure is not done too fast, as this may result in central apneas (CPAP emergent apneas). Figure 20.1 shows a hypnogram of a patient with OSA. Rapid titration resulted in CPAP-emergent central apneas.
- If the patient awakens and complains that the pressure is too high, the pressure should be restarted at a lower pressure chosen as one that the patient reports is comfortable enough to allow are turn to sleep.

Table 20.1 summarizes the titration steps and indications.

Table 20.1 The Titration Steps and Indications

<12 years of age	≥12 years of age
≥1 OA	≥2 OA
≥1 Hypopnea	≥3 Hypopneas
≥3 RERAs	≥5 RERAs
≥1 minute of loud unambiguous snoring	≥3 minutes of loud unambiguous snoring

OA: obstructive apnea; RERAs: respiratory effort-related arousals.

20.3 WHEN TO SWITCH FROM CPAP TO BPAP IN PATIENTS WITH OSA?

- If there are continued obstructive respiratory events at 20 cm H_2O of CPAP during the titration study, the patient may be switched to BPAP. Figure 20.2 shows a hypnogram of a patient who continues to have obstructive hypopneas on CPAP pressure of 20 cm H_2O. Upon shifting the patient to BPAP, the obstructive events disappeared.
- If the patient with OSA is uncomfortable or intolerant of high pressures on CPAP, he/she may be tried on BPAP.

20.4 BPAP TITRATION FOR OSA

- The recommended minimum starting EPAP should be 4 cm H_2O for pediatric and adult patients.
- The recommended minimum starting IPAP should be 8 cm H_2O for pediatric and adult patients.

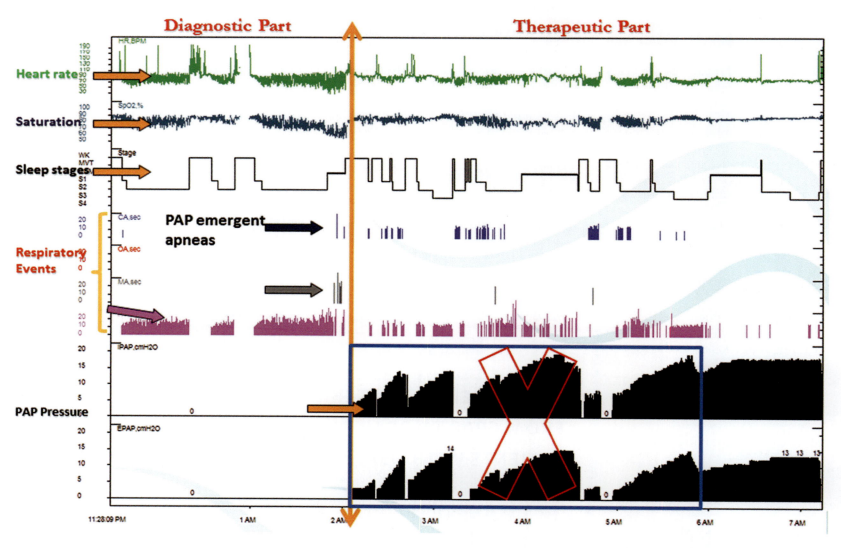

Figure 20.1 A hypnogram of a patient with OSA. Rapid titration resulted in PAP emergent central apneas.

Manual Titration with Positive Airway Pressure

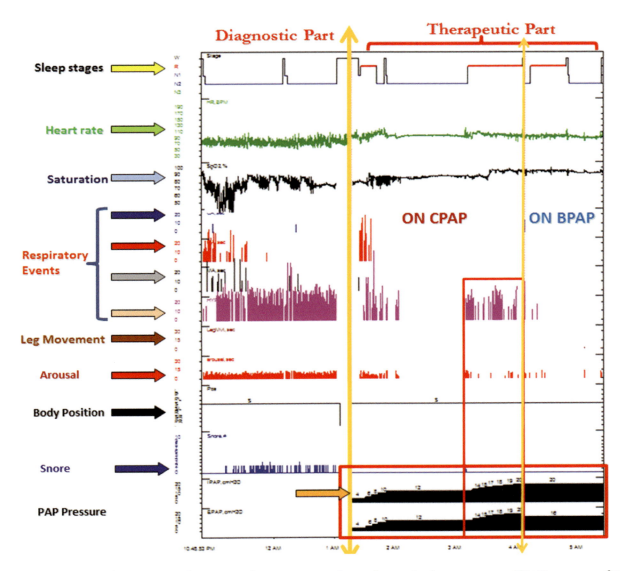

Figure 20.2 A hypnogram of a patient who continues to have obstructive hypopneas on CPAP pressure of 20 cm H_2O. Upon shifting the patient to BPAP, the obstructive events disappeared.

- The recommended maximum IPAP should be 20 cm H_2O for patients <12 years, and 30 cm H_2O for patients ≥12 years.
- The recommended minimum IPAP-EPAP differential (pressure support) is 4 cm H_2O.
- The recommended maximum IPAP-EPAP differential (pressure support) is 10 cm H_2O.
- Increase IPAP and EPAP pressure by at least 1 cm H_2O with an interval no shorter than 5 minutes.

The AASM has categorized the success of PAP titration into several categories:

- **Optimal titration:** It reduces the respiratory disturbance index (RDI) <5 for at least a 15-min duration and should include supine REM sleep at the selected pressure that is not continually interrupted by spontaneous arousals or awakenings (a consensus agreement).
- **Good titration:** It reduces RDI ≤10 or by 50% if the baseline RDI <15 and should include supine REM sleep that is not continually interrupted by spontaneous arousals or awakenings at the selected pressure (a consensus agreement).
- **Adequate titration:** It does not reduce the RDI ≤10 or one in which the titration grading criteria for optimal or good are met with the exception that supine REM sleep did not occur at the selected pressure (a consensus agreement).
- **Unacceptable titration:** It is one that does not meet any one of the above grades (a consensus agreement).
- **Repeat PAP titration study:** It should be considered if the initial titration does not achieve a grade of optimal or good and, if it is a split-night PSG study, it fails to meet the AASM criteria (i.e., titration duration should be >3 hr) (a consensus agreement).

FURTHER READING

1. Kushida, C. A., Chediak, A., Berry, R. B., Brown, L. K., Gozal, D., Iber, C., et al. (2008). Clinical guidelines for the manual titration of positive airway pressure in patients with obstructive sleep apnea. *Journal of Clinical Sleep Medicine. JCSM: official publication of the American Association of Sleep Technologists.* 4(2), 157.

REVIEW QUESTIONS

1. Titration with Auto-PAP is not recommended as:

A. It gives higher pressure

B. Algorithms of different PAPs are different

C. Optimal titration is difficult to achieve because of automatic change in pressure

D. Is not safe

2. BPAP is recommended when:

   A. Pressure tolerance is poor with CPAP

   B. More than 10 cm H_2O PAP is required

   C. Patient is having cardiac illness

   D. Patient had stroke earlier

3. During PAP titration, if pressure is increased at a faster pace:

   A. CSB appears

   B. OSA worsens

   C. Mixed apnea increases

   D. Treatment emergent central sleep apnea may occur

4. During PAP titration, starting pressure is:

   A. 4 cm H_2O always

   B. May be chosen according to patient's comfort and severity of illness

   C. Always close to 10 cm H_2O

   D. Randomly chosen

5. How many RERAs should appear in children <12 years before increasing the pressure by 1 cm H_2O:

   A. 1

   B. 2

   C. 3

   D. 4

ANSWER KEY

1. C   2. A   3. D   4. B   5. C

21
WRITING AN INFORMATIVE REPORT

LEARNING OBJECTIVES

After reading this chapter, the reader must be able to write an informative report that can be understood by any medical person.

CONTENTS

Contents of an Informative PSG Report 447

Review Questions 450

Answer Key 451

The organization of a sleep study report is dependent on the needs of the end user. Reports do come with superfluous amounts of data, which, fortunately, can be customized according to the needs of the users. For in-depth analysis and research purposes, a report can contain multiple tables, statistics, graphs, and charts. On the other hand, if the report is mainly for clinical use, it usually includes the patient's demographic data, main findings of the sleep study, sleep study hypnogram that reflects the summary of the sleep study, and the recommendations of the treating physician. Important sleep parameters that must appear in the report include a summary of sleep architecture, respiratory events, oxygen saturation levels, limb movements, arousals, and heart rate. If a PAP device was used, the report should summarize the subjective and objective response to therapy and the optimal setting of the device. The sleep technologists' comments are important and can add further insight into the study report as they stay around the patient during monitoring. Examples of important technologists' comments include an account of the patient's physical and emotional status during the study, unusual sleep-related behavior or atypical findings that are not evident in the monitored sleep parameters.

While writing the report, it must be considered that a physician who does not specialize in sleep medicine is able to understand the interpretation of the report. The physician should be able to provide a clear opinion and modify his/her own treatment, if required. Sleep study report generated in a standard performa often has multiple tables and a page that shows the graphical representation of data (Figure 21.1). Based on observations at night and data available in a tabulated and graphical format, one comprehensive summary should be provided that clearly indicates your opinion.

A good report would be like this:

The patient was referred for the diagnostic sleep study with the suspicion of obstructive sleep apnea due to his long history of systemic hypertension. The recording included three EEG channels—Frontal, Central and Occipital, which were referred to the contralateral mastoid. Two channels from EOG and two channels from the submentalis were included. Respiratory effort was measured using RIP belts from the thorax, abdomen, and RIP-sum. The respiratory flow was recorded using a thermistor and nasal cannula and overnight oximetry by oximeter placed at the right index finger with three seconds averaged value. Snore microphone was placed to diagnose snoring. $EtCO_2$ was recorded using a capnograph. Body position sensor was used to ascertain body position and leg EMG was recorded from two channels from anterior tibialis. Synchronized video was also recorded.

The patient was explained about the overnight sleep study in advance and was given time to get acclimatized to the sleep lab. The recording was started at his usual bed-time (or earlier or later to that) and terminated at his usual wake time (or earlier or later to that). The patient took (or was asked not to take) his medications at his usual time which included (mention here names as some medications may alter sleep-wake cycle and proportions of sleep stages). Morning report suggests that patient was comfortable (if not, mention

Writing an Informative Report

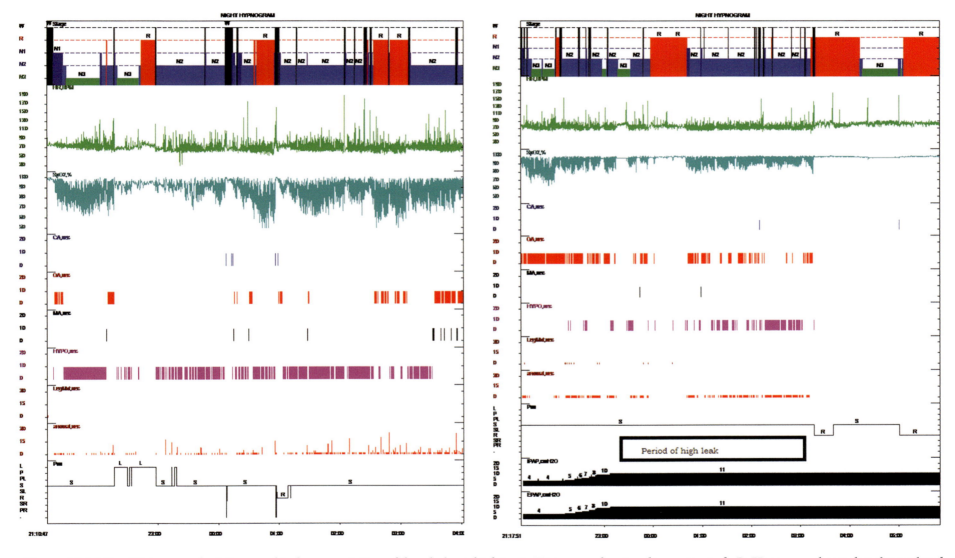

Figure 21.1A,B Histograms depicting graphical representation of the whole night data: A: Histogram showing diagnostic study B: Histogram shows that sleep related breathing events improved after correction of mask fitting at same pressure.

the reason for discomfort) during the study and that he had his usual kind of sleep. Data generated from the recording suggests total sleep time of 300 min with asleep efficiency of 80%. Sleep onset latency was 25 min that matches with sleep diary data, and hypnogram suggests that sleep was well-maintained except for 2–3 arousals to go to the washroom. However, he fell asleep soon after each arousal. Total arousal index was 25 which is higher than expected, and was mostly related to the sleep-disordered breathing events. Data suggests that proportion of stage 1 was higher (12%), while that of N3 lower (8%), again related to sleep-disordered breathing. Snoring started soon after sleep initiation and many events of the respiratory-event-related arousals, hypopnea and obstructive sleep apnea were observed with respiratory disturbance index (RDI) of 18.3. Average oxygen saturation was 92% with the minimum saturation of 79% and desaturation index of 22. Overnight oximetry showed sawtooth appearance. No effect of sleep stage and body position could be observed on RDI or desaturation index. Cheyne-Stokes breathing, periodic breathing, and sleep-related hypoventilation were not observed during the study. Frequent limb movements were observed with PLMS score of 24, however, they were not associated with arousals. EKG did not show any abnormality.

This report clearly indicated that:

1. Sleep latency was within normal limits and also he did not have difficulty in falling asleep after waking up for the need to go to the washroom, so insomnia can be ruled out.
2. Adequate time was given to diagnose the condition in question, that is, sleep apnea.
3. The patient usually did not get deep sleep (low N3).
4. Sleep-disordered breathing was responsible for the non-refreshing sleep (multiple arousals).
5. Sleep was interrupted by the urge to void, which may be related to the sleep-disordered breathing.
6. Moderately severe sleep apnea without any modification from body position and presence of systemic hypertension suggests that the patient needs treatment for sleep-disordered breathing.
7. High PLMS score could be an incidental finding as patient does not have RLS.

In summary, treatment of sleep-disordered breathing would take care of most of the complaints.

REVIEW QUESTIONS

1. If usual bed time and start time is not mentioned in the report:

A. inference of sleep onset latency may be mistaken
B. duration of sleep onset latency may be mistaken
C. sleep onset latency cannot be computed
D. sleep onset latency is overestimated

2. If PAP titration report mentions only the total RDI, but not the RDI achieved at the highest pressure:
   A. therapeutic pressure will be underestimated
   B. therapeutic pressure will be overestimated
   C. therapeutic pressure is difficult to be ascertained
   D. suggests that Auto-PAP will be better option

3. Minimum total sleep time required for a reliable report is:
   A. 60 minutes
   B. 120 minutes
   C. 240 minutes
   D. 360 minutes

4. PLMS should be mentioned as it is:
   A. helps is diagnosing RLS
   B. helps in diagnosing PLMD
   C. helps in diagnosing parasomnia
   D. helps in diagnosing seizure

5. Following respiratory parameter provides a reliable indication of severity of sleep apnea:
   A. central apnea index
   B. hypopnea index
   C. apnea-hypopnea index
   D. respiratory disturbance index

ANSWER KEY

1. A   2. C   3. C   4. B   5. D

22

GUIDELINES FOR SUPPLEMENTAL OXYGEN

LEARNING OBJECTIVES

After reading this chapter, reader should be able to:

1. Decide when to institute oxygen therapy during sleep study.
2. Provide oxygen therapy using the AASM guidelines.

CONTENTS

22.1	Introduction	453
22.2	Oxygen Delivery Devices and Polysomnography	454
22.3	Precautions	454
22.4	Devices for Oxygen Delivery	454
22.5	Steps Towards Institution of Oxygen Delivery	455
Further Reading		458
Review Questions		458
Answer Key		458

22.1 INTRODUCTION

Oxygen is a gaseous element that forms 21% of the inhaled air. Oxygen is considered a drug, and hence, it is administered in accordance with the institution policy and procedure. Medical oxygen is present in three common methods: compressed gas, liquid form, and oxygen concentrator.

In the sleep disorders center (SDC), desaturation events related to sleep-disordered breathing (SDB) are commonly seen. In most cases, the use of positive airway pressure is sufficient to maintain a patient's oxygen level within a normal range. However, in certain cases such as patients with co-existing lung or heart disease or patients with obesity hypoventilation syndrome, the need to increase FiO2 may arise and necessitate the addition of supplemental oxygen. Therefore, sleep technologists should be acquainted with the rules and regulations governing the administration of oxygen.

In general, there should be clearly written protocols and parameters to regulate oxygen administration in the SDC. As a general rule, the least amount of oxygen to bring about the desired therapeutic effect, which is often defined as a predetermined level of oxygen saturation.

22.2 OXYGEN DELIVERY DEVICES AND POLYSOMNOGRAPHY

During diagnostic polysomnography, airflow is usually monitored by using a thermistor, nasal pressure, and capnography. When oxygen is supplied to the patient via a nasal cannula, oxygen supplementation may interfere with airflow signal in the form of dampened flow signal. To overcome this problem, a cannula with a bifurcated access to each nostril can be used allowing both oxygen delivery and pressure monitoring.

22.3 PRECAUTIONS

Oxygen as a chemical substance has potential dangers. Although oxygen is not flammable by itself, it strongly enhances combustion. Therefore, care should be practiced to ensure that flammable substances used in the SDC such as collodion should not be used near oxygen. If oxygen cylinders are used, it should be realized that oxygen is stored in the cylinders under very high pressure; hence, cylinders must be secured from falling and must be stored away from heating sources.

22.4 DEVICES FOR OXYGEN DELIVERY

A. **High-flow devices:** These delivery systems deliver oxygen of precise concentration at flow higher than the patient's inspiratory flow. An example of these devices is the Venturi mask. High-flow oxygen is usually delivered through a facemask with humidification. High-flow systems are not often used in the sleep center.

B. **Low-flow devices:** This method provides 100% oxygen at a rate that is less than the peak inspiratory flow rate of the patient. It is recommended to use humidification with these systems. Examples of these systems include nasal cannulas and simple masks.

22.5 STEPS TOWARDS INSTITUTION OF OXYGEN DELIVERY

A. **Preparation for Oxygen Initiation in the Sleep Disorders Center**
- Confirm physician's order for supplemental nocturnal oxygen titration.
- Explain the oxygen titration procedure.
- Be aware of the patient's diagnosis and any history of CO_2 retention.
- Document SpO_2 of the patient while awake.

B. **During Sleep Study Recording**
- If obstructive respiratory events associated with SpO_2 drop below 88%, start PAP therapy and assure that all respiratory events are corrected. Figure 22.1 shows a zoomed 5 min epoch of a patient with obstructive respiratory events associated with oxygen desaturation. CPAP pressure should be increased to eliminate the obstructive events before adding oxygen.
- If SpO2 stays below 88% for >5 min after elimination of respiratory events, then oxygen supplementation is indicated (Figure 22.2).
- Administer oxygen at 1 liter/min, and increase it by 1 liter/min, with an interval no shorter than 15 min, until the SpO2 is between 88% and 94%.
- Oxygen flow should not exceed 3 liters/min unless ordered by a physician.
- In general, each sleep disorders center should have clear policies about oxygen administration.

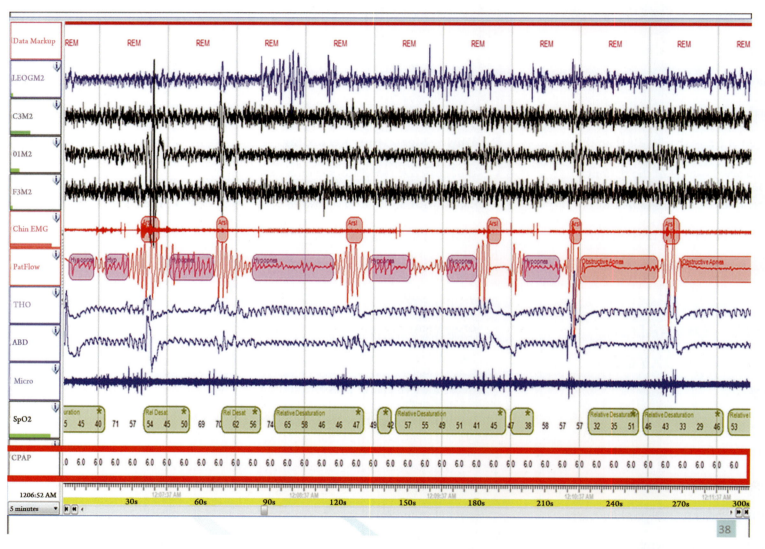

Figure 22.1 OSA must be treated before starting oxygen therapy: 5 min epoch of patients with obstructive respiratory events associated with oxygen desaturation. CPAP pressure should be increased to eliminate the obstructive events before adding oxygen.

Guidelines for Supplemental Oxygen 457

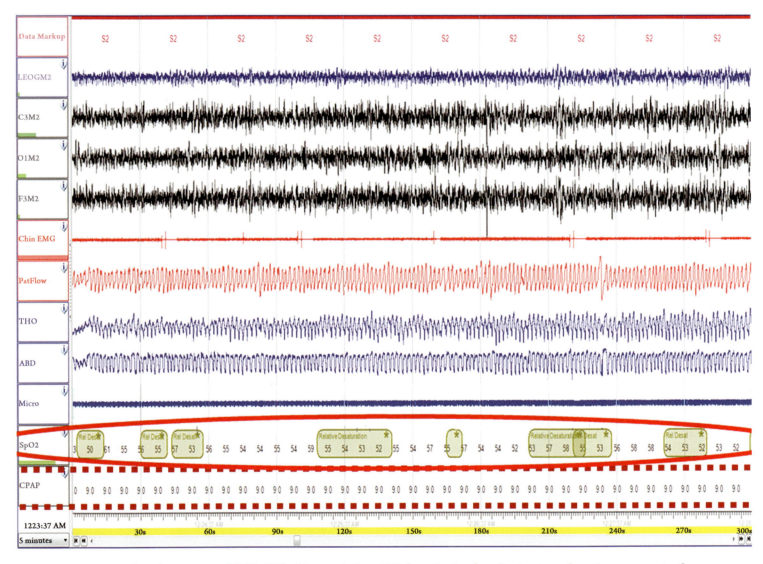

Figure 22.2 Oxygen saturation after elimination of OSA: If SpO2 stays below 88% for >5 min after elimination of respiratory events, then oxygen supplementation is indicated.

FURTHER READING

1. Kushida, C. A., Chediak, A., Berry, R. B., Brown, L. K., Gozal, D., Iber, C., et al. (2008). Clinical guidelines for the manual titration of positive airway pressure in patients with obstructive sleep apnea. *Journal of Clinical Sleep Medicine: JCSM*: official publication of the *American Academy of Sleep Medicine*, 4(2). 157–171. Epub 2008/05/13.

REVIEW QUESTIONS

1. Before starting oxygen flow, ensure that:
    A. all sleep disordered events are corrected
    B. arterial PCO_2 is below 45 mmHg
    C. patient is sleeping
    D. leg movements are absent

2. Oxygen delivery should be started at a rate of (L/min)
    A. 4
    B. 3
    C. 2
    D. 1

3. Minimum interval to increase the oxygen flow is:
    A. 10 minutes
    B. 15 minutes
    C. 20 minutes
    D. 25 minutes

4. Unnecessary oxygen delivery can interfere with:
    A. scoring of CSA
    B. scoring of OSA
    C. scoring of hypopnea
    D. scoring of MSA

ANSWER KEY

1. A   2. D   3. B   4. C

23
INFECTION CONTROL

LEARNING OBJECTIVES

After reading this chapter the learner should be:

1. Able to discuss the common sources of infection in sleep laboratory.
2. Able to take optimal precautions for control of infections in sleep laboratory.

CONTENTS

Importance of Infection Control 459

Review Questions 460

Answer Key 461

Infection control is an important issue during the sleep study. A patient has to spend two nights in a hospital room. In addition, a number of instruments are applied during the study, and these may transmit infection. In this chapter, we will discuss the common modes of infection transmission and their prevention measures.

The first source of infection could be the room itself. This is an example of airborne infection. If there is any seepage in the wall or in the bathroom, fungal spores can grow and be a source of infection. It is therefore prudent to inspect the room regularly and take corrective measures in case seepage is observed. It is also good to get the room fumigated before posting another patient for a sleep study. Besides, one patient may also be a source of infection for another patient. It is advisable not to take patients with an active respiratory infection for a sleep study. Even otherwise, a person with an active upper or lower respiratory infection may have spurious data so it is best to avoid such patients. Also, it is important that the room is well-lit and ventilated, and there is a regular exchange of air to prevent infection.

The second common source of infection is the linen. All linen used in a case study should be thoroughly washed and dried and each patient should be given fresh linen. This is because linen may be a source of infection in case it contains mites, ticks, lice, or any other arthropods.

The other common sources of infection include the nasal cannula, mask of PAP, and hosepipe of the PAP machine. It is essential that the nasal cannula is changed after every sleep study. Since the mask and the hosepipe are expensive items and cannot be discarded so often, they should be sent for autoclaving after each use. Mask straps can also transmit infection as they are applied directly to the patient's skin/hair. They should also be washed after each use.

The EEG and EMG electrodes may also be infrequent sources of infection, and, therefore, they should be thoroughly cleaned with alcohol after every sleep study. Alcohol not only removes the greasy conductive gel but also works as a disinfectant.

Respiratory belts seldom act as fomite as they are usually worn over the clothes. However, care must be taken that the patient does not accidently contaminate them. In case they are contaminated, they must be washed thoroughly with soap.

For further details, please refer to the infection control policy in Chapter 24.

REVIEW QUESTIONS

1. Electrodes are disinfected using:
   A. soap
   B. collioiden
   C. medical spirit
   D. water

Infection Control

2. PAP mask should be disinfected using:

    A. temperature

    B. ETO

    C. soap

    D. warm air

3. Bed linen may spread following infections:

    A. skin

    B. meningitis

    C. gastrointestinal

    D. urinary

4. Fumigation of sleep lab is necessary when:

    A. it is well ventilated

    B. it is having attached toilet

    C. floor is not tiled

    D. any of the wall has seepage

5. To prevent cross infection, following should be disposed after every use:

    A. electrode

    B. nasal cannula

    C. strap of mask

    D. bed linen

ANSWER KEY

1. C   2. B   3. A   4. D   5. B

24

SLEEP LAB MANAGEMENT

CONTENTS

24.1 Setting the Sleep Laboratory 464

24.2 Staff of the Sleep Laboratory and Their Responsibilities 466

24.3 Staff: Qualification, Regular Training, Number and Working Hours 468

24.4 Workflow From Sleep Clinic to Sleep Laboratory 470

24.5 Policy for Accounting and Budgeting 471

24.6 Policy for Quality Assurance 472

24.7 Policy for Maintaining the Records 473

24.8 Policy for Infection Control in the Sleep Lab 476

24.9 Policy in Case of Any Emergency During Sleep Study 477

24.10 Protocols 478

24.11 Forms, Questionnaires, and Templates 484

Further Reading 498

> **LEARNING OBJECTIVES**
>
> After reading this chapter, the reader must be able to:
> 1. Able to logically formulate various policies for a sleep-laboratory.

24.1 SETTING THE SLEEP LABORATORY

Sleep laboratory must be established in a calm and quiet place in the hospital. The place must be accessible by wheelchair.

Recommended infrastructure of the sleep laboratory is as follows:

1. Room and furniture for the sleep-lab
    a. Size of the room: At least 10 x 12 feet
    b. Should be properly ventilated
    c. Thick curtains must be drawn on the windows
    d. Must be sound attenuated
    e. Must be well lit
    f. Have an attached washroom with Geyser installed
    g. Must have air-conditioning and a television with remote control facility
    h. Uninterrupted power supply must be ensured
    i. Artwork
    j. Furniture required for the sleep lab
        i. King size bed with mattress, linen, pillows, blanket
        ii. Cub-board
        iii. Bed-side table
        iv. Platform for the PSG equipment
        v. Couch/bed for the attendant along with linen
    k. Dustbins for disposal of recyclable waste
    l. Disposal facility for the consumables in sleep lab and sharp objects, if and when used

One polysomonography can be carried out in one room. So the number of rooms required varies depending upon the number of polysomnography machines one wishes to establish.

2. Monitoring room
    a. Size of the room: At least 10 X 12 feet
    b. Should be properly ventilated
    c. Must be sound attenuated
    d. Must be well-lit
    e. Must have an attached washroom for sleep technologists
    f. Have uninterrupted power supply
    g. Must have an area with running water to clean reusable items, for example, mask and electrodes

h. Have a washroom close to the monitoring room
i. Must have a kitchenette for staff
j. Furniture required for the monitoring room
   (1) Computer table
   (2) Cupboard for storage of accessories
   (3) Revolving chairs for sleep technologists

Depending upon the number of beds in the facility, the storage room may be separated from the monitoring room.

3. Equipment for the sleep laboratory
   a. Level I digital PSG machine from a standard manufacturer with extended EEG channels
   b. Infrared video recording with maneuvering facility, synchronized to the PSG recording
   c. Real-time audio recording
   d. Intercom between the sleep laboratory and monitoring room
   e. CPAP/BiPAP instrument with remote control
   f. CPAP masks:
      i. Full face
      ii. Nasal mask
      iii. Nasal pillows
   g. Oxygen line in the sleep lab with humidifier
   h. Suction facility in the sleep laboratory
      i. Consumables:
      i. Neuroprep Gel
      ii. Cotton buds
      iii. Electro-conductive gel
      iv. Electrodes
      v. Cannula
      vi. Disinfectant
      vii. Cotton Gauges
      viii. Gloves and Mask
4. Equipment for the monitoring room:
   a. Computer system
   b. Printer with papers
   c. Intercom with amplifier connected to sleep laboratory
   d. Defibrillator
   e. Emergency kit: Medications used for emergency; measures to secure airway
5. Documents that must be available in a sleep lab
   a. Inventory register, duly completed till date
   b. Accounts and budget Register
   c. Appointment register
   d. Service logs for the equipments

e. Standard Operating Procedure (SOP) manual
   i. Equipments and staff of sleep laboratory
   ii. Organizational structure
   iii. Protocol for sleep study
   iv. Emergency Protocol Manual
   v. Quality Assurance Procedure Manual
   vi. Protocols for various sleep studies
   vii. Trouble-shooting guides
   viii. Emergency Contact Numbers
f. Practice Parameters/Guidelines: The most recent version should be available
g. AASM Scoring Manual: The latest version should be available

6. Facility to store case record forms

24.2 STAFF OF THE SLEEP LABORATORY AND THEIR RESPONSIBILITIES

1. Administration of the Sleep Clinic and Laboratory shall be as follows:
   i. Director/In-Charge
   ii. Sleep Physicians
   ii. Trainee Sleep Physicians
   iv. In-Charge Sleep Technologist
   v. Sleep Technologists
   vi. Trainee Sleep Technologists
   vii. Manager Sleep Laboratory
2. Responsibilities of the Staff:
   Sleep laboratory is fairly a busy place with lot of equipments and infrastructure. Hence, the responsibilities must be divided among the staff members.

   A. Director/In-Charge:
   a. Administrative
      (1) Development and revision of policies related to the functioning of the sleep clinic and sleep laboratory in collaboration with other members
      (2) Participation in the selection of the sleep technology
      (3) Review of the technical side of the instruments to be purchased
      (4) Developing the modules to improve the quality of work in the sleep clinic and sleep laboratory
      (5) Expansion of services at the appropriate time
      (6) Building the team for multi-disciplinary practice of sleep medicine

b. Training and Research:
   (1) Supervise the training programs of the sleep-technologists/sleep physicians/short-term trainee
   (2) Periodic revision of the syllabus with the other members of the team for trainees
   (3) Finding opportunities for the further training of team members
   (4) Organizing departmental workshops/training programs for the members of the team
   (5) Conducting and motivating the members of the team to conduct research
c. Networking and Patient Education:
   (1) Develop the material for marketing and patient education along with team members
   (2) Networking with other physicians in the geographical area where the clinic is situated
   (3) Running awareness campaigns for the medical fraternity and patients
d. Clinical Care:
   (1) Providing clinical care to the patients through sleep clinic/PAP clinic

B. Sleep Physicians/Trainee Sleep Physicians:
   (1) Providing clinical care to the patients attending sleep clinic
   (2) Scoring the sleep study
   (3) Conducting research in association with other members of the team
   (4) Participation in departmental educational programs
   (5) Participation in various administrative works along with director/in-charge as required
   (6) Completing the patients case sheet at the time of admission
   (7) Providing special instructions, if any, in the case sheet before the sleep study
   (8) Monitoring the prescribed medications and mentioning them in the case report sheet at the time of admission

C. Sleep Technologists Including Trainee Technologists:
   a. Administrative: One Sleep Technologist will be assigned the administrative work by the local administration. He/She will be responsible for the following duties:

(1) Looking after the administration of the sleep lab
(2) Developing the inventory of the sleep lab and regularly maintaining it
(3) Refilling the consumables in the sleep lab at appropriate intervals
(4) Keeping a record of the quality of data recorded in the sleep lab every day
(5) Ensuring that all the equipments are functional in the sleep
(6) Communication with the biomedical engineer, whenever required

b. Clinical:
(1) Recording and scoring the sleep study as mentioned in the protocol section
(2) Ensuring infection control in sleep laboratory, as mentioned in the protocol section/section of infection control
(3) Ensuring the quality of recording of the study
(4) Interacting with the sleep physicians for improving the services and patient care
(5) Participation in various activities (for improving quality of work, patient education) in the sleep lab

D. Manager - Sleep Laboratory:
a. Exploring new business opportunities
b. Managing billing-related issues

24.3 STAFF: QUALIFICATION, REGULAR TRAINING, NUMBER AND WORKING HOURS

24.3.1 Sleep Physicians Including Director/In-Charge

A. Qualifications:
   (i) Preferably certification in sleep medicine by any international/national board after graduate/post-graduate medical degree.
   (ii) In case, board-certified person is not available in the country, any physician having documented interest in sleep medicine may be opted. In that case, other minimum qualifications shall remain the same.
B. Training:

Sleep Lab Management

(i) Sleep physicians are expected to participate in in-house and other training programs at regular intervals to keep their knowledge updated.

C. Number of Sleep Physicians and Team Building:
   (i) At least one sleep physician should be attached with each sleep laboratory.
   (ii) Practice of sleep medicine requires multidisciplinary approach, and team comprises of a Sleep Physician, Psychiatrist, Neurologist, Pulmonologist, Pediatrician, ENT Surgeon, Dental Surgeon, and a general physician for providing optimal care to the patient. The Director/In-Charge should try to build the team.

D. Working Hours:
   (i) Sleep physicians shall be working in 8 hours day shifts, 6 days a week to provide optimal care to the patients attending Sleep Clinic/Sleep Laboratory.
   (ii) In case of any emergency during the sleep study, sleep physicians are expected to attend the patient at night as well.

24.3.2 Sleep Technologists

A. Qualifications:
   (i) Preferably certification in sleep technology by any international/national board after graduate degree.
   (ii) In case, board-certified person is not available in the country, any EEG technologist having documented interest in Sleep Medicine Technology may be opted. In that case, other minimum qualifications shall remain the same.
   (iii) In such case, he shall undergo 6 months intensive training under the supervision of Sleep Physician in the laboratory.

B. Training:
   (i) Sleep Technologists are expected to participate in in-house and other training programs at regular intervals to keep their knowledge updated.
   (ii) Must be trained in basic life support (BLS)

C. Number of Sleep Technologists
   (i) It will depend upon the number of beds in the sleep laboratory. In general, for a 1–2 bedded laboratory, at least 3 sleep technologists should be available: One for the night shift, the other for the day-shift, and the third, reliever, in case of emergency.

(ii) One sleep technologist will take care of two sleep studies at a time (that includes one titration study) as recommended by the American Academy of Sleep Medicine.

D. Number of Working Hours:

(i) Each night shift will be spanning 12 hours starting at 20.00 to 8.00 AM for 5 days. In usual circumstances, no study shall be recorded on Saturday and Sunday nights.

(ii) Sleep Technologists shall be given two days off after 5 working nights.

(iii) The daytime technicians would undertake tasks such as performing MSLT, MWT, Scoring, Teaching and Training, and other administrative duties.

24.4 WORKFLOW FROM SLEEP CLINIC TO SLEEP LABORATORY

1. Sleep study is an elective procedure; hence, patients may be scheduled depending upon the number of beds in the facility and patient load. However, at the same time, Director/In-Charge of the facility should try to reduce the waiting period as far as possible.

2. Patients may be given appointments after a Sleep Physician has examined them. In cases of patients referred from other physicians as well, it is best that the patients pay at least one visit to the Sleep Physician, before the sleep study.

3. During the visit in the Sleep Clinic (which is usually situated in an outpatient's department), the sleep physician is expected to thoroughly study the history, examine the patient and get any other investigations done, if required. Findings must be recorded and a provisional diagnosis should be made. Provisional diagnosis must include other comorbid illnesses as well.

4. Findings must be discussed with the patient and the reason for the sleep study should be informed.

5. Based on the provisional diagnosis, if any intervention is required before the sleep study, it may be done.

6. In case, based upon his/her clinical judgment, the sleep physician desires to discontinue some medications before the sleep study, the patient should be informed.

7. Appointment should be scheduled according to the local hospital policy (some hospitals may require part payment in advance for scheduling an appointment; in case of medical insurance, a process has to be initiated with the appropriate person of the facility).
8. Patient should preferably be given a pamphlet explaining what will be recorded, how the study will be done, and what is expected from him/her on the day of study.
9. In general, two nights should be given to the patient: First for the diagnosis and, second, for the intervention, if required. However, the sleep physician may use his/her clinical judgment for the split-night studies.
10. On the day of the sleep study, the patient should be called at least 6–8 hours before his/her usual bedtime and should be allowed to spend time in the sleep lab. This will help in acclimatization.
11. Before admitting the patient in the Sleep Laboratory, it is mandatory to take informed consent in writing to conduct the sleep study.
12. After the diagnostic study, the patient should be requested to fill the information in the post-test form.
13. Before the CPAP study, the patient should be handed over a clean and autoclaved mask in the morning by the sleep technologist. The patient should be instructed to practice wearing it during the day so as to facilitate acclimatization.
14. Take the informed consent in writing for the PAP study.
15. After the sleep study, information must be obtained from the patient using appropriate questionnaires.
16. All records must be kept as per the policy of the facility.
17. In case of voluntary termination of a sleep study, appropriate protocol must be followed.
18. Report should be provided to the patient within a week of the sleep study.

24.5 POLICY FOR ACCOUNTING AND BUDGETING

1. This should be done according to the policy of the facility, where the sleep laboratory is situated.
2. Sleep Physicians and sleep technologists musts obtain professional indemnity insurance according to the policy of the facility.

Table 24.1 Format for the Accounting in a Sleep Laboratory

Date	Name of patient/purchased article	Receipt Number	Income	Expenditure
Net income from the sleep lab: Total income – Total expenditure.					

3. In-Charge Sleep technologist of the facility should maintain monthly record in the following format (Table 24.1). This must be get signed by the Director/In-charge of the facility at the end of the month.

24.6 POLICY FOR QUALITY ASSURANCE

1. Assuring the quality of data is an important parameter for the success of the sleep disorders center. A center should try to achieve the highest standards of the quality so as to best serve the patients. To achieve this, the following must be ensured:

   a. By the In-Charge Sleep Technologist:
      i. All the measures to prevent infection control are taken in the sleep lab.
      ii. All t parts of the equipments are working and consumables are refilled before their exhaustion.
      iii. Hygiene is maintained in the sleep laboratory and monitoring room.
      iv. Patients are providing information regarding their experience before and after the test, as their feedback may help in improving the standards.

   b. Sleep Technologists:
      i. They must ensure that they possess adequate knowledge of their subject and that they are engaged in advancement of their knowledge by attending periodic trainings.
      ii. They must ensure that they stick to the protocols developed by the facility.
      iii. If there is any mal-function in the equipments or if there is any patient-related issue, they should bring it to the notice of In-Charge Sleep Technologist and/or Director/In-Charge of the sleep clinic at the earliest.
      iv. They must ensure that their behavior with the patients is professional.

Sleep Lab Management

v. They must encourage the patient to provide post-test information in appropriate form.

vi. They should score the data as per the latest guidelines from the American Academy of Sleep Medicine.

c. Sleep Physicians Including Director/In-Charge:

i. They are responsible for the overall functioning of the sleep clinic and sleep lab. They must ensure that they get regular feedbacks from the staff.

ii. In case some action for rectification is required, they must ensure that they take a prompt action.

iii. They must ensure that the recordings are assessed daily for the quality and that the sleep technologists get the feedback (Table 24.2).

iv. They must ensure that they conduct a monthly meeting with all staff members to rectify all issues related to patients, equipments, quality of recordings, and overall services of the facility.

The American Academy of Sleep Medicine has proposed some indicators for the quality assurance. These are tabulated in Table 24.3.

24.7 POLICY FOR MAINTAINING THE RECORDS

Records in the Sleep Clinic and Sleep Laboratory should be maintained for a minimum duration as advised by the local administration of the law of land. Two kinds of records are required in a sleep

Table 24.2 Template for Maintaining Quality of PSG Recordings

Date	Hospital ID of Patient	Type of Study	EEG	EMG Chin	EOG	LEG EMG	EKG	Respiratory Flow	Respiratory effort	SpO_2	Capnograph	Body position	Audio and Video	PAP, if applicable	Sign: Sleep Technologist	QC Checked by.........

Table 24.3 Indicators of Quality Assurance in a Sleep Laboratory

Indicator	Frequency of reporting	Tolerance	Source of Information	Description
Patient Care Indicators				
Patient injury/death	Daily and summarized monthly	Zero	Patient record, PSG record	Review next day and in monthly meetings. Take corrective actions
Patient Complaint	Daily and summarized monthly	Variable	Patient record, post-test questionnaires	Review next day and in monthly meetings. Take corrective actions
PSG terminations against medical advise	Daily and summarized monthly	Zero	Patient record, Sleep technologist's notes	Review next day and in monthly meetings. Take corrective actions
Refusal to CPAP	Daily and summarized monthly	10%	Patient record, Sleep technologist's notes	Reattempt the PAP titration after longer mask-acclimatization
Patient Satisfaction	Daily and summarized monthly	Variable	Patient Satisfaction Questionnaire	Review next day and in monthly meetings. Take corrective actions, if required
Procedure Indicators				
PSG Recording Quality	Daily by In-Charge Sleep Technologist/ Sleep Physician and summarized monthly	Zero	Recording quality work-sheet	Take corrective action, if required
PAP titration quality	Daily by In-Charge Sleep Technologist/ Sleep Physician and summarized monthly	Zero	Recording quality work-sheet	Take corrective action, if required
Adherence to standards during MSLT	Daily by In-Charge Sleep Technologist/ Sleep Physician and summarized monthly	Zero	Recording quality work-sheet	Take corrective action, if required
Reliability of Sleep Stage Scoring	Daily by In-Charge Sleep Technologist/ Sleep Physician and summarized monthly	Zero	Recording quality work-sheet	Take corrective action, if required
Reliability of Respiratory data Scoring	Daily by In-Charge Sleep Technologist/ Sleep Physician and summarized monthly	Zero	Recording quality work-sheet	Take corrective action, if required
Reliability of Leg Movement Scoring	Daily by In-Charge Sleep Technologist/ Sleep Physician and summarized monthly	Zero	Recording quality work-sheet	Take corrective action, if required
Administrative Functions				
Scheduling Appointment	Weekly monitoring, summarized monthly	—	Appointment logs	Monitor time between initial work-up and scheduled appointment
Scheduling Follow-ups	Weekly monitoring, summarized monthly	—	Appointment logs	Monitor follow-ups and drop-outs

Sleep Lab Management

laboratory—one for administrative purpose and the others that are related to the patients.

Administrative records include inventory register (Table 24.4), appointment register (Table 24.5), service logs of the equipments (Table 24.6), and accounts and budget register (Table 24.1). Basic templates for these registers are provided here, although they may be modified for individual facility.

Patient-related records are important not only for future references but also for the medicolegal purposes. These records should be maintained for a minimum period as per the local policy.

A. Case record form must be completed at the time of admission after obtaining written informed consent. Case record form must bear at least the following:
- Patient ID number

Table 24.4 Format for the Inventory Register

S. No.	Date	Manufacturer	Equipment/Part number	Number of articles/parts available	Number of articles/parts procured	Number of articles/parts discarded	Signature of Sleep tech	Sign of director/In-Charge of the facility

Table 24.5 Template for Appointment Register

Patient ID	Name	Age/Gender	Contact Number	Diagnosis	Type of Sleep Study required	Referred by	Special Instructions for Sleep Tech, if any

Table 24.6 Template for Service log

Date	Problem in the equipment/part	Equipment/part inspected by	Recommendation made	Equipment/component: serviced/Changed	Serviced by	Signatures of Sleep Tech

- Demographic data
- Contact number
- Informed consent for the sleep study and PAP
- Medical history
- Examination
- Findings of other laboratory investigations, if required
- Complete diagnosis
- Prescription of medication during the admission in sleep-lab
- Necessary and clear instructions to the sleep technologists
- Technician notes of sleep study as defined in various protocols
- Pre-test and post-test questionnaires
- Copy of report of the sleep study which is given to the patient at the time of discharge

B. Raw data of sleep study: Raw data may be stored for variable duration, depending upon the local policy. Some facilities may choose to store the data for training and research purpose as well.

24.8 POLICY FOR INFECTION CONTROL IN THE SLEEP LAB

A. Sleep Physician must ensure that the following patients should not be taken for the sleep study:
   a. Patients with active upper airway/lower airway infections
   b. Patients with skin infections that may be transmitted through electrodes/belts/bed linen

B. To prevent the spread of infection from one patient to another, the sleep technician must ensure that:

   a. Gold Cup Electrodes are properly cleaned after every study.
      They must be cleaned using mild soap and water. A toothbrush with soft bristles may be used to remove paste and then disinfect using bleach in water.
   b. Sticking electrodes/Button electrodes are to be disposed after every use.
   c. Cannula used to measure respiratory flow is changed after every use.
   d. Respiratory belts are wiped with alcohol damp cloth after every use.
   e. Oximeter probe must be cleaned with alcohol damp cloth after every use.
   f. Thermister are cleaned with alcohol damp cloth after every use.
   g. Chin straps are washed with warm water after every use.

Sleep Lab Management

h. Mask of the CPAP, hosepipe and straps are cleaned with lukewarm water after every study and then sent for Ethylene Oxide Sterilization.

i. Do not apply the electrodes on the parts of skin where there is a breach in continuity.

C. Maintenance of Hygiene in the Sleep Lab:

   a. Sleep lab technician along with the nursing in-charge of the ward will ensure that:

   b. The bed linen/pillow-cover is changed after every study.

   c. Area where machines are kept in the sleep lab is clean and free of dust.

   d. There is free circulation of air after every study.

   e. The washroom attached to the sleep lab is clean and hygienic.

E. Precautions to be taken by the Sleep Technologists to Prevent Infection:

   i. They must wear clean clothes.

   ii. They must wear gloves while applying electrodes and other sensors on the patients.

   iii. If they have any active respiratory infection, they must seek medical advice at the earliest and start treatment. If they have been allowed to handle patients by the physician, they must wear a mask while in the sleep lab.

24.9 POLICY IN CASE OF ANY EMERGENCY DURING SLEEP STUDY

1. Common emergency situations that can arise during a sleep study:

   A. Cardiac arrhythmias
   B. Precordial chest pain
   C. Seizures
   D. Dyspnea

2. Policy to handle such situations:

   A. Sleep technologists must possess adequate knowledge and skills to recognize and handle such events.

   B. In countries where sleep technology certification is not available, sleep technologists must be trained to recognize and handle such events.

   C. Such situations may require different levels of experience, knowledge, and skills. Hence, sleep technologist must involve the nursing staff, trainee-physicians and sleep physician of the facility while handling such a situation.

   D. Course of action in such cases:

In such cases, the sleep technologist is expected to initiate the following course of action:

i. Inform the nursing staff in the facility

ii. If dyspnea occurs during a titration study, remove the mask and turn off the PAP machine.

iii. Provide basic life support to the patient.

In such a case, nursing staff is expected to initiate the following course of action:

i. Call the trainee-physician/physician on duty immediately.

ii. Immediately inform the Sleep Physician through the appropriate person.

iii. Help the Sleep Technologist to provide basic life support to the patient.

In such cases, the trainee-physician is expected to take the following course of action:

i. Assess the gravity of the situation.

ii. Provide optimal care to the patient.

iii. Work with the sleep technologist and attending nursing staff in the benefit of the patient.

iv. Call the physicians/surgeons from other specialties, if required.

In such cases, the sleep physician on call is expected to take the following course of action:

i. Attend to the patient immediately.

ii. Provide optimal care to the patient.

iii. Work with sleep technologist and attending nursing staff in the benefit of the patient.

iv. Call the physicians/surgeons from other specialties, if required.

v. Shift the patient to another facility, if required.

E. Once the patient is stabilized, sleep physicians and sleep technologists must review the event to uncover its possible reasons by examining the case record forms. The discussion must be recorded and preventive actions should be ensured for future, if possible.

24.10 PROTOCOLS

24.10.1 Protocol for the Diagnostic Sleep Study

1. **Admission of the patient:**

   a. Patient has to be admitted after work-up and a case record sheet will be filled.

b. Please check the case record sheet for special instructions, if any.
2. **Preparation of the patient:**
   a. Please ensure that hair and face of the patient are free of oil and grease.
   b. In cases of males, please ensure that the patient is shaven, unless he wears a beard for personal or religious practice. In that case, please ensure that the beard is free of greasy material.
   c. Please ensure that the patient is comfortable and is wearing proper clothes.
   d. Ensure that the air-conditioner is working properly and the temperature is set to 22°C.
   e. Blanket/sheet is to be made available to the patient.
   f. Ensure that the patient has taken medications, as prescribed by the physicians and approved by the sleep physician.
3. **Placing the electrodes on the patient's body:**
   a. Start placing electrodes (EEG/EOG/EMG/ECG) 45 min before the patient's usual bed-time.
   b. Please clean the area and prepare it with Neuroprep gel.
   c. Use adequate amount of the conductive gel.
   d. Electrodes are to be fixed as per the AASM manual ver. 2.3 released in 2015.
   e. Please fix electrodes so as to minimize the chances of their falling off at night.
4. **Placement of respiratory sensors:**
   a. RIP belts are to be placed on the thorax and abdomen. Please adjust the belts so that the unstretched belt covers two-third of the circumference for improving the signals and longevity of the belts.
   b. Both thermister as well as cannula should be placed and fixed properly.
   c. Make sure the cannula is attached to the Capnograph at the appropriate time to ensure $EtCO_2$ signals.
   d. Place the oximeter properly so that it does not fall off at night.
5. **Place the body position sensor over the chest belt in the middle of the chest.**
6. **Recording the study:**
   a. Please fill the name, age, gender, and ID of the patient.

b. Please fill the anthropometric data (height/weight).

c. Please fill the name of the referring physician.

d. Please choose the appropriate montage (Diagnostic/Seizure/Parasomnia).

7. **Checking the signals and biocalibration before lights off:**

   a. Please check the impedance from electrodes and make sure that it is below 5 K Ohms in the EEG/EMG/EOG leads.

   b. Please check that the EEG/EOG/EMG leads have good quality signals and are free from artifacts.

   c. Please check that the ECG signals are of good quality.

   d. Please check that the respiratory flow signals are appearing properly.

   e. Please check that the oximeter and capnograph signals are proper.

   f. Please check that the body position is showing appropriate signals.

   g. Please check that the video and audio signals are working properly.

   h. If any of these signals are not appropriate, fix the electrode using appropriate measures, as conveyed during your training.

   i. Do the bio-calibration before every study. Put a note in the patient's record sheet.

8. **If all the signals are adequate and have good quality, then start the recording and ask the attendants to turn off the lights. Place a marker in the sleep recording.**

9. **During the study:**

   a. Please check all the signals periodically and enter the signal quality at an interval of 30 min in the patient's file, till the end of the study.

   b. Audio and video signals from the sleep lab are to be kept turned-on during the whole study.

   c. Please make yourself available in case of any assistance required by the patient during the study.

   d. If any unusual behavior is seen, mention it in the patient's file along with the time when it was observed.

10. **Terminating the study:**

    a. Study to be terminated at the usual wake time of patient or at 7.00 AM.

    b. Please ensure proper closure of the study.

    c. Ensure that data has been transferred to the computer system before turning off the computer system.

Sleep Lab Management

11. **After the study:**
    a. Please remove the electrodes so as to cause minimum pain to the patient.
    b. Electrodes are to be cleaned after the study.
    c. Place all sensors at an appropriate place so as to minimize damage.
    d. If data of a diagnostic study suggest that titration is required, please provide mask to the patient with instruction to wear it during the day for acclimatization.

24.10.2 Protocol for the Titration Sleep Study

1. **Preparation of the patient:**
    a. Please ensure that hair and face of the patient are free of oil and grease.
    b. In cases of males, please ensure that patient is shaven, unless he wears a beard for personal or religious practice. In that case, please ensure that the beard is free of greasy material.
    c. Please ensure that the patient is comfortable and is wearing proper clothes.
    d. Ensure that the air-conditioner is working properly and the temperature is set at 22°C
    e. Blanket/sheet is to be made available to the patient.

2. **Placing the electrodes on the patient's body:**
    a. Start placing electrodes (EEG/EOG/EMG/ECG) 45 min before the patient's usual bedtime.
    b. Please clean the area and prepare it with Neuroprep gel.
    c. Use adequate amount of the conductive gel.
    d. Electrodes are to fixed as per the AASM manual ver. 2.3 released in 2015.
    e. Please fix the electrodes so as to minimize the chances of their falling off at night.

3. **Placement of respiratory sensors:**
    a. RIP belts are to be placed on the thorax and abdomen. Please adjust the belts so that the unstretched belt covers two-third of the circumference for improving the signals and longevity of the belts.
    b. Place fix the mask properly and make it comfortable for the patient. Attach it to the hosepipe.
    c. Place the oximeter properly so that it does not fall off at night.

4. **Place the body position sensor over the chest belt in the middle of the chest**

5. **Recording the study:**

a. Please fill the name, age, gender and ID of the patient
b. Please fill the antropometric data (Height/weight)
c. Please fill the name of referring Physician
d. Please choose appropriate montage (Titration)

6. **Checking the signals before lights off:**
   a. Please check the impedance from electrodes and make sure that it is below 5 K Ohms in the EEG/EMG/EOG leads.
   b. Please check that the EEG/EOG/EMG leads have good quality signals and are free from artifacts.
   c. Please check that the ECG signals are of good quality.
   d. Please check that the respiratory flow signals are appearing properly.
   e. Please check that the oximeter signals are proper.
   f. Please check that the body position is showing appropriate signals.
   g. Please check that the video and audio signals are working properly.
   h. If any of these signals are not appropriate, fix the electrode using appropriate measures as conveyed during your training.
   i. Perform bio-calibration before every study. Put a note in the patient's record sheet.

7. **If all the signals are adequate and have good quality, then start the recording and ask the attendants to turn off the lights.**

8. **During the study:**
   a. Titrate the pressure of PAP as per guidelines of the AASM provided to you.
   b. Please check all the signals periodically and enter the signal quality at an interval of 30 min in the patient's file, till the end of the study.
   c. Please mention the PAP pressure in the file at 30 min interval along with number of hypopneas and apneas seen during the preceding 30 min.
   d. Audio and video signals from the sleep lab are to be kept turned on during the whole study.
   e. Please make yourself available in case of any assistance required by the patient during the study.
   f. If any unusual behavior is seen, mention it in the patient's file along with time when it was observed.

9. **Terminating the study:**

a. Study to be terminated at the usual wake time of the patient or at 7.00 a.m.
b. Please ensure proper closure of the study.
c. Ensure that data has been transferred to the computer system before turning off the computer system.

10. After the study:

a. Please remove the electrodes so as to cause minimum pain to the patient.
b. Electrodes are to be cleaned after the study.
c. Place all sensors at the appropriate place so as to minimize damage.
d. Please hand-over the mask to the in-charge of the ward for autoclaving.

24.10.3 Protocol for the Multiple Sleep Latency Test (MSLT)

A. Precautions to be obtained before MSLT:

a. This should be done after the diagnostic study. During the diagnostic study, sleep apnea and any other cause that may lead to sleep disruption should be ruled out.
b. Total sleep time during the previous night should be at least 6 hours.
c. Get the sleep log for at least one week prior to the MSLT.
d. Stimulant medications or medications that suppress REM should be withdrawn at least two weeks ahead of the test. Physician consultation should be done to avoid the influence of hypnotic medication on the day of the test. Vigorous physical activity, exposure to bright light, smoking, and caffeine should be avoided on the day of the test, as they may be counterproductive to the sleep. Any other measure to promote wakefulness should be terminated at least 15 min prior to every nap opportunity. Drug screening may be performed in selected cases.

B. Starting the test:

a. The test is started in the morning at least two hours after waking up. The patient should be given a light breakfast. Light lunch should be provided during noon.
b. Five opportunities for the sleep are provided, 2 hours apart during the day.

C. Preparing the patient and the lab:

a. The sleep lab should be sufficiently dark and quiet to promote sleep. The temperature of the lab should be optimal.

b. Make sure that the scalp and face of the patient are free of any grease/oil.

c. Patient's clothes must be comfortable and should not be synthetic.

d. Make sure that the patient's sleep is not disrupted because of any other reason, for example, going to the washroom.

e. Apply the EEG electrodes, EOG electrodes, chin EMG, and EKG for the test.

D. **Starting data recording:**

a. Perform bio-calibration to assess alpha waves in the EEG, eye movement, and chin EMG. Preferably, this should be done in every nap.

b. Instruct the patient "Please close your eyes and try to fall asleep" before the nap and request him/her to lie down and be comfortable in bed.

E. **Duration of the each recording during the test:**

a. Recording should be done for 20 min. If the patient does not fall asleep during 20 min, record sleep latency as 20 min.

b. If the patient falls asleep during this 20 min, from the first epoch-defining sleep, another 15 min should be provided for REM sleep to appear. For example, if the epoch-defining sleep is seen at the 10th min, the test should be terminated at the 25th min (10 min+ 15 min) instead of the 20th min.

c. After each nap, the patient should be asked and information should be filled in the "MSLT questionnaire."

d. Report "start and stop time," "sleep onset latency from lights out," for each recording, and "Mean sleep latency" and "number of sleep onset REM (SOREM) out of five naps." SOREM is defined as REM appearing within 15 min of the sleep onset during a recording.

24.11 FORMS, QUESTIONNAIRES, AND TEMPLATES

24.11.1 Forms, Questionnaires, and Templates

ADULT CONSENT FORM FOR UNDERGOING SLEEP STUDY

I, _____, Resident of _____, hereby declare that I have been explained in detail of the procedure, which I will undergo on _____

I confirm that:

1. I have not hidden any information that could be vital to my diagnosis and management. If any unforeseen medical emergency arises during unforeseen condition because of this reason, the facility shall not be held responsible.
2. Considering my medical condition, I have been advised to undergo the following type of sleep study (Please check what is appropriate):
   - Overnight Diagnostic Sleep Study with Synchronized Video-Audio Recording
   - Overnight Manual Titration with PAP with Synchronized Video-Audio Recording
   - Split-Night Study
   - MSLT/MWT
3. I have been explained regarding the potential adverse effects of the above-mentioned procedures.
4. The cost that I have been explained covers only the polysomnography. In case of any unforeseen emergency, if any other medical intervention is required, the facility will add the expenses to my account.
5. I have been explained that the video-audio signals will be recorded for better understanding of my medical condition.

I hereby give my consent that:

1. Facility may carry out sleep study as mentioned above along with video-audio recording.
2. Facility may/may not use the data arising out of my investigation for the research/teaching purpose provided my identity is kept anonymous.
3. Facility may/may not use the audio-video data arising out of my investigation for the research/teaching purpose provided my identity is kept anonymous.

_____   _____   _____
(Signature of patient)   (Signature of next of kin)   (Signature of witness)

CONSENT FOR VOLUNTARY TERMINATION OF SLEEP STUDY

I, _____ hereby declare that I have been explained in detail regarding my medical condition by Dr. _____ on date.

I confirm that:

1. I have been advised to undergo sleep study/sleep study with PAP titration by my treating Physician.

2. I am not able to undergo the diagnostic sleep study/sleep study with PAP titration for the following reason: _____ .

3. I am aware that not undergoing the sleep study/PAP titration may have adverse consequences on my health.

4. Despite knowing the possible adverse consequences, I am requesting the Sleep Technologist to terminate the sleep study.

5. If any untoward medical condition emerges in future, because of this reason (termination of sleep study on my request), I shall myself be responsible for that.

_____   _____   _____

(Signature of patient)   (Signature of next of kin)   (Signature of witness)

Sleep Lab Management

Template for Appointment Register

Patient ID	Name	Age/Gender	Contact Number	Diagnosis	Type of Sleep Study required	Referred by	Special Instructions for Sleep Tech, if any

Information to be filled by Sleep technologist during diagnostic sleep study at interval of 30 minutes

Study Started at _____	All information filled as mentioned in protocol: Yes/No	Patient's pre and posttest information received and filed? Yes/No	Any other information you wish to provide _____

	Quality of signals												Auxiliary channels, if any	Any reportable behavior?
Time	EEG	EMG Chin	EOG	LEG EMG	EKG	Respiratory Flow	Respiratory effort	SpO_2	Capnograph	Body position	Audio and Video			

Sleep Lab Management

Information to be filled by Sleep technologist during titration sleep study at interval of 30 minutes

Name:							Gender: M/F
Age: …. years							UHID: ………
Date of testing: ………							

Time	PAP pressure		Leak	Apneas/Hypopneas observed	Oxygen rate	Sleep Stage	Body position
	EPAP	IPAP					

Template for Quality Assurance of PSG Record

Date	Hospital ID of Patient	Type of Study	EEG	EMG Chin	EOG	LEG EMG	EKG	Respi-ratory Flow	Respi-ratory effort	SpO$_2$	Capno-graph	Body posi-tion	Audio and Video	PAP, if appli-cable	Sign: Sleep Technolo-gist	QC Checked by………

Template for Inventory Register

S.N.	Date	Manufacturer	Equipment/Part number	Number of articles/parts available	Number of articles/parts procured	Number of articles/parts discarded	Signature of Sleep tech	Sign of director/In-Charge of the facility

Template for Accounting and Budgeting of a Sleep Lab

Date	Name of patient/purchased article/ expenditure towards service	Receipt Number	Income	Expenditure

Total Income: … … … … … … … … … .

Total Expenditure: … … … … … … … …

Net Income: … … … … … … … … … … …

Template for maintaining Service Logs of Equipments

Date	Problem in the equipment/part	Equipment/part inspected by	Recommendation made	Equipment/component: serviced/Changed	Serviced by	Signatures of Sleep Tech

Template for Pre-Sleep Questionnaire

	Name:	Gender: M/F
	Age: …. years	UHID: ……………
	Date of testing: ……….	
1.	Do you take any medication regularly?	Yes/No
2.	If yes, please mention their names below: … … … … … … … … … … … … … … … … … …	
3.	Have you taken them today?	Yes/No
4.	If not, why? A. Suggested by Sleep Physician B. Not available with me	
5.	Have you taken a nap today?	Yes/No
6.	If yes: How many hours? At what time?	… … … … Hours Between … … … and … ……
7.	Do you regularly take alcohol?	Yes/No
8.	If yes, when did you last take it?	… … … … … … ….
9.	Do you smoke/chew tobacco?	Yes/No
10.	If yes, when did you last take it?	… … … … … ….
11.	Have you prepared yourself for the sleep study as per the instructions are given in flyer?	Yes/No
12.	If no, why? … … … … … … … … … … … …	
13.	Are you feeling comfortable here?	Yes/No
14.	If no, why? … … … … … … … …	
15.	How was your sleep during past week?	Good Bad Too bad
If you think that you have any concerns regarding the test, please feel free to discuss it with Mr./Ms.… … … … … … …who is a Sleep technician.		

Sleep Lab Management

Template for Post-Diagnostic Night Questionnaire

	Name:	Gender: M/F
	Age: …. years	UHID: …………
	Date of testing: ………..	
1.	How was your sleep last night?	Better than usual As usual Worse Too bad
2.	If it was not good, what was the reason? …………………………… …………………………… …………………………… ……………………………	
3.	Have you had the problem that you have come for, last night as well?	Yes/No
4.	At what time did you go to bed?	………… p.m.
5.	How much time did it take to fall asleep?	…………. min
6.	Did you wake up at night?	Yes/No
7.	If yes, how many times?	………
8.	Were you able to fall asleep easily after awakenings?	Yes/No
9.	If not, why? …………………………….. ……………………………	
10.	Overall, how do you rate your experience in a sleep lab on a scale of 0–10? 0 being worst and 10 being satisfying	………………..
If you think that you have any concerns regarding the test, please feel free to discuss it with Mr./Ms.………………… who is a Sleep technician.		

Template for Post CPAP Night Questionnaire

	Name:	Gender: M/F
	Age: …. years	UHID: …………
	Date of testing: ……….	
1.	How was your sleep last night?	Better than usual As usual Worse Too bad
2.	If it was not good, what was the reason? … … … … … … … … … … … … … … … … … …	
3.	How did you feel with the PAP at night? … … … … … … … … … … … … … …. … … … … … … … … … … … … … ….	
4.	How are feeling today?	Better than usual As usual Worse Too bad
5.	How much time did it take to fall asleep?	… … …. min
6.	Did you wake up at night?	Yes/No
7.	If yes, how many times?	… ….
8.	Were you able to fall asleep easily after awakenings?	Yes/No
9.	If not, why? … … … … … … … …..	
10.	Overall, how do you rate your experience with PAP on a scale of 0–10? 0 being worst and 10 being satisfying	… … … … … ….
11.	Would you consider using CPAP at home? Yes/No	
12.	If not, why? … … … … … … … … … ….	
If you think that you have any concerns regarding the test, please feel free to discuss it with Mr./Ms.… … … … … … …who is a Sleep technician.		

Template for MSLT Questionnaire

	Name:			Gender: M/F		
	Age: …. years			UHID: …………		
	Date of testing: ……….					
	Item	Nap 1	Nap 2	Nap 3	Nap 4	Nap 5
1.	At what time was the test started?	……	……	……	……	……
2.	At what time did the test finish?	……	……	……	……	……
3.	Did you fall asleep during this test?	Yes/No	Yes/No	Yes/No	Yes/No	Yes/No
4.	If yes, how much time did it take to fall asleep?	……..min	……..min	……..min	……..min	……..min
5.	Did you dream during the nap?	Yes/No	Yes/No	Yes/No	Yes/No	Yes/No
6.	Did you fall asleep between two tests?	Yes/No	Yes/No	Yes/No	Yes/No	Yes/No
7.	If yes, for how long?	……..min	……..min	……..min	……..min	……..min
8.	Did you dream during that nap?	Yes/No	Yes/No	Yes/No	Yes/No	Yes/No

FURTHER READING

1. George, C. F. (1996). Standards for polysomnography in Canada. The Standards Committees of the Canadian Sleep Society and the Canadian Thoracic Society. CMAJ. 155(12), 1673–1678.

2. Fischer, J., Dogas, Z., Bassetti, C. L, Berg, S., Grote, L., Jennum, P., et al., (2012). Executive Committee (EC) of the Assembly of the National Sleep Societies (ANSS). Board of the European Sleep Research Society (ESRS), Regensburg, Germany. Standard procedures for adults in accredited sleep medicine centers in Europe. J. Sleep Res. 21, 357–368.

3. Chesson, A. L., Ferber, R. A., Fry, J. M., Grigg-Damberger, M., Hartse, K. M., Hurwitz, T. D., Johnson, S., Kader, G. A., Littner, M., Rosen, G., Sangal, R. B., Schmidt-Nowara, W., Sher, A. Practice parameters for the indications for polysomnography and related procedures. Polysomnography Task Force, American Sleep Disorders Association Standards of Practice Committee. Sleep. 1997, 20, 406–422.

4. American Academy of Sleep Medicine. Sleep Centre Management: A comprehensive manual. American Academy of Sleep Medicine, Weschester, IL, 2007.

25

FINANCIAL VIABILITY FOR A SLEEP LAB

LEARNING OBJECTIVES

After reading this chapter, reader should be able to:

1. Compute the financial viability of a sleep laboratory.

CONTENTS

25.1 Revenue for Two Beds 500

25.2 Hospital-Owned/Academic/Medical College Programs 504

A good Sleep Center should have state-of-the-art sleep suites designed and decorated for optimum patient comfort. Each suite should have its own flat-panel television, queen-sized bed, and private bathroom. The manpower team should have experienced technologists who are trained in comprehensive testing for a full range of sleep disorders, and certified physicians/specialists in sleep medicine to evaluate the test results, and prescribe an effective course of treatment so you can sleep, and feel, better.

There are three basic business structures for physician-based sleep labs:

1. Hospital-owned;
2. Independent diagnostic and testing facility; and
3. Academic/medical college sleep programs.

A single trained technician can take care of three sleep labs at a time. The cost would be more viable if there are two beds started at a time per center. It would be very difficult to break even in a single bed center considering the expenditure incurred.

For the above-mentioned three business structures, the basis of set-up is almost the same. The capital investment and the recurring expenditure are explained in detail. The below is only a guide and does not include the cost of land, which is highly variable (Tables 25.1–25.3).

25.1 REVENUE FOR TWO BEDS

It is assumed that 300 patients can be accommodated for one Sleep Lab (two beds, where one patient requires two nights of study; this calculation does not include split-night study) per year when managed optimally. Per Patient Revenue for year 1 is considered at Rs. 15,000. A nominal increase of Rs. 1,500 per year in the procedural revenue is considered.

The breakeven for one sleep lab with the assumed cost and revenue will happen in 4 years (Tables 25.4 and 25.5).

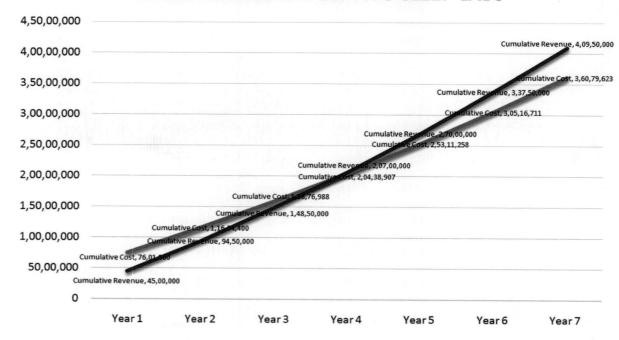

Financial Viability for a Sleep Lab

Table 25.1 Capital Cost of a Sleep Laboratory

S.No.	Item	Description	Cost for one bed (in INR)	Cost for two beds (in INR)
1	Patient room	Furniture, bathroom, camera, speaker	2,00,000	4,00,000
2	Diagnostic sleep system	Computer, jack boxes, servers and remote access	15,00,000	30,00,000
3	Pulse oximetry	Induced with PSG equip	25,000	25,000
4	CPAP, BPAP		65,000	1,30,000
5	EEG/EMG/EOG setup (electrodes).	Electrodes Kit	40,000	80,000
6	Airflow Thermistor		40,000	80,000
7	Snoring Microphones & Snore Sensor		15,000	30,000
8	Respiratory Effort Belts		40,000	80,000
	Total		19,25,000	38,50,000

Table 25.2 Recurring Cost Per Year

S. No	Description	Remarks	Cost for one bed per year	Cost for two beds per year
1	Sleep Physician	1 for 2 beds – INR 1,50,000 Per Month	18,00,000	18,00,000
2	Nurse Practitioner	1 for 2 beds – INR 35,000 Per month	4,20,000	4,20,000
3	Sleep Technician	2 for 2 beds (including Reliever) – INR 40,000 per month for 2.	4,80,000	4,80,000
4	Secretary	1 for 2 beds – INR 20,000 per month	2,40,000	2,40,000
5	House Keeper	1 for 2 beds – INR 15,000 per month	1,80,000	1,80,000
6	Electricity	500 Units per month @ INR 10 per unit for one room	60,000	1,20,000
7	Consumables	INR 500 per patient – 150 patients per year per bed	75,000	1,50,000
8	Annual Maintenance Contract	7% of the Capital cost per annum excluding furniture	1,20,750	2,41,500
9	General Maintenance	INR 5000 lumpsum per month	60,000	1,20,000
	Total		34,35,750	37,51,500

Table 25.3 Recurring Cost Year-Wise for Two Beds

Item	Remarks (% increment YoY)	Year 1	Year 2	Year 3	Year 4	Year 5	Year 6	Year 7
Sleep Physician	7 %	1,800,000	1,926,000	2,060,820	2,205,077	2,359,433	2,524,593	2,701,315
Nurse Practitioner	7 %	420,000	449,400	480,858	514,518	550,534	589,072	630,307
Sleep Technician	7 %	480,000	513,600	549,552	588,021	629,182	673,225	720,351
Secretary	7 %	240,000	256,800	274,776	294,010	314,591	336,612	360,175
House Keeper	7 %	180,000	192,600	206,082	220,508	235,943	252,459	270,131
Electricity	5%	120,000	126,000	132,300	138,915	145,861	153,154	160,811
Consumables	10%	150,000	165,000	181,500	199,650	219,615	241,577	265,734
AMC	7 % (of Capital cost excluding furniture)	241,500	241,500	241,500	241,500	241,500	241,500	241,500
General Maintenance	10%	120,000	132,000	145,200	159,720	175,692	193,261	212,587
Total		37,51,500	4,002,900	4,272,588	4,561,919	4,872,351	5,205,453	5,562,912

Table 25.4 Calculating the Break-Even for the Sleep Lab

Expected Revenue	Revenue Per Year	Cumulative Revenue
Total Patients in First Year (300 patients x Rs. 15,000)	45,00,000	4,500,000
Total Patients in Second Year (300 patients x Rs. 16,500)	4,950,000	9,450,000
Total Patients in Third Year (300 Patients x Rs. 18,000)	5,400,000	14,850,000
Total Patients in Fourth Year (300 patients x Rs. 19,500)	5,850,000	20,700,000
Total Patients in Fifth Year (300 patients x Rs. 21,000)	6,300,000	27,000,000
Total Patients in sixth Year (300 patients x Rs. 22,500)	6,750,000	33,750,000
Total Patients in seventh Year (300 patients x Rs. 24,000)	7,200,000	40,950,000
Total	4,09,50,000	

Table 25.5 Break-Even for Two Beds

Year	Fixed Cost	Recurring Cost	Cumulative Cost	Revenue per year	Cumulative Revenue
Year 1	3,850,000	3,751,500	7,601,500	4500000	4,500,000
Year 2		4,002,900	11,604,400	4950000	9,450,000
Year 3		4,272,588	15,876,988	5400000	14,850,000
Year 4		4,561,919	20,438,907	5850000	20,700,000
Year 5		4,872,351	25,311,258	6300000	27,000,000
Year 6		5,205,453	30,516,711	6750000	33,750,000
Year 7		5,562,912	36,079,623	7200000	40,950,000

25.2 HOSPITAL OWNED/ ACADEMIC/MEDICAL COLLEGE PROGRAMS

25.2.1 Raising Awareness of the "Unrecognized Benefits" of Having a Sleep Center

Given the current evolving environment of health care, it is difficult to predict the level of profitability of sleep programs. As such, it becomes imperative for academic sleep centers to impress upon institutional leaders the vital role and "unrecognized benefits" of having an academic sleep center as part of their operations. Additional outcomes research should be targeted to quantify benefits and cost savings. Examples might include:

- Identifying and treating sleep disorders saves health care costs: Numerous studies have shown that once sleep apnea is identified and adequately treated, health costs and health-care utilization decrease.
- Identifying and treating sleep disorders improves comorbidities: Studies have indicated benefit from identification and treatment of sleep apnea, on cardiovascular disease (hypertension, arrhythmias, heart failure, and stroke) and diabetes. Addressing sleep apnea also likely improves symptoms associated with psychiatric illness. These improvements can decrease long-term, health-care costs.
- Identifying and treating sleep disorders improves quality of life: This too has been shown in a number of randomized trials.
- Identifying and treating sleep disorders may help to reduce readmission rates: Prevention of unnecessary readmission is a major priority for hospitals. If institutional leaders can be convinced that treatment of patients with sleep apnea prevents readmission, they may be more willing to support sleep center programs. For example, patients with decompensated heart failure have both high risk for readmission and high frequency of undiagnosed sleep apnea; for such patients, Sleep Centers may be able to improve patient health and hospital readmission rates simultaneously.
- Identifying and treating sleep apnea is increasingly important to surgical services: Growing data highlight the risk patients with sleep apnea face when undergoing surgery, especially if sleep apnea is unexpected or undiagnosed.

- Hospitals employ a large number of individuals with sleep disorders: Identifying and treating these disorders could improve employee performance and safety.
- Hospitals employ a large number of individuals who perform shift work: An onsite sleep center is needed to help optimize shift working employee›s productivity and quality of life.

Another consideration is to approach the hospital and university leadership to request that institutions support the vital role of academic sleep centers by reimbursing them based on such outcomes as:

- Decreased health-care costs resulting from their effective management of patients with sleep disorders.
- Referrals to high revenue generating departments (e.g., surgical specialties) for management of patients with sleep disorders. This idea could require somewhat of a paradigm shift in how resources are allocated at a given academic institution/hospital.

GLOSSARY

Active wakefulness: it is seen as mixed frequency EEG along with darting eye movements and high muscle tone during a sleep study.

Apnea: cessation of airflow or (≥90%) decrease in apnea sensor excursions compared to baseline of a minimum duration of ten seconds in adults. Apneas are classified as obstructive, mixed, or central based on the pattern of respiratory effort.

Apnea hypopnea index: the apnea hypopnea index (AHI) is the total number of apneas and hypopneas per hour of sleep.

Apnea index: the apnea index (AI) is the total number of apneas per hour of sleep.

Arousal index: the arousal index (ArI) is the total number of arousals per hour of sleep.

Central apnea: it occurs when both airflow and ventilator effort are absent.

Cheyne-Stokes breathing (CSB): a breathing rhythm with a specified crescendo and decrescendo change in breathing amplitude separating central apneas or hypopneas.

Desaturation: oxygen desaturation is a frequent consequence of apnea and hypopnea. Several measures are used to quantify the severity of desaturation.

Hypopnea: it is a reduction in airflow with the minimum amplitude and duration as specified in the hypopnea rules for adults. The reduction in airflow must be accompanied by a ≥3% desaturation or an arousal or a ≥4% desaturation.

Hypoventilation: a specified period of increased P_aCO_2 of >50 mmHg in children or >55 mmHg in adults, or a rise of P_aCO_2 during sleep of ≥10 mmHg that exceeds 50 mmHg for a specified period of time in adults.

Mixed apnea: it occurs when there is an interval during which there is no respiratory effort (i.e., central apnea pattern) followed by an interval during which there is obstructed respiratory efforts.

Obstructive apnea: it occurs when airflow is absent or nearly absent, in the presence of respiratory effort.

$PaCO_2$: it is a partial pressure of carbon dioxide in arterial blood.

Posterior dominant rhythm: it is seen in the occipital channels of EEG during quiet wakefulness. In most of the adults it is alpha.

Quiet wakefulness: it appears as slow eye movement and alpha in the EEG and lowering of chin tone in the polysomnography data.

Respiratory disturbance index: the respiratory disturbance index (RDI) is the total number of events (e.g., apneas, hypopneas, and RERAs) per hour of sleep. The RDI is generally higher than the AHI, because the RDI includes the frequency of RERAs, while the AHI does not.

Respiratory effort related arousals: they are the sequences of breaths characterized by increasing respiratory effort (esophageal manometry); inspiratory flattening in the nasal pressure or positive airway pressure (PAP) device flow channel; or an increase in end-tidal partial pressure of carbon dioxide (PCO_2) (children) leading to an arousal from sleep. Respiratory effort related arousals do not meet criteria for hypopnea and have a minimum duration of ≥10 seconds in adults or the duration of at least two breaths in children. RERAs (>5 events per hour) associated with daytime sleepiness were previously called upper airway resistance syndrome (UARS), which was considered a subtype of obstructive sleep apnea (OSA). These patients have abnormal sleep and cardiorespiratory changes typically found in OSA.

Respiratory events: the respiratory events are defined as breathing abnormalities during sleep.

INDEX

A

AASM scoring manual, 388
Abnormal sleep, 508
Acceptable EOG derivations, 136
Acetylcholine, 15, 56, 58
Action potential, 13, 53, 55–58, 60, 85
Active
　respiratory infection, 460, 477
　wakefulness, 199, 271, 507
　　darting eye movements, 89, 507
　　high muscle tone, 507
Acute
　hypercapnic respiratory failure, 27
　pulmonary edema. See PAP
Adaptive servo-ventilation, 25, 26
Adolescence, 2, 405, 406
Adrenaline, 56
Advancement of the machines, 45
Airborne infection, 460
Air-conditioning, 116
Airflow measures, 98
Alcohol, 30, 425, 460, 476, 494
Alice 6 from Philips Respironics, 44

Alpha
　rhythms, 14
　waves during wakefulness, 200
Alternate leg muscle activity, 209
Alternating current, 60, 61, 99, 107
American Academy of Sleep Medicine, 23, 24, 26, 41, 43, 45, 46, 77, 111, 121, 122, 198, 270, 388, 434, 440, 444, 453, 466, 470, 473, 479, 481, 482
Amplitude, 9, 12, 39, 61, 62, 67, 69, 73, 77, 79, 84, 87, 93, 94, 99, 100, 110, 186, 244, 284, 313, 315, 320, 340, 341, 343, 344, 352, 354, 355, 363, 406–408, 411–414, 507
Annual maintenance contract (AMC), 502
Anterior tibialis muscle, 131, 132
Anteroposteriorly, 73
Antihistamines, 30
Apnea, 16, 22–24, 26, 38, 40, 41, 43, 46, 47, 99, 101, 198, 209, 212, 214, 219, 247, 266, 340, 342, 345, 346–351, 357, 358, 363, 388, 394, 397, 401, 408, 419, 440, 441, 448, 450, 483, 504, 507, 508
　classifications
　　central apnea, 340, 408, 507
　　central based on the pattern of respiratory effort, 507
　　mixed, 507
　　obstructive, 22, 27, 40, 41, 43, 47, 101, 198, 209, 214, 219, 340, 350, 351, 358, 394, 397, 400, 440, 441, 443, 448, 450, 455, 456, 507, 508
　hypopnea index, 23, 25, 26, 350, 440, 507, 508
Arousal, 1, 270, 284–86, 290–292, 295, 297, 325, 330, 331, 333, 350, 352, 354, 355, 363, 365, 408, 411–413, 440, 450, 507, 508
　index, 440, 450, 507
Arrhythmia, 22, 209
Arterial blood, 508
　gas analysis (ABG), 26–28
Artifacts, 47, 48, 67, 68, 70, 77, 88, 116, 134, 179, 192, 194, 199, 253–262, 264–266, 280, 315, 388, 432, 480, 482
　cardioballistic artifact, 254
　EKG artifact, 254, 267
　electrical artifact, 254, 267
　electrode pop artifact, 254, 255, 267
　localized artifacts, 254
　movement artifact, 254, 267
　muscle artifact, 254, 267
　physiological activities of orofacial structures, 254, 267
　respiration artifact
　　sweating artifact, 254

sweating artifact, 267
Atonia, 10, 11, 16, 93, 96, 204, 370, 388, 406
    anterior horn cells, 16
    during REM sleep, 16, 38, 97
    glycine-mediated (glycineric) inhibition, 16
    hypoglossal nerve, 16
    sign of narcolepsy, 16
Atrial fibrillation, 24, 381
Atrioventricular, 93, 375, 380
Auricular electrode, 72
Axon, 56–60
    dendrite, 60

B

Basal forebrain, 3
Berlin questionnaire, 22
Bi-level positive airway pressure, 23, 439
Biocalibration, 179, 480
    utility, 194
Bipolar montage, 73, 227, 229
Blood pressure, 183, 191, 270
Body
    movements, 38, 315
    position, 110, 140, 214
        sensor, 140
Bradycardia, 378
Brain, 3, 4, 12, 13, 16, 39, 40, 41, 55, 67, 85, 87–89, 122, 204, 253, 405
    electrical activity, 12, 14, 16, 38, 54, 60, 67, 69, 71, 85, 88, 93, 214, 215, 271
Brainstem, 13–16
Break-Even, 502, 503

sleep lab, 502
two beds, 503
Breathing, 22–27, 38, 43, 47, 101, 106, 110, 111, 116, 183, 194, 209, 229, 236, 247, 340, 342–344, 348, 350, 357, 358, 363, 364, 408, 435, 439, 449, 450, 454, 507, 508
    abnormalities, 508
    pattern, 350, 358
    respiratory events, 508
Bruxism, 219, 267, 363, 366, 367

C

Cadwell, 42, 44, 145, 169
Caffeine, 30, 419, 483
Calcium ion (Ca++), 56–58, 60, 85
Calibration, 77, 78, 151, 152, 179, 183, 184, 191, 214, 418, 420, 421, 480, 482, 484
    command in Easy III, 182
    alpha, 187
    microphone, 189
    blinking, 185
    blood pressure, 191
    eye movement, 186
    leg EMG signals, 190
    respiratory signals, 188
    teeth clenching, 192
    voice signals in microphone, 193
Cannula, 45, 98, 100, 110, 111, 137, 138, 139, 209, 211, 214, 340, 341, 352, 353, 419, 435, 448, 454, 460, 479
Capnograph, 43, 137, 183, 214, 216, 448, 480
Carbon dioxide (CO_2), 43, 54, 111, 137, 214, 216, 350, 455, 508

Cardiac
    activity, 12, 39
    myocytes, 55
Cardiorespiratory changes, 508
Cardiovascular complications, 22, 23
    cerebrovascular complications, 23
Cardiovascular disease, 504
Cataplexy, 16, 28, 29
    serotonin-norepinephrine reuptake inhibitors (SNRI) venlafaxine, 29
    serotonin reuptake inhibitors (SSRI) fluoxetine, 29
    sodium oxybate (xyrem), 29
Central sleep apnea, 24, 26, 43, 47, 101, 209, 247, 266, 340, 349, 350, 351, 357, 401, 408, 507
Chest
    abdominal movements, 101
    imaging, 26
    wall
        disorder, 26
        movement, 38
Cheyne-Stokes breathing, 24, 25, 47, 209, 236, 245–247, 340, 350, 358, 450, 507
    diagnosis, 24
    management, 24
        oxygen, 25
        PAP support, 25
        pharmacological therapy, 24
Children, 22, 23, 43, 118, 405, 406, 408, 411–414, 439, 507, 508
Chin EMG, 17, 189, 194, 230, 325, 363, 366, 369, 420, 421, 434, 484
Chloride ion, 54, 56, 85
Chronic obstructive pulmonary disease, 40, 46

Index

Circadian
  clock, 4
  process, 3
  rhythm, 3, 5, 6, 30
Coma, 2
Common mode rejection (CMR), 69, 71, 88
  ratio (CMRR), 69
Common sleep disorders, 21. See Obstructive sleep apnea
Concepts of aliasing, filter setting, sampling rate, 107
  polarity, 99
Concepts related to digitalized recordings, 77
Continuous positive airway pressure, 23, 25, 27, 28, 394, 397, 401, 403, 419, 427, 439–441, 443, 455, 456, 465, 471, 474, 477, 496, 501
Contraction, 59, 60, 67, 88, 93, 94, 192, 219, 267
Coronal plane, 127
Cortex, 4, 13–16, 85–88, 253
  association fibers, 13
  glial cells, 13, 86
  interneurons, 13
  neurons, 13
  projection fibers, 13
  pyramidal cells, 13, 14, 16
Cortical pyramidal neurons, 13
Crescendo-decrescendo breathing pattern, 350

D

Daytime sleepiness, 22, 26, 28, 29, 508
Decreased memory, 23
Deep sleep, N3, 2, 3, 6, 9, 11, 12, 16, 96, 122, 198, 270, 276, 284, 306, 313, 314, 315, 316, 317, 394, 396, 397, 406, 407, 408, 450

Delta waves, 6, 9, 16, 39, 89, 135, 198, 204, 206, 208, 306, 313, 314, 316
  frontal derivations, 91, 185, 186, 205, 206, 208, 289, 313, 407
Deoxygenated blood, 107
Deoxyhemoglobin, 107
Depolarization, 13, 15, 56, 58, 87, 93
Depression, 23, 394, 398
Desaturation, 23, 27, 109, 340, 343–345, 348, 355, 394, 396, 397, 399–401, 403, 408, 414, 450, 454–456, 507
Devices for oxygen delivery, 454
Diabetes mellitus, 30
Diagnosis, 23, 24, 26–30, 43, 48, 49, 197, 209, 214, 267, 387, 388, 424, 455, 470, 471, 475, 476, 485, 487
Difference of polarity between cornea and retina, 89
Dipole, 61
Direct current, 60, 61, 99, 101
Dopamine agonists, 30, 31
  carbidopa-levodopa (sinemet), 30
  pramipexole, 30
  ropinirole, 30
Dorsomedial hypothalamus, 4, 5

E

Easy III device, 44
Efferents, 14
Electrical calibration, 184
Electrical concepts, 60
  amplifiers, 61, 69, 84
    amplification, 61
    differential discrimination, 61
  basic concepts of electricity, 60
    dipole, 53, 57, 61, 86, 90, 94, 263

filters, 71
ground, 71
polarity, 70
Electrical potential, 54, 61, 88, 93, 94, 124, 263, 432, 433
Electrical potentials of neurons, muscles, and heart, 54
  action potential of the neuron, 55
  communication between nerves and muscles, 58
  potential generation across cardiac muscles, 60
  resting membrane potential, 54
Electrical signals, 54, 74, 85, 87, 118, 124, 221
Electrocardiogram, 41, 45, 93, 132, 209, 215, 233
  ECG, 26, 41, 54, 67, 68, 85, 88, 93, 94, 98, 132, 137, 183, 254, 372–381, 387, 394, 397, 479, 480, 481, 482
  EKG, 54, 67, 117, 118, 223, 236, 254, 259–261, 267, 382, 420, 421, 450, 473, 484, 488, 490
Electrocardiography, 26, 38
Electrodes, 13, 38, 41, 43, 60, 70, 71, 85–89, 90, 93, 94, 116, 118, 122, 124, 126, 128–132, 136, 137, 204, 221, 226, 227, 253, 254, 261–263, 363, 406, 420, 421, 432, 433, 434, 460, 464, 476, 477, 479–484, 501
Electroencephalogram, 1, 6–8, 11–13, 15–17, 38, 39, 41, 45, 54, 57, 62, 67, 68, 70, 77, 80, 82, 85–87, 89, 117, 118, 122, 124–130, 133, 134, 183, 192, 194, 198, 199, 204, 209, 218, 230, 235, 236, 253, 258, –260, 262, 263, 270, 271, 273, 275–282, 284, 285, 287, 293, 295, 301–303, 308, 315, 321, 322, 325, 336, 359, 388, 405–407, 409, 410, 420, 421, 432, 448, 460, 465, 469, 473, 479–482, 484, 488, 490, 501, 507, 508
Electromagnetic field, 117
  stray capacitance, 117
  stray inductance, 117
Electromyogram (EMG), 17, 38, 39, 41, 43, 54, 59, 60, 67, 68, 85, 89, 93, 94, 95, 117, 118, 131, 136, 189, 190, 192, 194, 204,

210, 219, 230, 276, 321, 322, 325, 350, 359, 363, 366, 367, 369, 370, 406, 420, 421, 434, 448, 460, 473, 479, 480–482, 484, 488, 490, 501

Electrooculogram (EOG), 17, 38, 39, 41, 54, 85, 88–90, 129–131, 134, 136, 186, 194, 199, 204, 230, 263, 276, 289, 308, 321, 406, 420, 421, 433, 448, 473, 479–482, 484, 488, 490, 501

Electrophysiological, 14, 221
Electrophysiology aspects of sleep, 2
Endocardium, 93
End tidal CO2 (ETCO2), 26, 54, 111, 216, 350
Epochs, 235, 236, 273, 292, 293, 308, 311, 315, 322, 323, 336, 337, 355, 363, 371, 406, 421
   concepts, 235
   EEG signals using different filters, 80
Espohageal manometery, 214
Excessive
   daytime sleepiness (EDS), 22, 23, 26, 28
   fragmentary myoclonus (EFM), 209
   salivation during sleep, 22
   sweating, 22
Excitatory postsynaptic potential, 53, 56, 57, 85, 124
Extended
   montage for seizures, 218
   parameters for special circumstances, 219
External stimuli, 2, 12
   auditory, 2
   tactile in nature, 2
Eye movements during
   REM-opposite phase, 134
   wake state-opposite phase, 133

F

Filters, 71
Fomite, 460
Forms, questionnaires, and templates, 484

G

GABA
   γ-aminobutyric acid, 4, 56
GABAergic neurons, 3, 14, 15
Gastroesophageal reflux disease, 43, 214
Gastrointestinal (tract), 40
General maintenance, 501, 502
Glutamate (Glu), 56
Grade 1 AV block, 380
Guidelines
   supplemental oxygen, 453

H

Head circumference, 128, 129
Heart
   failure, 23–27, 46, 107, 504
   rate, 38, 39, 41, 270, 378, 394, 448
   rate variability, 394
      bradycardia, 378, 394, 397
      tachycardia, 374, 379, 394, 397
Hemispheres, 71, 204, 406
Hertz; a measure of frequency (cycles per second), 9, 12, 16, 60, 71, 77, 99, 101–103, 107, 254, 258, 267, 271, 276, 284, 287, 288, 313, 315, 320, 339, 363, 406–410
High-frequency filter, 77, 81–83, 99, 101, 103, 107
High threshold bursting (HTB), 15

Histogram, 391, 393–395, 403
Homeostatic process, 3, 4, 16
   brain, 3
   hypothalamus, 3–5
Home sleep testing (HST), 23, 46, 49, 198
Hosepipe, 460, 477, 481
Hospital owned/academic/medical college programs, 504
House Keeper, 501, 502
Human scalp, 38
Hyperpolarization, 13, 16, 87
Hypertension, 23, 27, 448, 450, 504
Hypnagogic
   hypersynchrony, 408
   foot tremor, 209
   hypersynchrony, 409, 410
Hypnogram, 11, 270, 391, 392, 394, 402, 403, 441–443, 447, 450
Hypocretin, 3
Hypopnea, 16, 23, 24, 40, 99–101, 209, 245, 247, 340, 342–347, 350, 352, 355, 357, 394, 397, 399, 408, 411–414, 440, 441, 450, 507, 508
Hypothyroidism, 26
Hypoventilation, 26–28, 43, 111, 214, 340, 350, 450, 454, 507
Hypoxemia, 22, 28, 40, 109, 214, 340

I

Idiopathic central alveolar hypoventilation syndrome, 26
Idiopathic hypersomnia, 40
Impedance, 61, 69, 70, 71, 74, 85, 183, 186, 253, 254, 267, 418, 432–434, 480, 482
Infant, 2
Infection, 116, 459, 460, 468, 472, 476, 477

Index

control, 459, 476
Informative report writing, 447
Inhibitory postsynaptic potential, 56, 57, 85
Inion, 124
Insomnia, 22, 24, 40, 198, 402, 450
Instructions to the patients before sleep study, 118
Insulin resistance, 23
Internal stimuli, 2
    anxiety, 2, 116, 117
    dyspnea, 2, 24, 478
    pain, 2, 424, 477, 481, 483
Interpretation of histograms, 391
Iron deficiency, 30, 31
Ischemic heart disease, 23

J

Jaw size, 23
Junction, 15, 132

K

K complexes in frontal derivations, 205
Kleine-Levine syndrome (KLS), 40

L

Laboratory tests, 26
Leg movements, 248–250, 350, 359, 360, 364, 365
Lenz's law, 101
Level I polysomnography, 54
Limb movement, 40, 198, 204, 350, 361, 362
Lithium, 30
Locus coeruleus, 3

Low-frequency filter (LFF) (high pass filter), 77, 81–83, 99, 102, 107, 258, 267
Lung parenchymal, 26

M

Maintenance of wakefulness test (MWT), 48, 417, 420, 470, 485
Malfunction of one electrode, 123
Management, 24, 27, 29, 463
    OHS, 27
Manual titration with PAP, 418, 439, 440, 485
Mastoid, 41, 43, 71, 88, 122, 124, 129, 131, 136, 198, 226, 267, 448
Medication use, 26
Methylphenidate, 29
Microsleep, 273, 274
Mid-adolescence, 2
Middle-aged women, 22
Mixed apnea, 351, 408, 507
Modafinil, 29
Montage, 72, 73, 116, 125–130, 148, 164, 218, 221–224, 226–229, 480, 482
    bipolar montage, 73, 227, 229
    diagnostic study, 224, 235, 350, 395, 398, 400, 417–419, 434, 449, 471, 481, 483
    referential montage, 226, 228
    scoring
        electrocardiogram, 233
        leg movement, 232
        respiratory data, 231
        sleep stages, 204, 230, 276

titration study, 40, 223, 225, 350, 419, 434, 439–441, 444, 470, 478
Morning headache, 22
Multiple sleep latency test, 29, 48, 417, 419, 420, 428, 470, 474, 483–485, 497
    questionnaire, 428, 497
Muscle
    contraction, 59, 60
    tone, 6, 12, 28, 39, 96, 270, 315, 507
    weakness, 26
Myelination, 405
Myocytes, 55, 58, 60, 93
Mysteries of sleep, 37

N

Narcolepsy, 3, 16, 28, 29, 40, 394, 398
    cataplexy, 28
    hypnagogic hallucination, 28
    interrupted fragmented sleep, 29
    irresistible attacks of sleep, 28
    sleep paralysis, 28
Nasal
    airflow, 137
    cannula, 45, 98, 100, 111, 138, 139, 214, 340, 352, 353, 419, 448, 454, 460
    oral airway, 98
Nasion, 124
Neuro-biological mechanisms, 122
Neurobiology, 3
    basal forebrain, 3
    characteristics, 3
    GABAergic neurons, 3, 14, 15

glutaminergic neurons, 3
hypocretin, 3
laterodorsal tegmental, 3
monoaminergic nuclei, 3
   cholinergic neurons, 3, 14
   histaminergic neurons, 3
   noradrenergic neurons, 3
   serotonergic neurons, 3
pedunculopontine, 3
reticular activating system, 3, 4, 13, 86, 87
sleep processes
   circadian, 3–6, 16, 30
   homeostatic, 3, 4, 16
ventrolateral preoptic nucleus, 3
Neurological disorders, 30
Neurologic disorder, 26
Neuromuscular disorders, 26
Neuronal membrane potential, 56
Neurons, 3, 4, 13–16, 54–57, 85–87
   thalamocortical system, 15
Neurophysiology of EEG rhythms from wakefulness to sleep, 12
Neurotransmission, 56
Neurotransmitter, 56, 57, 85, 87
Nicotine, 30
Nipples, 137
Nocturia, 22
Nocturnal
   heart burn, 22
   palpitation, 22
   seizures, 88, 387
Non-rapid-eye-movement sleep (NREM), 2, 6, 11–13, 24, 27, 38, 39, 89, 270, 276, 284, 406

N1 sleep, 2, 6, 7, 12, 15, 24, 122, 270, 273–279, 282–286, 291, 292, 295, 297, 298, 305, 315, 324, 326, 327, 333, 334, 337, 338, 406–408, 421
   30 seconds epoch, 7–10
N2 sleep, 2, 6, 8, 12, 15, 16, 24, 80, 122, 270, 276, 284, 287–297, 299–313, 315, 323, 334, 335, 406–408
   30 seconds epoch, 7–10
N3 sleep, 2, 3, 6, 9, 11, 12, 16, 96, 122, 270, 276, 284, 306, 313, 314, 315, 316, 317, 394, 396, 397, 406, 407, 408, 450
   30 seconds epoch, 7–10
Norepinephrine, 29
Normal sleep, 40, 393
Nostrils, 137
Nurse practitioner, 501, 502

O

Obesity, 22, 26
   hypoventilation syndrome, 26–28, 43, 454
Obstructive sleep apnea, 22–24, 26, 27, 40, 41, 43, 46, 47, 49, 101, 109, 116, 198, 209, 214, 219, 241, 244, 247, 340, 348, 351, 358, 394, 396–401, 403, 408, 417, 439–442, 448, 450, 456, 457, 508
   complications, 23
   diagnosis, 23
   risk factors, 22
   severity, 23
   symptoms, 22
   treatment, 23
Obstructive sleep apnea syndrome, 22
Ohms, 60, 61, 418, 480, 482
Optional parameters, 214
Oro-nasal airflow, 38

Out-of-center, 46, 49
Oxygen, 25, 28, 88, 107, 108, 122, 127–129, 209, 213, 256, 341, 429, 453–455, 457, 465, 489
   delivery, 454, 455
      devices and polysomnography, 454
   exchange in lungs, 108
   saturation, 25, 41, 54, 107, 209, 213, 340, 394, 396, 448, 450, 454
   supplementation, 25, 454, 455, 457
   therapy, 25, 28, 109, 453, 456
Oxyhemoglobin, 27, 107
Oxymeter, 139

P

Parasomnia, 41, 45, 48, 110, 125–130, 198, 214, 219, 223, 228, 229, 387, 388, 424
Parkinson's disease, 30, 198
Partial pressure of CO_2, 26, 27, 214, 350, 507, 508
Partial pressure of O_2, 27
Pedunculopontine tegmental nuclei, 3
Perception, 2
Periodic limb movement, 40, 131, 198, 204, 209, 249, 350, 450
Peripheral
   nervous system, 13, 55
   sensory system, 13
Pharmacological therapy, 24
Pharyngeal pH, 214
Phase delay, 2
Philips Respironics, 42, 44, 46, 145
Photodiode, 107
Physiological changes, 1, 12, 40
   during normal sleep, 12

Index

Physiology and recording of electrical potentials, 85
Pickwickian syndrome, 38
Piezoelectric principle, 99
Piezo technology, 101, 106
Placement of
    body position sensor, 142
    ECG electrodes, 137
    EEG electrodes according to 10–20 systems, 124
    EOG electrodes, 130, 131
    Measures of Respiration, 137
    nasal cannula, 138
    oximeter, 142
    RIP belts, 140
    thermistor, 138
Plethysmography, 101, 408
Pleural pathology, 26
Polarity, 70, 99, 104
Policy for
    accounting and budgeting, 471
    case of any emergency during sleep study, 477
    infection control in the sleep lab, 476
    maintaining the records, 473
    quality assurance, 472
Polysomnographic
    data, 270
    recording, 38, 40, 405
Polysomnography (PSG), 12, 23, 24, 27, 29, 30, 37, 38, 39, 40, 41, 46, 49, 53, 54, 67, 87, 110, 111, 115, 117, 118, 128, 145, 172, 191, 197, 221, 223, 236, 253, 254, 270, 276, 363, 387, 394, 417, 418, 420, 424, 431–434, 436, 440, 444, 454, 464, 465, 473, 474, 485, 490, 501, 508
    concepts, 53
    scoring rules in children, 405
Position dependent OSA, 399
Positive airway pressure (PAP)/titration, 23–25, 27, 28, 40, 43, 48, 49, 109–111, 209, 223, 350, 394, 395, 403, 417–419, 424, 427, 429, 434, 435, 439, 440, 442, 444, 448, 454, 455, 460, 467, 471, 473, 474, 476, 478, 482, 485, 486, 489, 490, 496, 508
    adequate titration, 444
    bi-level PAP, 23, 27, 28, 439–441, 443, 501
    good titration, 444
    optimal titration, 444
    PAP support, 25
    PAP therapy, 27
    questionnaire, 429
    repeat PAP titration study, 444
    unacceptable titration, 444
Positive deflection, 88, 90, 98, 284, 289
Post CPAP night questionnaire, 427, 496
Post-diagnostic night questionnaire, 426, 495
Posterior dominant rhythm, 276, 279, 406, 407, 508
Potassium ion, 54, 56, 85
Potential of hydrogen (pH), 43, 214
Pre-auricular point, 126
Precautions, 454, 477, 483
Pregnancy, 31
Preparing the machines, 116
Preparing the patient, 117
Pre-sleep questionnaire, 425, 494
Pressure
    limits, 440
    transducer, 41, 98–100, 110, 137, 194, 209, 214, 340, 343,–345, 348, 434
Protocols, 417, 466, 478
    diagnostic sleep study, 418, 478
    maintenance of wakefulness Test, 420
    manual titration with PAP, 418
    multiple sleep latency Test, 419, 483
    titration sleep study, 481
Pulmonary
    diseases, 26, 27
    function tests, 26, 27
    vascular pathology, 26
Pumping process, 93
Pyramidal neurons, 13, 14, 15

Q

Quality assurance indicators, 474
Quality of life, 24, 25, 27, 49, 198, 504, 505
Quartz, 99
Quiet wakefulness, 272, 508

R

Rapid eye movements (REM), 2, 6, 10–12, 15, 16, 22, 27–29, 38–40, 89, 92, 93, 96, 97, 122, 133, 134, 187, 198, 199, 203, 204, 219, 269–271, 276, 284, 307–312, 315, 317–321, 322–325, 328, 330–339, 363, 369, 370, 387, 388, 394, 396, 397, 398, 400, 406, –408, 419, 420, 424, 444, 483, 484
Real time EMG data, 95
Recording, 116
    eye movements, 88
Record of snoring, 110
Recurring cost for two beds, 502
Red blood cells (RBC), 107, 108

Refractory period, 13
Relationship of process C and S, 6
Relaxation, 93
REM sleep
  30 seconds epoch, 7–10, 238, 242, 248, 271, 272, 275, 371
REM sleep behavior disorder, 11, 16, 40, 93, 204, 219, 269, 363, 387
Renal failure (uremia), 30
Resistance, 23, 60, 61, 99, 117, 118, 508
Respiration, 12, 38, 39, 54, 99, 101, 214, 240, 242–244, 256, 270, 340, 341, 354, 406
Respiratory data, 98, 340
Respiratory disturbance index (RDI), 40, 444, 450, 508
  apneas, 22–24, 27, 209, 394, 397, 399–401, 403, 408, 424, 440–442, 482, 507, 508
  hypopneas, 23, 24, 27, 99, 109, 209, 340, 350, 394, 397, 400, 424, 440, 441, 443, 482, 507, 508
Respiratory effort related arousal (RERAs), 350, 356, 363, 394, 397, 408, 440, 441, 508
  abnormal sleep, 508
  cardiorespiratory changes, 508
  obstructive sleep apnea, 22, 40, 41, 43, 47, 101, 198, 209, 214, 219, 340, 351, 358, 448, 450, 508
  partial pressure of carbon dioxide, 508
  pressure or positive airway pressure (PAP), 508
  upper airway resistance syndrome, 508
Respiratory efforts, 41, 54, 137, 188, 209, 212, 340, 350, 408, 440, 441, 501, 507, 508
Respiratory events, 209, 248, 269, 364, 394, 408, 440, 441, 448, 455–457, 508
  breathing abnormalities, 508
Respiratory flow, 41, 45, 54, 99, 104, 105, 107, 209, 211, 340, 350, 448, 476, 480, 482
Respiratory inductance plethysmography (RIP), 101, 106, 137, 140, 141, 194, 209, 212, 340, 341, 345, 348, 408, 434, 435, 448, 479, 481
Respiratory waveform, 101, 102, 104, 257
Resting membrane potential, 53–56, 58, 93, 94, 98
Restless legs syndrome, 29–31, 40, 198, 387, 450
  creepy-crawly sensations, 29
  diagnosis, 23, 24, 26, 29, 30, 475, 487
  International restless legs syndrome study group, 29
  medical conditions associated with RLS, 30
  prevalence of RLS, 30
  treatment, 30
Reticular activating system, 3, 4, 13, 86, 87
Reticular neurons, 14
Retinohypothalamic tract, 4
Reversibility, 2
Rheumatologic diseases, 30
Rhythmic movement disorder, 363
Risk factors for OSA, 22

S

Saw tooth waves, 203, 320
  central derivations, 203
SCOPER system, 45
Scoring of
  EKG data, 382
  respiratory data, 340
  sleep stage, 270
Seizure, 40, 88, 110, 125–130, 214, 219, 227, 228, 363
Sensation
sensory stimulus, 2
Sensory stimulation, 13
Serotonin, 29
Serotonin-norepinephrine reuptake inhibitors, 29
Setting HFF to 1 Hz removes snoring, 103
Sino-atrial, 93
Skeletal myocytes, 55
Skeletal restriction, 26
Sleep across age, 2
Sleep and wake promoting areas in brain, 4
Sleep architecture, 6, 22, 391, 394, 396, 448
Sleep disordered breathing (SDB), 23, 109, 363
Sleep disorders, 6, 21, 38, 40, 43, 46, 48, 49, 54, 198, 391, 454, 455, 472, 499, 504, 505
Sleepiness, 3, 22, 26, 28, 29, 40, 48, 508
Sleep lab
  financial viability, 499
Sleep laboratory, 464, 466
Sleep laboratory setting, 116, 464, 466, 468, 469, 470–474, 501
Sleep latency and sleep onset REM, 29, 394, 398, 420, 484
Sleep medicine, 23, 121, 122, 270, 469, 470, 473
Sleep onset rapid eye movement (SOREM), 29, 394, 398, 420, 484
Sleep onset REM, 398
Sleep physician, 46, 49, 118, 421, 424, 425, 466–471, 473, 474, 476–479, 494, 501, 502
Sleep-related seizure, 219
Sleep spindles, 6, 16, 39, 198, 284, 287, 288, 290, 293, 314, 315, 323, 406, 407
Sleep stages, 1, 2, 3, 11, 16, 39, 41, 88, 96, 122, 204, 230, 236, 270, 276, 391, 394, 406, 421, 448
See, Non-rapid-eye-movement sleep (NREM)

rapid-eye-movement, 2
Sleep studies
  guidelines, 46
  types and their utility, 47
Sleep studies, advantages, and limitations, 45
Sleep studies/monitoring devices types, 40
  level I, 43
  level II, 41, 43, 254
  level III, 41
  level IV, 40
Sleep technician, 501, 502
Sleep technologists, 43, 46, 49, 423, 432, 448, 454, 464–467, 469–473, 476–478
Slow transition, 270
Snore microphone, 140
Snoring embedded in respiratory waveform, 102
Snoring signals in microphone, 217
  crescendo-decrescendo signals, 189, 217
Sodium ion, 54, 56, 58, 60, 85
Somnomedics, 42, 44, 145
Somnoscreen device from Somnomedics, 44
Source of electrical signals, 85
Spinal cord, 13, 16
Spindle activity in central derivations, 202
Split night study, 395, 401
Spontaneous movements, 12
Staff, 466, 468
Standard duration of an epoch, 236
STOP-Bang questionnaire, 22
Stroke, 23, 504
Supplemental oxygen guidelines, 453
Suprachiasmatic nucleus, 4, 5

T

Template, 473, 475, 487, 490–496, 497
  appointment register, 475, 487
  inventory register, 475, 491
  MSLT questionnaire, 428, 497
  post CPAP night questionnaire, 427, 496
  quality assurance, 466, 472, 490
Ten–twenty (electrode placement system), 124, 125
Test protocols, 417
Thalamocortical, 13–16, 87
Thalamus, 13, 16, 86
  thalamic relay neurons, 13
    tonic firing vs. burst firing, 14
  thalamic reticular neurons, 14
Thermistor, 41, 98, 99, 100, 137, 138, 194, 209, 211, 340, 341, 343–348, 419, 434, 436, 448, 454
  thermocouple, 98
Thermoregulation, 12
Theta
  rhythms, 15
  waves, 6, 83, 284
Thorax, 212, 340, 448, 479, 481
Time constant, 79
Titration, 40, 43, 48, 110, 111, 209, 223, 225, 350, 395, 403, 417–419, 424, 434, 439–442, 444, 455, 470, 474, 478, 481, 482, 485, 486, 489. See Positive airway pressure (PAP)/titration
  BPAP titration for OSA, 441
  CPAP titration for OSA, 440
  pressure limits, 440
  steps and indications, 441

switch from CPAP to BPAP in patients with OSA, 441
Transcutaneous CO_2, 350
Transthoracic echocardiogram, 26
Treatment, 23, 28
Tricyclics, 30
Troubleshooting, 431, 432
True eye movement, 207
T-tubules, 60

U

Unrecognized benefits of sleep center, 504
Unrefreshing sleep, 22
Upper airway resistance syndrome, 99, 508
Uremia, 30, 31

V

Ventilation, 24, 25, 27, 41, 116
Ventricles, 93, 94
Ventricular tachycardia, 374
Ventrolateral preoptic nucleus, 3
  VLPO, 3, 4, 5
Vertex sharp waves, 275, 406
Vertex waves, 7, 39, 198, 276
  central derivations, 201–203, 275, 277, 278, 284, 285, 287, 288, 315, 320, 407, 409, 410
  wavesin N1, 277
Video data, 110
Video polysomnography, 387
Video recording, 43, 54, 110, 214, 387, 388, 436, 465
Volts, 60, 67

W

Wakefulness, 3, 4, 11–17, 26, 27, 28, 38–40, 204, 235, 270–274, 276, 279, 284, 286, 296, 315, 324, 406, 417, 419, 420, 483, 507, 508
Waveforms upon signal capturing rate, 84
Waves during eye movement, 132
Waves generated through calibration, 78
Weight loss/reduction, 27
Workflow from sleep clinic to sleep laboratory, 470

Y

Young adults, 22

Z

Zeitgebers, 3
Zig-zag, 106
Z-RIP-module, 435